THE
INTERNATIONAL SERIES
OF
MONOGRAPHS ON PHYSICS

GENERAL EDITORS
W. MARSHALL   D. H. WILKINSON

THE
PHYSICS OF
LIQUID CRYSTALS

BY

P. G. de GENNES

CLARENDON PRESS · OXFORD
1974

Oxford University Press, Ely House, London W.I

GLASGOW   NEW YORK   TORONTO   MELBOURNE   WELLINGTON
CAPE TOWN   IBADAN   NAIROBI   DAR ES SALAAM   LUSAKA   ADDIS ABABA
DELHI   BOMBAY   CALCUTTA   MADRAS   KARACHI   LAHORE   DACCA
KUALA LUMPUR   SINGAPORE   HONG KONG   TOKYO

ISBN 0 19 851285 6

© OXFORD UNIVERSITY PRESS 1974

PRINTED IN NORTHERN IRELAND AT THE UNIVERSITIES PRESS, BELFAST

PREFACE

LIQUID crystals are beautiful and mysterious; I am fond of them for both reasons. My hope is that some readers of this book will feel the same attraction, help to solve the mysteries, and raise new questions.

We have known of the existence of liquid crystals for a comparatively long time—eighty years—yet many experiments which could have been done thirty years ago are only now being performed. The importance of potential applications to thermography and electro-optic display was realized only ten years ago (mainly through the work of Fergason and Hellmeier), but the apparent lack of applications in earlier times cannot, by itself, explain certain delays. More fundamentally, the study of liquid crystals is complicated because it involves several different scientific disciplines: chemistry, optics, and mechanics, more specialized tools such as nuclear magnetic resonance, and also a certain sense of vision in three-dimensional space in order to visualize complex molecular arrangements. A semi-theorist like myself is not very well trained in any of these techniques. For this reason, this book is very incomplete; certain aspects (and particularly the chemical aspects) are reduced to a bare minimum. On the other hand, what a theorist can and should systematically introduce is comparisons with other fields. In the present context, comparisons with magnetic systems are often useful and will be mentioned. Comparisons between the so-called smectic phases and the superfluid helium II and superconductors are also highly instructive. However, they imply a certain familiarity with low-temperature physics, which I did not want to impose as a pre-requisite; for this reason, the references to superfluids are kept short.

As do all theorists, I relish exact calculations, but in the present case I have tried to insert qualitative discussions rather than equations whenever possible. Two parts remain particularly heavy and unpleasant however: these are the sections of Chapters 3 and 5 concerned with the hydrostatics and hydrodynamics of nematics. This subject has been source of some controversy between the classical mechanics group at Johns Hopkins and the theoretical physics group at Harvard. The two groups use very different languages; however (fortunately), their essential results are the same. I have tried to show this at the expense of a rather long and dull analysis, but, apart from this particular topic,

the theoretical speculations have usually been condensed and replaced by one or two useful references.

I must emphasize that the references quoted (both for experimental data and for theoretical work) do not claim to be a complete list and do not refer to historical priority. For instance, I do *not* mention the books of Lehmann and of Schenk because an average reader will get a much more clear presentation of their main observations through the later reviews by Friedel or Saupe†. Similarly, in the discussion of nematic order at the molecular level, I have omitted many theoretical papers which are interesting but still very tentative.

Even as regards the choice of topics, the present work is incomplete; in particular, the lyotropic materials are not discussed. I believe that we still need a few years to understand them better. The same remark holds for the 'exotic' smectic phases D, E, F, G... and for the 'blue phase' of certain cholesterol esters. These phases are extremely interesting, but (in my opinion at least) we grasp them too poorly to discuss them in this book; my purpose is clarification more than compilation. Having taught the corresponding material during the past three years, I know that the present version is still very far from achieving this purpose, but I hope that, in spite of its obvious defects, it will help 'liquid crystallers' to reach a universally common language.

My debt to a number of colleagues and friends is enormous. I would like to mention first G. Durand and M. Veyssié, who bravely started experiments on liquid crystals at Orsay during the autumn of 1968. Now, owing to their efforts and to the participation of other teams we have established a very active group; to a large extent, the present book represents its common views, based on constant discussions. The mistakes are my mistakes but the spirit is their spirit. I would also like to mention friends from outside the Orsay unit: J. Billard, who showed me the first liquid crystals that I ever saw and who, more recently, taught me a number of fundamental features of mesomorphic phases; R. B. Meyer, with whom I have been in constant correspondence and whom I hoped (in vain) to convince to write some chapters; our visitors from last year D. Litster, P. Pershan, and especially P. Martin, whose advice and positive criticism has been of constant help; F. C. Frank who, together with J. Friedel and M. Kleman, patiently instructed me on the structure of some difficult defects; G. Sarma and N. Boccara, with whom I have been exchanging ideas for a long time; S.

† The history of liquid crystals has been analysed recently by H. KELKER (*Molecular crystals and liquid crystals*, **21**, 1 (1973)).

Alexander, D. Martire, and J. Vieillard Baron, who coached me on the hard-rod problem. A special mention is due to J. L. Ericksen, H. Gruber, W. Marshall, A. Rapini, and Y. R. Shen for their close scrutiny of the initial manuscript, to C. Williams, who prepared the list of elastic constants, and to L. Leger, who did the same for the friction constants. Most of the typing work—particularly painful in view of the unpredictable changes of mood of the author—was done, with charming patience, by Marie France Jestin. M. Crasson and R. Seveste have succeeded in producing figures and photographs from my vague indications and scribblings.

Last, but not least, I want to thank my wife Anne Marie for her co-operation: so many sunny weekends have been sacrificed to this book. Now, looking at the result, I am not entirely convinced that it was worthwhile, and I certainly do not dare to dedicate the book to her.

'Well, now that we have seen each other,' said the unicorn 'If you believe in me, I'll believe in you. Is that a bargain?'
'Yes, if you like,' said Alice.

P. G. de G.

Orsay, December 1972

CONTENTS

1

ANISOTROPIC FLUIDS: MAIN TYPES AND PROPERTIES

'Que m'a donné le monde que ce mouvement d'herbes?'
SAINT JOHN PERSE

1.1. Introduction

DURING our years in high school, we have all been taught that matter exists only in three states: solid, liquid, and gas. However, this is not quite correct: in particular, certain organic materials do not show a single transition from solid to liquid, but rather a cascade of transitions involving new phases; the mechanical properties and the symmetry properties of these phases are intermediate between those of a liquid and those of a crystal. For this reason, they have often been called *liquid crystals*. A more proper name is 'mesomorphic phases' (mesomorphic: of intermediate form).

To understand the significance of these new states of matter, it may be useful to recall first the distinction between a *crystal* and a *liquid*.

In the crystal, the components (molecules, or groups of molecules) are regularly stacked. The centres of gravity of the various groups are located on a three-dimensional periodic lattice. In the liquid, the centres of gravity are not ordered in this sense. These two states of matter differ most obviously by their mechanical properties; a liquid flows easily. More fundamentally, the crystal is distinguished from the liquid by its X-ray diffraction pattern, showing sharp Bragg reflections characteristic of the lattice.

With this fundamental distinction present in our minds, we see that mesophases can be obtained in two different ways:

(1) Imposing positional order in one or two dimensions rather than in three dimensions. This does happen in nature. In the main practical case, we have positional order in one direction only; the system can be viewed as a set of two dimensional liquid layers stacked on each other with a well-defined spacing; the corresponding phases are called *smectic*.

(2) Introducing degrees of freedom which are distinct from the localization of the centres of gravity. For non-spherical molecules, the *orientation* of the molecule is the most natural candidate. Orientational transitions may take place in a crystal, or in a liquid (or even in a smectic) phase:

(a) Many crystals show a transition from a strongly-ordered state to a phase where each molecule commutes between several equivalent orientations. The high-temperature phase is positionally ordered, but orientationally disordered. It is sometimes (loosely) called a *plastic crystal*. Examples of such rotational transitions are: solid hydrogen, ammonium halides, and also certain types of organic molecules.

(b) Certain organic *liquids* show a low temperature phase where the molecules are aligned preferentially along one direction: these anisotropic liquids are called *nematics*. They are positionally disordered, but orientationally ordered. At higher temperatures, they undergo a transition to a conventional (isotropic) liquid phase.

The denomination 'liquid crystals' is commonly applied to both smectics and nematics. These two types are found only when the constituent molecules, or groups of molecules (the 'building blocks,' as we shall call them) are *strongly elongated*.

The situation is completely opposite for the 'plastic crystals,' where the molecules are (usually) *nearly spherical* in shape. For this reason, the two fields are quite distinct; in the present book, we shall be concerned only with liquid crystals.

Depending upon the nature of the building blocks, and upon the external parameters (temperature, solvents, etc) we can observe a wide variety of phenomena and transitions amongst liquid crystals. This chapter therefore starts off by discussing the building blocks, which give rise to the known types of liquid crystals, and then goes on to give a broad classification of nematics and smectics.

1.2. The building blocks

As explained above, to generate a liquid crystal, one must use elongated objects. At the present time, we know at least three different ways to achieve this: with small organic molecules; with long helical rods which either occur in nature or can be made up artificially; and with more complex units which are really associated structures of molecules and ions.

We shall discuss these three examples in turn.

1.2.1. Small organic molecules

The classical example is *p*-azoxyanisole (PAA) with the formula

$$CH_3-O-\bigcirc-N{=}N-\bigcirc-O-CH_3$$
$$\downarrow$$
$$O$$

From a (rough) steric point of view this is a rigid rod of length $\sim$20 Å and width $\sim$5 Å. (The two benzene rings are nearly coplanar).

Another example of practical interest is *N*-(*p*-methoxybenzylidene)-*p*-butylaniline (MBBA) with the formula

$$CH_3$$
$$\diagdown$$
$$O-\bigcirc-CH{=}N-\bigcirc-CH_2 \qquad CH_2$$
$$CH_2 \qquad CH_3$$

Both PAA and MBBA are 'nematogens.' This word means that they give rise to the nematic type of mesophase which will be discussed in Section 1.3.1. However, for PAA the nematic state is found only at high temperatures (between 116°C and 135°C at atmospheric pressure), while MBBA is nematic from $\sim$20°C to 47°C, thus allowing much easier experimentation.

A broad class of organic molecules with the same general pattern,

$$R-\bigcirc-A{=}B-\bigcirc-R'$$

(i.e. two *para*-substituted aromatic rings rigidly linked by a double, or triple, bond A—B) also give mesophases. In the above formula R and R' are short, partly flexible chains. The major types inside this scheme are listed in Table 1.1.

Empirical rules describing the influence of R and R' on the phase diagram, and related problems, are reviewed in reference [1].

TABLE 1.1

Main nematogenic types together with references describing their chemical synthesis

A—B	
—CH=N—	H. KELKER and B. SCHEURLE (1969). *Angew. Chem. (Int. ed.)* **8**, 884.
	B. FLANNERY and W. HAAS (1970). *J. Phys. Chem.* **74**, 3611.
—N=N—	H. KELKER *et al.* (1970). *Angew Chem. (Int. ed.)* **9**, 962.
	J. VAN DER WEEN *et al.* (1972). *Mol. Cryst.* **17**, 291.
—N=N— ↓ O	R. STEIN STRASSER and L. POHL (1971). *Tetrahedron Lett.* **22**, 1921.
—CH=N— ↓ O	W. R. YOUNG, I. HALLER, and A. AVIRAM (1971). *IBM Journal* **15**, 41.
—CH=CH—	W. R. YOUNG, I. HALLER, and A. AVIRAM (1972). *Mol. Cryst.* **15**, 311.
—CH=C— | Cl	W. R. YOUNG, A. AVIRAM, and R. J. COX (1971). *Angew. Chem. (Int. ed.)* **10**, 410.
—C≡C—	J. MALTHETE, M. LEDLERCQ, J. GABARD, J. BILLARD, and J. JACQUES (1971). *C.r. hebd. Séanc. Acad. Sci., Paris* **C273**, 265; *Mol. Cryst.* (to be published).

Another favourable class is obtained with *cholesterol esters*, of general formula:

(we use a simplified convention in which the hydrogens are not explicitly shown). Note that the rings are not aromatic, and the structure is not planar. However the ring system is rigid, while the saturated chain C and the radical R (when it is not too short) behave like two somewhat more flexible tails attached to the rigid part; thus there is some steric similarity to the preceding group.† The three-dimensional structure of a cholesterol ester is shown on Fig. 1.8.

---

† Cholesterol itself has H instead of R—CO—, and does not give mesophases, probably because the OH group creates strong hydrogen bonds between different molecules.

We observe finally that, in all the pure systems which have been discussed here (PAA, cholesterol esters, etc.) the only simple way to induce a transition is to vary the temperature. For this reason, such systems are commonly called *thermotropic*.

1.2.2. Long helical rods

A number of *synthetic polypeptides*, in suitable solvents, have a rod-like conformation with typical rod lengths of order 300 Å, and widths 20 Å [2]. In concentrated solutions these systems give mesophases [3]. Similar phases are found also with deoxyribonucleic acids (DNA) and with certain viruses [4]; the standard example is *tobacco mosaic virus* (TMV), with length 3000 Å, and width ∼200 Å. One definite advantage of such viruses, from the point of view of physical experiments, is that all rods from one virus species are exactly the same size.

Finally, on a still larger scale, model systems made with glass or plastic *fibres* (diameters ∼10 μm, lengths ∼100 μm or more) floating in water, might be of great interest, and it is my personal hope that such systems will be prepared and studied.†

Note that, for all the systems listed in this section, the transitions are induced most easily by changing the concentration of rods rather than the temperature; for this reason they are commonly called *lyotropic*.

1.2.3. Associated structures

Typical examples of such structures are found in the soap-water systems. Here we have an aliphatic anion $CH_3-[CH_2]_{n-2}-CO_2^-$ (with n in the range 12–20) plus a positive ion (Na^+, K^+, NH_4^+, or others). The polar head of the acid (i.e., the $-CO_2^-$ group) tends to be in close contact with water molecules, while the apolar aliphatic chain avoids the water. These two opposite requirements are typical of 'amphiphilic' materials. A single chain in solution cannot satisfy both of them, but a cluster of chains can, as is shown in Fig. 1.1(a). The resulting objects (e.g. rods or leaflets) may become the building units of larger meso-morphic structures [5]. Other examples of amphiphilic chains leading to similar geometries are the 'block copolymers' shown on Fig. 1.1(b).

To summarize: we have found three types of building blocks.

---

† The two main difficulties associated with these systems are as follows. (1) The possibility of flocculation of the rods due to van der Waals attractions. This can be counteracted by using charged systems, detergents, etc. (2) The slow dynamical response of long objects will tend to make the system highly viscous. The optimal (theoretical) solution would be to have charged rods floating in superfluid helium.

(1) Pure organic molecules such as PAA. In such systems the phase transitions are induced most naturally by a change of temperature. They are commonly designated as *thermotropic*.

(2) Rods in a liquid substrate. Here the temperature effects are difficult to control (very often an increase in T will rapidly destroy the individual rods), and the natural parameter which we can adjust to induce phase transitions is the *concentration* of the rods. Such systems are called *lyotropic*.

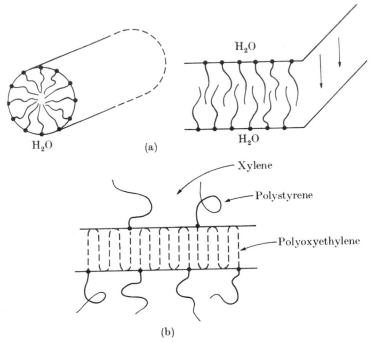

FIG. 1.1. A few typical building blocks for amphiphilic materials (a) rods and sheets for the system fatty acids–water (the polar head of the fatty acid is represented by an arrow). (b) Sheet structures for a copolymer: each chain has one soluble part (here polystyrene, the solvent being xylene) and one unsoluble part (here polyoxyethylene).

(3) Amphiphilic compounds. These may give rise to associations and to mesomorphic behaviour, either in the presence of a selective solvent (eg. water in soaps) or as a pure phase (in particular, the latter is the case for certain block copolymers). Thus, depending on which of the above conditions hold, amphiphilic compounds may be lyotropic or thermotropic.

In the present book, we shall be concerned only with thermotropic materials. Having in mind these various species, we can now start to describe the unusual thermodynamic phases to which they give rise. The classification of mesophases (first clearly set out by G. Friedel in 1922 [6]) is essentially based on their symmetry. There are two major classes: nematics and smectics, which we shall now discuss.

1.3. Nematics and cholesterics

1.3.1. Nematics proper

A schematic representation of the order in a 'nematic'† phase is shown in Fig. 1.2. The main features are as follows.

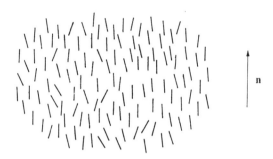

FIG 1.2. The arrangement of molecules in the *nematic* mesophase.

(1) The centres of gravity of the molecules have *no long-range order*, and consequently there is no Bragg peak in the X-ray diffraction pattern. The correlations in position between the centres of gravity of neighbouring molecules are similar to those existing in a conventional liquid. In fact, nematics do flow like liquids. For a typical nematic such as PAA the viscosities are of order 0·1 Poise.‡

(2) There is some order however in the *direction* of the molecules; they tend to be parallel to some common axis, labelled by a unit vector (or 'director') **n**. This is reflected in all macroscopic tensor properties: for instance, optically, a nematic is a uniaxial medium with the optical axis along **n**. (The difference between refractive indices measured with polarizations parallel or normal to **n** is quite large: typically 0·2 for PAA.) In all known cases, there appears to be complete rotational symmetry around the axis **n**.

† The word 'nematic' was invented by G. Friedel. It comes from the Greek νημα = thread, and refers to certain thread-like defects which are commonly observed in these materials. The physical nature of these defects ('disclination lines') will be discussed in Chapter 4.

‡ For comparison, the viscosity of water at room temperature is ∼10^{-2} Poise.

(3) The direction of **n** in arbitrary in space; in practice it is imposed by minor forces (such as the guiding effect of the walls of the container). This is a situation of broken rotational symmetry,' reminiscent of a Heisenberg ferromagnet [7], where all spins tend to be parallel but where the energy is independent of the direction of the total moment **M**.

(4) The states of director **n** and −**n** are indistinguishable. For instance, if the individual molecules carry a permanent electric dipole, as in Fig. 1.3, there are just as many dipoles 'up' than there are dipoles 'down'' and the system is not ferroelectric.

F<small>IG</small>. 1.3. In a nematic single crystal, if the molecules carry a dipole (represented by an arrow), there are as many dipoles 'up' than there are dipoles 'down.'

(5) Nematic phases occur only with materials which do not distinguish between right and left; either each constituent molecule must be identical to its mirror image (achiral) or, if it is not, the system must be a 'racemic' (1:1) mixture of the right- and left-handed species (we shall come back to this point in section 1.2.2 below). From a crystallographic point of view, the properties (2), (4), and (5) may be summarized by the symbol $D_{\infty h}$ in the Schonflies notation.

The dual aspects of a nematic phase (liquid-like but uniaxial) are exhibited most spectacularly in the *nuclear magnetic resonance spectrum*; the uniaxial symmetry causes certain line splittings (which are absent in the conventional isotropic liquid phase). On the other hand, the lines are relatively narrow; this implies rapid molecular motions and is a natural consequence of the fluidity. Some applications of these n.m.r. measurements will be discussed in Chapters 2 and 5.

1.3.2. The cholesterics: a distorted form of the nematic phase

If we dissolve in a nematic liquid a molecule which is *chiral* (i.e. different from its mirror image), we find that the structure undergoes a helical distortion. The same distortion is also found with pure cholesterol esters (which are also chiral). For this reason, the helical phase is called *cholesteric*.

1.3.2.1. The helical structure.

Locally, a cholesteric is very similar to a nematic material. Again the centres of gravity have no long-range

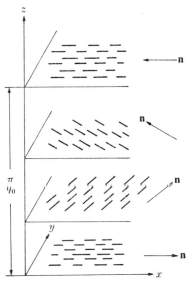

Fɪɢ. 1.4. The arrangement of molecules in the *cholesteric* mesophase. (The successive planes have been drawn for convenience, but do not have any specific physical meaning.)

order, and the molecular orientation shows a preferred axis labelled by a director $\mathbf{n}$. However $\mathbf{n}$ is not constant in space. The preferred conformation, shown in Fig. 1.4, is helical. If we call the z-axis the helical axis, we have the following structure for $\mathbf{n}$:

$$\left.\begin{array}{l} n_x = \cos(q_0 z + \phi) \\ n_y = \sin(q_0 z + \phi) \\ n_z = 0. \end{array}\right\} \qquad (1.1)$$

Both the helical axis (z) and the value of ϕ are arbitrary; we see here another type of broken symmetry. The structure is periodic along z, and (since the states $\mathbf{n}$ and $-\mathbf{n}$ are again equivalent) the spatial period L is equal to one half of the pitch:

$$L = \frac{\pi}{|q_0|} \qquad (1.2)$$

Typical value of L are in the 3000 Å range, i.e. much larger than the molecular dimensions. Since L is comparable to an optical wavelength, the periodicity results in Bragg scattering of light beams. We shall discuss these optical effects in Chapter 6.

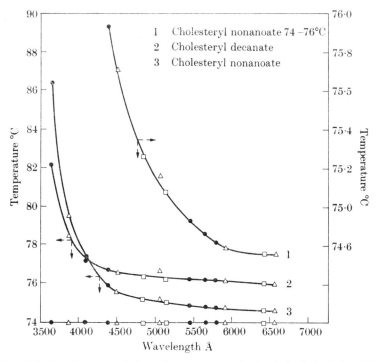

FIG. 1.5. Relation between pitch and temperature in typical cholesterics. What is plotted horizontally is the optical wavelength for Bragg reflection on the helical structure, which is equal to the pitch multiplied by the refractive index $\bar{n} \sim 1·5$ (Fergason J., Goldberg N., Nadalin R., *Molecular Crystals* 1, 315, 1966).

Both the magnitude and sign of q_0 are meaningful. The sign distinguishes between right- and left-handed helices; a given sample at a given temperature always produces helices of the same sign. If we change the temperature T, q_0 changes (Fig. 1.5). In some particular cases $q_0(T)$ may even change sign at a particular temperature T^*. This case is interesting: (1) For $T = T^*$, the material is found to behave like a conventional nematic. (2) When we cross the temperature T^*, we find that the physical properties such as specific heats, etc. remain quite smooth. The properties (1) and (2) show that the local molecular arrangement is indeed very similar in the nematic and in the cholesteric state, as was first noted by G. Friedel [6].

In the published literature, the cholesteric helix is sometimes visualized as in Fig. 1.6, i.e. as a stacking of flat strips in the xy-planes. In this model the local optical symmetry is not uniaxial but biaxial (with axes ξ, η, and z). However, this would imply very strong differences on the molecular scale between the nematic and the cholesteric state, and is in contradiction with the preceding paragraph; thus Fig. 1.6 is wrong.†

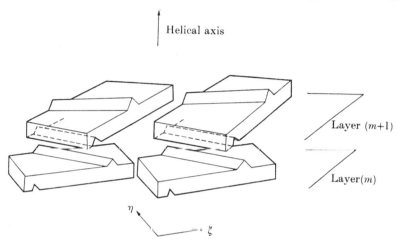

F IG. 1.6. An incorrect model for the cholesteric mesophase. The model assumes nearly flat molecules which pile up on each other and build up a helical structure. Central to the model is the assumption that the molecules form well defined *layers* normal to the helical axis; in reality, the planes of the molecules are completely free to rotate around the local optical axis (the ζ axis of the figure).

1.3.2.2. Cholesterics occur only with non-racemic systems. Let us consider a general twisted structure $n_x = \cos\theta(z)$ $n_y = \sin\theta(z)$ and find the form of the free energy F (per unit volume) as a function of the twist

$$q = \frac{\partial\theta}{\partial z}$$

(1) In materials which do not distinguish between the right and the left the plot of $F(q)$ must be symmetrical $F(-q) = F(q)$. Then there are two possibilities (Fig. 1.7a).

The minimum value of F occurs at $q = 0$; this corresponds to nematics.

† More precisely, the difference in refractive index between the ξ and η directions is at most of order $(q_0 a)^2$, where a is a molecular dimension, i.e. of order 10^{-4}.

Alternatively, F has two symmetrical minima at $q = \pm q_1$. This would be the analogue of the helimagnetic structures observed in certain rare-earth metals [8]. However, we know from this case that such lateral minima can occur only if the interactions extend up to *second neighbours*. This is not likely to occur with our molecules where the interactions (contact repulsion and van der Waals) are short range. Indeed (at the present stage) no helices of this type have been found in liquid crystals.

(2) If the constituent molecules differ from their mirror image, the plot of $F(q)$ is not symmetrical (Fig. 1.7b): the minimum cannot fall

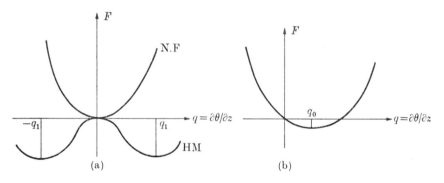

Fɪɢ. 1.7. Variation of the free energy with twist for various physical systems. (a) systems which do not distinguish right and left. The curve labelled N, F applies for a nematic or for a ferromagnet (minimum of energy at O twist). The curve labelled HM applies for a helimagnet. (b) Systems distinguishing right from left such as actual cholesterics.

at $q = 0$, and the optimum twist q_0 is non-zero.† This corresponds to the actual situation in cholesterics.

Thus, nematics and cholesterics appear as two subclasses of the same family, with the correspondence rules

Racemic or achiral system $\rightarrow$ nematic N
System different from its mirror image $\rightarrow$ cholesteric N*

Let us conclude with a remark concerning orders of magnitude; experimentally, the natural twist q_0 is always small on a molecular scale ($q_0 a \sim 10^{-2}$ or 10^{-3}, where a is the length of the molecules). This means that the overall steric difference between the constituent molecule and

---

† Except at 'accidental points' such as $T = T^*$ in our earlier discussion.

its mirror image is rather small. For instance if one looks at a three-dimensional model of a cholesterol ester (Fig. 1.8; see facing page 132), one finds a shape which is not very far from a simple rod, and thus not very different from its mirror image.

1.4. Smectics

Smectic (from the Greek $\sigma\mu\eta\gamma\mu\alpha$ = soap) is the name coined by G. Friedel for certain mesophases with mechanical properties reminiscent of soaps. From a structural point of view, all smectics are *layered structures*, with a well-defined interlayer spacing, which can be measured by X-ray diffraction.† Smectics are thus more ordered than nematics. For a given material, the smectic phases always occur at temperatures below the nematic domain.

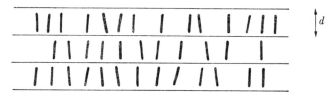

Fig. 1.9. The arrangement of molecules in a smectic A.

G. Friedel recognised only one type of smectics—the type which is now called smectic A. However, starting with some very early work by Vorlander, it became progressively clear that there are many different types of smectics, giving rise to different macroscopic textures, readily recognized by optical observation. This has led to a very useful classification, which is due to the East German school at Halle [9]. The three main types in this classification are defined by the letters A, B, and C.

1.4.1. Smectics A

A picture of the molecular arrangement for smectic A is shown in Fig. 1.9. The major characteristics are as follows.

(1) A layer structure (with layer thicknesses close to the full length of the constituent molecules).

(2) Inside each layer, the centres of gravity show no long range order; each layer is a two-dimensional liquid.

The properties (1) and (2) together define a remarkable type of one-dimensional ordering, which is in fact quite singular (see Chapter 7).

† The first X-ray evidence for the layers was obtained by E. Friedel (the son of G. Friedel), *C.r. hebd. Séanc. Acad. Sci.*, *Paris*, **180**, 269 (1925).

(3) The system is *optically uniaxial*, the optical axis being the normal Oz to the plane of the layers. It is consistent with all known data to assume that there is complete rotational symmetry around Oz.

(4) The directions z and $-z$ are equivalent: at least in all clear-cut cases which are known at present.

Properties (3) and (4) lead to a symmetry (D_∞) in the Schonflies notation. Note the difference between nematic $(D_{\infty h})$ and smectic A (D_∞). We have seen earlier that, if we try to set up a nematic phase with a material which differs from its mirror image, it will in fact distort into a cholesteric. No similar distortion is found for smectics A. As we shall see in Chapter 7 (page 299) the requirement of constant interlayer thickness imposes the condition curl $\mathbf{n} = 0$ for all macroscopic deformations of smectics. The helical arrangement of equation (11) has curl $\mathbf{n} = -q_0\mathbf{n} \neq 0$ and is thus forbidden. Many cholesterol esters are in fact smectics A below their cholesteric domain.

1.4.2. Smectics C

The structure of a smectic C is defined as follows:

(1) Each layer is still a two-dimensional liquid.

(2) The material is optically biaxial [10].

The most natural (although not unique) interpretation of these features amounts to assuming that, in a smectic C, the long molecular axis is tilted with respect to the normal z of the layers (Fig. 1.10). This interpretation is substantiated by a number of X-ray experiments, which give a layer thickness $d = l \cos \omega$, where l is the length of the molecule, and ω the tilt angle. Note that, if the molecules are tilted in the (x, z) phase, the principal axes of the dielectric

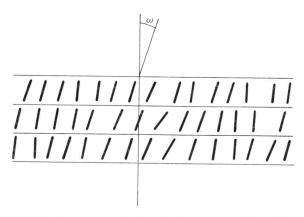

FIG. 1.10. The tilted arrangement of molecules in a smectic C (tilt angle ω).

tensor are two orthogonal directions in the xz-plane plus the y direction.

(3) The simple smectic C structure described in points (1) and (2) above is obtained only when the constituent molecules are optically inactive (or with a racemic mixture). If we add optically-active molecules to a smectic C, the structure distorts; the direction of tilt precesses around the z-axis and a helical configuration C* is obtained (see Fig. 1.11) [11].

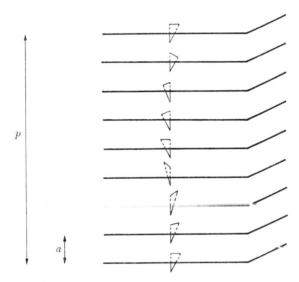

FIG. 1.11. Twist induced in a smectic C by the presence of a chiral agent. The pitch p is much larger than the interlayer distance (typically $p/a \sim 10^3$). Note that p need not be an exact multiple of a.

Again assuming that no ferroelectricity is present, we see that the symmetry elements of a smectic C are a two-fold axis (y) and a plane normal to it (xz), corresponding to the point group C_{2h}.

1.4.3. Smectics B

In both the A and the C type of smectics, each layer behaves as a two-dimensional liquid. In the B type, however, the layers appear to have the periodicity and rigidity of a two-dimensional solid. X-ray reflections, corresponding to an order *inside* each layer, are observed. The layers are not very flexible; under the microscope the texture of the B phase (the so-called 'mosaic' texture) shows domains inside which the layers are quite flat (see Fig. 7.5). This is to be contrasted with the A and

C smectics, where most observed textures involve a strong curvature of the layers.

Thus, the B phase appears as the most ordered of the three major smectic phases A, B, C. Indeed, if one material is able to display all three phases, the sequence in increasing temperatures is always

$$S \rightleftharpoons B \rightleftharpoons C \rightleftharpoons A,$$
$$\xrightarrow{\quad\quad}$$
$$T \text{ increasing}$$

where S stands for solid. A typical material showing all the phases listed above is terephthal-bis (-p-butylaniline) (TBBA), with formula

giving the following set of transition temperatures (in °C)

$$S \underset{113}{\rightleftharpoons} B \underset{144}{\rightleftharpoons} C \underset{172}{\rightleftharpoons} A \underset{200}{\rightleftharpoons} N \underset{236}{\rightleftharpoons} I.$$

The difference, at the microscopic level, between a smectic B phase and a solid is not yet clearly understood. Various models have been proposed and will be discussed in Chapter 7. As regards symmetry, note that there are at least two subgroups of the B type, which might be called B_A (with molecules normal to the layers) and B_C (with tilted molecules). Examples of both subgroups are found in nature.

1.5. Other mesomorphic phases

We have restricted our attention to the major types of thermotropic mesophases. However, it is clear that symmetry consideration alone would allow for many more phases. A recent discussion of more general types, based on group-theoretical arguments, is given in references [12] and [13]. From a more empirical point of view, it is important to quote at least the following families.

1.5.1. 'Exotic' smectics

Apart from the simple A, B, and C types described above, the thermotropic smectics may include more complex types, which have been classified by the Halle school as D, E, F, G. A further type H had been introduced by de Vries [14] on the basis of X-ray data, but has been found to be miscible with smectics B [15].

The E phase has some amount of positional order inside each layer [16] and may be a variant of the B phase.

The F phase shows textures reminiscent of the smectics C, and the G phase shows 'mosaics' quite similar to a smectic B. The reason which induced Demus et al. [17] to classify e.g. G separately from B is essentially based on a systematic lack of miscibility. To appreciate this point, let us recall the operational rule used for classification.

When two materials X and Y give the same texture, and are miscible in all proportions maintaining this texture, they are classified in the same group. When they do not mix continuously, no conclusion can be drawn, in principle. However, if a subgroup Y_1, Y_2 ... Y_P of materials is found, all of them showing the B texture, all of them intermiscible, but none of them miscible with standard B compounds, it is tempting to give them a new label G. However, the distinction which is defined in this way may reflect a difference in molecular size (or shape) between the G group and the B group, which is enough to prohibit miscibility, rather than a fundamental difference in symmetry between the B and G phases with the same texture. In fact, these size or shape effects are expected to be particularly important in the strongly-ordered phases, just as they are in ordinary solids (this point has been stressed in particular by Schott†). Thus, it is not yet quite sure that the F and G groups must be considered as true novel phases.

The D phase is quite different: it has an over-all cubic symmetry. Some X-ray data has been obtained for poly-domain samples [16]. One possible model for the D phase is the following: start by bending the layers of a planar smectic obtaining, for instance, concentric cylinders or concentric spheres. Using the resulting rods or balls as building blocks, set up a cubic packing. Arrangements of this sort have indeed been observed in the cubic phases of soap–water systems [5]. The D phase might be their counterpart for thermotropic systems.

1.5.2. Long-range order in a system of long rods

In solutions of large, rod-like polymers [3] [4], and also for certain soap phases [5], we find a set of X-ray reflections which can be indexed in terms of a hexagonal packing of rods. This corresponds to two-dimensional order (Fig. 1.12). F. C. Frank has proposed the name *canonic* (from the greek κανων = rod) for these phases. Further

---

† Verbal remark at the meeting on the physics of liquid crystals, Pont à Mousson (1971).

understanding of these systems is limited by the difficulty of producing single crystals.

A useful short list of mesomorphic transitions, giving the nature of the phases involved, the transition temperatures and the latent heats, is tabulated in the review by A. Saupe, *Angewandte Chemie* (English ed.) **7**, 97–112 (1968). A compilation of all mesogens as discovered before 1960, was established by W. Kast: *Landolt Bornstein* (6th ed.), vol. II, part 2a, p. 266 (1960).

(a)   (b)

FIG. 1.12. Two examples of hexagonal rod systems: (a) lipid in water; (b) water in lipid.

1.6. Remarkable features of liquid crystals

Liquid crystals have unusual optical properties. Nematics and smectics A are uniaxial. Cholesterics (because of their periodic structure) give rise to Bragg reflections at optical wavelengths. In nematics and cholesterics, these properties are carried by a fluid, flexible substrate; thus they are extremely sensitive to *weak external perturbations*.

The pitch of a cholesteric and hence the wave length of the Bragg-reflected light depends on the temperature T. Thus the colour of the material can change drastically in a temperature interval of a few degrees. This leads to a number of applications: detection of hot points in microcircuits [18], localization of fractures and tumors in humans [19], conversion of infrared images etc. [20]. Reference [21] provides a useful introduction to this field.

The pitch is also sensitive to other agents such as pressure, chemical contaminants, etc. For instance, a method of display for ultraviolet images based on a photochemical reaction and the resulting change in pitch has been worked out [22].

Both nematics and cholesterics are very sensitive to external fields; the first magnetic-field effects were shown long ago by the Russian groups of Frederiks and Tsvetkov [23]. But a variety of new magnetic phenomena have been discovered quite recently. Electric field effects are more complex (because their action is influenced by impurity carriers and electrochemical phenomena), but very spectacular, and important for applications—in particular for *display systems*. Liquid-crystal films are inexpensive, work under low voltage and low power; also they can often operate in the presence of sunlight (because they modulate the *reflected* light including the sunlight itself and thus maintain a good

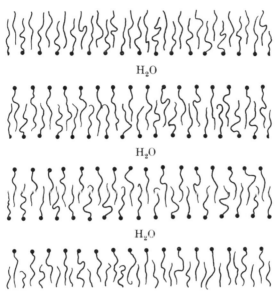

H_2O

H_2O

H_2O

FIG. 1.13. The lamellar phase of soaps ('neat soap'). The particular represented here corresponds to 'liquid-like' aliphatic chains, and also assumes that the chains are normal to the layers, on the average.

contrast). A few references on liquid-crystal display devices are listed under [24].

For all these reasons interest in nematics and cholesterics has grown rapidly since 1965. On the other hand, because of their higher viscosities the smectic mesophases have attracted less industrial interest. However, the amount of work on smectics is also increasing steadily, with various directions of applications.

(1) Physics of detergents. As already-mentioned, soaps and non-ionic detergents show a number of remarkable mesophases. The so-called 'neat' soaps, for instance, correspond to a lamellar phase with successive sheets of water and lipid (Fig. 1.13).

(2) Membrane biophysics. Biological membranes are thin (80 Å) sheets made of lipids and proteins. They play a crucial role in many living processes, but very little is known about their structure. Most physical techniques (e.g. n.m.r.) cannot be used on a single membrane, because the amount of matter available is too small. But it is possible to set up model systems with lipids and water (or even with lipid + protein + water [25]) which have a lamellar structure: it is hoped that each individual sheet will have some analogy with a membrane. With a bulk phase of this sort, one can use sample volumes large enough to allow for accurate physical studies.†

(3) It has been discovered recently that smectics A are quite sensitive to thermal or mechanical perturbations: this may lead to some interesting applications in the future.

(4) Furthermore, the textures and defects of the smectic phase are remarkable, and only partly understood.

Thus, on the whole, liquid crystals will probably find some remarkable applications in the next decade. Also, from a more fundamental standpoint the very existence of the mesophases raises a number of fascinating questions in statistical mechanics and in hydrodynamics. We shall try and discuss some of these in the following chapters.

† In a similar way, the hexagonal phases of certain rod-like molecules (e.g. nucleic acids) may be useful to carry out experiments on oriented molecules, while still retaining enough water to preserve the biological functions.

REFERENCES

CHAPTER 1

[1] GRAY, G. W. *Molecular structure and the properties of liquid crystals*, Academic Press, London (1962); see also *Mol. Cryst. liquid Cryst.*, **7**, 127 (1969); **21**, 161 (1973); the review by BROWN, G. H. and SHAW, G. H. *Chem. Rev.*, **57**, 1049 (1957) and the list of mesomorphs established by KAST, W. *Landolt–Bornstein*, 6th edn., Vol. II, part 2a, p. 266. Springer, Berlin (1960).

[2] For a review on physical determinations of the size of polypeptide molecules, see BENOIT, H. FREUND, L., and SPACH, G. *Poly-α-amino- acids* (G. Fasman ed.) Vol. **1**, p. 105. Dekker, New York (1967).

[3] ROBINSON, C. (a) *Discuss. Faraday Soc.*, **25**, 29 (1958); (b) *Proceedings of the Kent conference on liquid crystals* (1965) (G. Brown, G. Dienes, M. Labes eds.) p. 147. Gordon and Breach, New York (1966); SALUDJAN, P. and LUZZATI, V. *Poly-α-amino acids* (G. Fasman, ed.), Vol. 1, **p.** 157 Dekker, New York (1967).

[4] LUZZATI, V. *Prog. nucl. Acid Res.*, **1**, 347 (1963).

[5] LUZZATI, V. in *Biological membranes* (Chapman, ed.), Academic Press, New York (1968); SKOULIOS, A. *Adv. Colloid Interface Sci.*, **1**, 79 (1967).

[6] FRIEDEL, G. *Anns. Phys.* **18**, 273 (1922); see also CHISTIAKOV, I. G., *Sov. Phys. Usp.*, **9**, 551 (1967); BROWN, G. H., DOANE, J. W., and NEFF, V. D., *C.R.C. Crit. Rev. solid-state Sci.*, **1**, 303 (1970).

[7] For a discussion of Heisenberg ferromagnetism see, for instance, MATTIS, D. C. *The theory of magnetism* Harper, New York (1966).

[8] A good review of helimagnetism is found in HERPIN, A. *Théorie au magnétisme*, Presses Universitaires de France, Paris (1968).

[9] SACKMANN, H. and DEMUS, D. (a) *Mol. Cryst.* **2**, 81 (1966); (b) *Fortsch. chem. Forsch.*, **12**, 394 (1969).

[10] TAYLOR, T. R., FERGASON, J., and ARORA, S. L. *Phys. Rev. Lett.* **24**, 359 (1970); **25**, 722 (1970).

[11] HELFRICH, W., and OH, C. S. *Mol. Cryst. liquid Cryst.* **14**, 289 (1971); URBACH, W., and BILLARD, J. *C.r. hebd. Séanc. Acad. Sci., Paris.* **274**, 1287 (1972).

[12] BOCCARA, N. *Ann. Phys.* **76**, 72 (1973).

[13] GOSHEN S., MUKAMEL D., SHTRINKMAN, S. (to be published).

[14] DE VRIES, A. and FISHEL, D. L. *Mol. Cryst. liquid Cryst.* **16**, 311 (1972).

[15] BILLARD, J. and URBACH, W. *Mol. Cryst. liquid Cryst.* (to be published).

[16] DIELE S., BRAND, P., and SACKMAN, H. *Mol. Cryst. liquid Cryst.* **17**, 163 (1972).

[17] DEMUS, D., DIELE, S., KLAPPERSTRUCK, M., LINK, V., and ZASCHKE, H., *Mol. Cryst. liquid Cryst.* **15**, 161 (1971).

[18] KOPP, U. *Prakt. Metallogr.* **9**, 370 (1972).

[19] See, for instance, GAUTHERIE, M. *J. Phys. (Fr.)* **30**, (Suppl. C4) 122 (1969).

[20] ENNULAT, R. and FERGASON, J. *Mol. Cryst. liquid Cryst.* **13**, 149 (1971).

[21] FERGASON, J. *Scient. Am.* **211,** 77 (1964); *Mol. Cryst.* **1,** 309 (1966).

[22] HAAS, W., ADAMS, J., and WYSOCKI, J. *Mol. Cryst. liquid Cryst.* **7,** 371 (1969).

[23] FREDERIKS, V. K. and ZOLINA, V. *J. R.F.* (*Kharkov*) (*Phys.*) **62,** 457 (1969); TSVETKOV, V. N. and SOSNOVSKII, A. *Acta Phys.-chim. URSS*, **18,** 358 (1943).

[24] WILLIAMS, R. *J. chem. Phys.* **39,** 384 (1963); HELLMEIER, G. H., ZANONI, J., and BARTON, L. *Proc. Inst. elect. electron. Engrs.,* **56,** 1162 (1968); *Trans.* **17,** 22 (1970); VAN RAULTE J., *Proc. Inst. elect. electron. Engrs.* **56,** 2146 (1968); ORSAY GROUP on liquid crystals, *La Recherche,* **12,** 433 (1971).

[25] GULIK-KRZYWICKI, T., SCHECHTER, E., LUZZATI, V., and FAURE, M. *Biochemistry and biophysics of mithocondrial membranes* (G. F. Arzone *et al.*, eds.), p. 241. Academic Press, New York (1972).

LONG- AND SHORT-RANGE ORDER
IN NEMATICS

*'Quand tu penses, ne sens-tu pas que tu déranges secrètement quelque
chose?'*

PAUL VALÉRY

2.1. Definition of an order parameter

THE nematic phase has a lower symmetry than the high-temperature
isotropic liquid. We express this qualitatively by saying that the
nematic phase is 'more ordered.' To put this on a quantitative basis, we
need to define an order parameter which is non-zero in the nematic
phase but which vanishes, for symmetry reasons, in the isotropic phase.
In some physical systems an adequate choice of the order parameter is
obvious. For instance in a ferromagnet, the order parameter is the
magnetization $\mathbf{M}$; this is a vector with three independent components
M_α. In a nematic phase the choice is less trivial and we shall have to
proceed in successive steps.

2.1.1. Microscopic approach

2.1.1.1. Simple rods. Rigid rods are the simplest type of objects
which allow nematic behaviour. The axis of one rod will be labelled by a
unit vector $\mathbf{a}$. The rod is assumed to have complete cylindrical
symmetry about $\mathbf{a}$. The direction of the nematic axis $\mathbf{n}$ (i.e., the average
direction of alignment of the molecules) will be taken as the z-axis of the
(x, y, z) laboratory frame. We shall define $\mathbf{a}$ by its polar angles θ and ϕ,
where

$$a_x = \sin \theta \cos \phi$$

$$a_y = \sin \theta \sin \phi,$$

and

$$a_z = \cos \theta.$$

The state of alignment of the rods can be described by a distribution
function $f(\theta, \phi) \, d\Omega$ (giving the probability of finding rods in a small
solid angle $d\Omega = \sin \theta \, d\theta \, d\phi$ around the direction (θ, ϕ)).

From the discussion in Chapter 1, we know that, in conventional
nematics,

(1) $f(\theta, \phi)$ is independent of ϕ (the phase has complete cylindrical symmetry about **n**);

(2) $f(\theta) \equiv f(\pi - \theta)$ (the directions **n** and $-$**n** are equivalent). (2.1) The general appearance of $f(\theta)$ is shown in Fig. 2.1.

Now we wish to characterize the alignment not through the full function $f(\theta)$, but preferably by one related numerical parameter. The first idea would be to use the average

$$\langle \cos \theta \rangle = \langle \mathbf{a} \cdot \mathbf{n} \rangle = \int f(\theta) \cos \theta \, d\Omega,$$

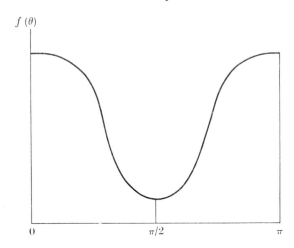

FIG. 2.1. The distribution function $f(\theta)$ for a system of rods in a nematic phase. $f(\theta)$ is large around $\theta = 0$ or π (i.e. for molecules parallel to the optical axis) and is small for $\theta \simeq \pi/2$.

but this vanishes identically as a result of the property (2), and there is no average dipole. Thus we must resort to higher multipoles. The first multipole giving a non-trivial answer is the *quadrupole;* this is defined as

$$S = \tfrac{1}{2}\langle (3 \cos^2\theta - 1) \rangle = \int f(\theta)\tfrac{1}{2}(3 \cos^2\theta - 1)/d\Omega \qquad (2.2)$$

For instance, if $f(\theta)$ is strongly peaked around $\theta = 0$ and $\theta = \pi$ (parallel alignment), $\cos \theta = \pm 1$ and $S = 1$. If, on the other hand, $f(\theta)$ peaked around $\theta = \pi/2$ (perpendicular alignment), we would have $S = -\tfrac{1}{2}$.† Finally, if the orientation was entirely random ($f(\theta)$ independent of θ) we would have $\langle \cos^2\theta \rangle = \tfrac{1}{3}$ and $S = 0$. Thus S is a measure of the alignment.

† However, on physical grounds, it is hard to invent a system of rods which would prefer to have the perpendicular alignment.

2.1.1.2. *Relation to n.m.r. spectra.* The quantity S can be extracted from n.m.r. data. To understand this, consider the following (simplified) example: the rod contains two (and only two) protons of spins I_1 and I_2 ($I_1 = I_2 = \frac{1}{2}$). An external field H is applied. As we shall see later, nematics usually tend to line up in the direction of H. Thus, in our notation, H is parallel to the z-axis. Each spin is coupled to the external field H and the dipolar field created by its partner. The spin Hamiltonian $\mathscr{H}$ describing this situation has the structure

$$\mathscr{H} = -\hbar\gamma H(I_{1z}+I_{2z})- \frac{(\hbar\gamma)^2}{d^3} \{3(\mathbf{I}_1 . \mathbf{a})(\mathbf{I}_2 . \mathbf{a})-\mathbf{I}_1 . \mathbf{I}_2\} \qquad (2.3)$$

$$= \mathscr{H}_{\text{Zeeman}}+\mathscr{H}_{\text{dipolar}} = \mathscr{H}_z+\mathscr{H}_D,$$

where $\hbar\gamma\mathbf{I}_1$, $\hbar\gamma\mathbf{I}_2$ are the magnetic moments associated with the spins $\mathbf{I}_1$ $\mathbf{I}_2$, $\mathbf{a}$ is the unit vector along the rod, and d is the distance between the protons.

The dipolar fields are conveniently measured in terms of

$$H_L = \hbar\gamma/d^3.$$

H_L is of order 1 G and the corresponding precession frequency γH_L is of order 10^4 s^{-1}. In our liquids, the direction of the long molecular axis (the vector $\mathbf{a}$) changes with time on a much faster scale (typically 10^{-9} s). In this limit of rapid motion, $\mathscr{H}_{\text{dipolar}}$ may be replaced by its average over the orientations of $\mathbf{a}$, which we shall call $\mathscr{H}_D$. To derive $\mathscr{H}_D$ we use the averages:

$$\langle a_z^2 \rangle = \tfrac{1}{3}+\tfrac{2}{3}S,$$

$$\langle a_x^2 \rangle = \langle a_y^2 \rangle = \tfrac{1}{3}-\tfrac{1}{3}S,$$

and

$$\langle a_x a_z \rangle = \langle a_y a_z \rangle = \langle a_x a_y \rangle = 0,$$

This leads to the following average Hamiltonian

$$\mathscr{H}_D = \Delta(-2I_{1z}I_{2z}+I_{1x}I_{2x}+I_{1y}I_{2y}), \qquad (2.4)$$

where, for brevity, we have put

$$\Delta = \hbar\gamma SH_L.$$

Since protons are of spin $\frac{1}{2}$ we have:

$$I_{1z}^2 = I_{1x}^2 = I_{1y}^2 = \tfrac{1}{4}, \text{ etc.}$$

This allows us to write the Hamiltonian (2.4) very simply in terms of the total spin $\mathbf{I} = \mathbf{I}_1+\mathbf{I}_2$. Using the equality

$$I_z^2 = (I_{1z}+I_{2z})^2 = I_{1z}^2+I_{2z}^2+2I_{1z}I_{2z} = \tfrac{1}{4}+\tfrac{1}{4}+2I_{1z}I_{2z}$$

and other similar ones, we obtain

$$\mathscr{H}_{\mathrm{D}} = \tfrac{1}{2}\Delta(-2I_z^2 + I_x^2 + I_y^2) + \text{constant}.$$

The constant terms do not contribute to the energy *intervals* with which we are concerned, and may be omitted. The spin Hamiltonian (2.3) is thus reduced to

$$\overline{\mathscr{H}} = -\hbar\gamma H I_z + \tfrac{1}{2}\Delta(-3I_z^2 + \mathbf{I}^2). \tag{2.5}$$

$\overline{\mathscr{H}}$ may be shown to commute with the operators $\mathbf{I}^2 = I_x^2 + I_y^2 + I_z^2$ and I_z. Thus the levels of $\overline{\mathscr{H}}$ may be indexed by two quantum numbers: a number I such that $\mathbf{I}^2 = I(I+1)$ and a number I_z running from $-I$ to $+I$ by unit steps.

For two spins $\tfrac{1}{2}$, the total spin I may take only two values $I = 0$ (singlet) or $I = 1$ (triplet). The singlet state may be shown to be unobservable in the resonance experiment. Thus we are left with the triplet state, with three levels corresponding to $I_z = -1, 0, 1$. The term involving $\mathbf{I}^2$ in eqn (2.5) is the same for these three levels and may again be omitted in discussion of the intervals. Thus we have the further reduction

$$\overline{\mathscr{H}} \to -\hbar\gamma H I_z - \tfrac{3}{2}\Delta I_z^2.$$

The corresponding levels are represented below:

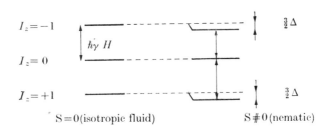

Also shown on this plot are the two allowed transitions. (A third transition, from $I_z = +1$ to $I_z = -1$, is forbidden by a general spectroscopic rule.) They correspond to the frequencies

$$\omega = \gamma(H \pm \tfrac{3}{2}H_{\mathrm{L}}S). \tag{2.6}$$

Thus, when going from the isotropic ($S = 0$) to the nematic phase ($S \neq 0$) the resonance line splits by an amount $3\gamma H_{\mathrm{L}}S$.

If the distance d between the protons is fixed (rigid molecule) and is known, H_{L} is known, and S can be extracted from the experimental splitting. Experiments of this type were performed very early; the first data on p-azoxyanisole were obtained in about 1953 [2]. Here we have two types of protons. (1) The 6 methyl-group protons at both ends of the molecule have rapid rotations and do not give rise to any interesting splitting. (2) The 8 protons linked to the aromatic ring fall into four pairs; each pair is nearly uncoupled to the others. The following diagram of PAA shows the pairs.

In a first approximation we may neglect the angle between **a** and the long axis **u** then we recover the simple model described above. The distance d is 2·45 Å and $H_L \sim 2\cdot9$ G. Typical values of S are in the range 0·4 (at the highest temperatures) to 0·6 (at the lowest temperatures). Detailed measurements of this sort have been performed by Saupe and his coworkers [3]. The date are accurate because each line is narrow: the 'motional narrowing' familiar in liquid phases [1] is also found in nematics. For general reviews on this subject see references [4], [5], [6].

2.1.1.3. Rigid molecules of arbitrary shape [7] [8].   Let **a, b, c** be three orthogonal unit vectors linked to the molecule; the degree of alignment may be defined by a natural generalization of eqn (2.2) through the quantities

$$S_{ij}^{\alpha\beta} = \tfrac{1}{2}\langle 3i_\alpha j_\beta - \delta_{\alpha\beta}\delta_{ij}\rangle, \tag{2.7}$$

where $\alpha, \beta = x, y, z$ are indices referring to the laboratory frame, while $i, j = a, b, c$, and $\delta_{\alpha\beta}$ and δ_{ij} are Kronecker symbols. The brackets $\langle\ \rangle$ represent a thermal average. $S_{ij}^{\alpha\beta}$ is symmetric in ij and in $\alpha\beta$; it is also a traceless tensor in respect of either pair

$$S_{ij}^{\alpha\alpha} = 0 \quad \text{and} \quad S_{i}^{\alpha\beta} = 0. \tag{2.8}$$

(We use the usual notation where summations are implied by repeated indices $S^{\alpha\alpha} = S^{11} + S^{22} + S^{33}$). We have seen, in Chapter 1, that the usual nematic structure has complete rotational symmetry about the optical axis; let us again take this as the z-axis. This property implies the following equalities:

$$S_{ij}^{xx} = S_{ij}^{yy} \quad \text{and} \quad S_{ij}^{xy} = 0. \tag{2.9}$$

Furthermore, the xy-plane is a plane of reflection for the structure; from this we derive

$$S_{ji}^{zx} = S_{ij}^{zy} = 0. \tag{2.10}$$

Thus, in the usual nematic structure, the only non-zero components of $S_{ij}^{\alpha\beta}$ are

$$S_{ij}^{zz} = -2S_{ij}^{xx} = -2S_{ij}^{yy} = S_{ij}, \tag{2.11}$$

and the state of alignment of a rigid molecule is described by a (3×3) matrix S_{ij}, which is symmetric and of zero trace.

The average dipolar interaction $\overline{\mathscr{H}}_{12}$ between two arbitrary spins I_1 and I_2, carried by the molecule, can be written entirely in terms of the S_{ij}s:

$$\left. \begin{aligned} \mathscr{H}_{12} &= -\frac{\hbar^2 \gamma_1 \gamma_2}{d^3} S_{uu}(2I_{1z}I_{2z} - I_{1x}I_{2x} - I_{1z}I_{2z}), \\ S_{uu} &= \tfrac{1}{2}\langle 3u_z^2 - 1 \rangle = \sum_{ij} S_{ij}u_iu_j, \end{aligned} \right\} \tag{2.12}$$

where $\mathbf{u}$ is the unit vector of the direction linking both spins and the u_is are its components in the molecular frame. Conversely, from detailed high-resolution n.m.r. spectra of a rigid molecule in the nematic phase, we can, in principle, reconstruct the matrix S_{ij}. In practice, however, this is often somewhat complicated, because the nematic molecules carry a number of nuclear spins and the spectra are complex.

To avoid the complexities inherent in a heavy molecule, it is sometimes more convenient to study a well-chosen *solute* in a nematic phase; if the solute is non-spherical in shape, it will be aligned by the neighbouring nematic molecules, and this alignment may also be characterized by a matrix $\tilde{S}_{ij}$. There is no direct way to relate $\tilde{S}_{ij}$ (for the solute) to S_{ij} (for the nematic solvent); nevertheless $\tilde{S}_{ij}$ gives some indication of the amount or order. A good example of such a solute is 1,3,5-trichloro-benzene [9]:

This is particularly simple; there are only three (equivalent) protons in the molecule. If we define the molecular axes as in the above diagram, the $\tilde{S}$ matrix is diagonal. Furthermore, $\tilde{S}_{11} = \tilde{S}_{22}$ ($= -\frac{1}{2}\tilde{S}_{33}$) because of the trigonal symmetry around the axis [3]. If $\mathbf{u}$ is any one of the three proton–proton vectors we also have $S_{uu} = \tilde{S}_{11} = \tilde{S}_{22}$. Let us call this single parameter $\tilde{S}$. In terms of $\tilde{S}$, the average spin Hamiltonian of the three protons, including dipolar coupling, is

$$\overline{\mathscr{H}} = -\gamma H(I_{1z}+I_{2z}+I_{3z}) +$$
$$+\Delta\{\mathbf{I}_1 . \mathbf{I}_2+\mathbf{I}_2 . \mathbf{I}_3+\mathbf{I}_3 . \mathbf{I}_1 - 3(I_{1z}I_{2z}+I_{2z}I_{3z}+I_{3z}I_{1z})\},$$
$$\Delta = \gamma H_{\mathrm{L}}\tilde{S}.$$

Introducing the total spin $\mathbf{I} = \mathbf{I}_1+\mathbf{I}_2+\mathbf{I}_3$, we have the very simple form

$$\overline{\mathscr{H}} = -\gamma HI_z+\tfrac{1}{2}\Delta(I^2-3I_z^2)+\text{constant}.$$

A complete analysis of the levels and of the allowed transitions shows that there are three resonance frequencies in the nematic phase, and the splitting directly measures $\tilde{S}$. Experimentally, in a variety of nematic solvents $\tilde{S}$ ranges from $+0\cdot08$ to $+0\cdot10$ (depending on the temperature). A recent review on n.m.r. studies of solutes has been written by Diehl and Khetrapal [6].

2.1.1.4. Other determinations of the order by magnetic resonance. Apart from these dipolar effects in n.m.r., there are also other resonance methods which permit a determination of S_{ij}: studies of n.m.r. quadrupole splittings using nuclei with spin $I \geqslant 1$ such as ^{14}N [10] or deuterium [11]; and studies of the anisotropy of the Zeeman and hyperfine couplings in the electron spin resonance (e.s.r.) of dissolved free-radicals [12]. The latter method has one advantage: the high signal intensities make it possible to apply the method to very small samples. However, it also has some drawbacks. First, it is always concerned with the alignment of a *solute* and not of the nematic matrix itself. Secondly, the rapid-motion limit which applies for n.m.r. (reducing $\mathscr{H}$ to $\overline{\mathscr{H}}$) is not always obtained here; the characteristic electron frequencies can be comparable to $1/\tau_{\mathrm{rotation}}$, for which we obtain a broad spectrum, and the determination of S_{ij} becomes inaccurate.

2.1.2. Macroscopic approach

2.1.2.1. Flexible molecules. It often happens that the molecules of interest have flexible parts. For instance, in the typical pattern for nematic molecules,

$$R-\langle\bigcirc\rangle-A{=}B-\langle\bigcirc\rangle-R'$$

R and R' may be long alkyl chains with non-negligible flexibility. Such chains also occur in cholesterol esters:

$$\text{Aliphatic chain}\qquad\qquad \text{Aliphatic chain}$$
$$\text{Cholesterol skeleton}$$
$$\text{Ester function}$$

Finally, in lyophilic liquid crystals such as hydrated soaps, (at least for the high-temperature phases) there is essentially no rigid part in the constituent molecules. In all such cases a microscopic description of the alignment, adequate to interpret high-resolution n.m.r. spectra, cannot be reduced to one (3×3) matrix S_{ij}; we need more parameters.

Consider, for instance, an aliphatic chain:

$$—CH_2—CH_2—CH_2—$$
$$(n-1)\ \ (n)\ \ (n+1)$$

belonging to the nematogen. Assume for simplicity that the only important spin–spin couplings are between the two protons linked to the same carbon. Then we may hope to extract from the n.m.r. data a parameter $S^{(n)}$ for the nth CH_2 group ($S^{(n)}$ specifying the alignment of the H–H direction with respect to the nematic axis as in eqn (2.2). For the examples described above, the chain is rather short and tied at one end of a rigid molecular portion. Then the successive $S^{(1)}$, $S^{(2)}$, ... for CH_2 groups of increasing distance from the attachment point will be different. This provides, in principle, a very interesting means of probing the rôle of flexible chains in nematic conformation and stability.†

However, for *defining* an order parameter, a collection of many quantities such as $S^{(1)}$ ... $S^{(n)}$ ... in the above example is an *embarras de richesse.*' This suggests that we return to a more macroscopic definition of the order, which will be applicable independently of any assumption on molecular rigidity.

2.1.2.2. Tensor order parameter.   A typical difference between the high-temperature isotropic liquid and the nematic mesophase is found

† Deuterated chains give better resolutions, and are beginning to be studied (J. Charvolin, B. Deloche, to be published).

in the measurement of all macroscopic tensor properties. For instance, the relation between the magnetic moment $\mathbf{M}$ (due to the molecular diamagnetism) and the field $\mathbf{H}$ has the form

$$M_\alpha = \chi_{\alpha\beta} H_\beta, \tag{2.13}$$

where $\alpha, \beta = x, y, z$. When the field $\mathbf{H}$ is static, the tensor $\chi_{\alpha\beta}$ is symmetric ($\chi_{\beta\alpha} = \chi_{\alpha\beta}$). In the isotropic liquid, we have

$$\chi_{\alpha\beta} = \chi \delta_{\alpha\beta},$$

while in the uniaxial nematic phase (always choosing the z-axis parallel to the nematic axis).

$$\chi_{\alpha\beta} = \begin{vmatrix} \chi_\perp & 0 & 0 \\ 0 & \chi_\perp & 0 \\ 0 & 0 & \chi_\parallel \end{vmatrix} \tag{2.14}$$

Recent data on $\chi_\parallel$ and the average $\chi = \frac{1}{3}\chi_\parallel + \frac{2}{3}\chi_\perp$ have been collected by the Bordeaux group [13].

To define an order parameter which vanishes in the isotropic phase we extract the anisotropic part $Q_{\alpha\beta}$ of the magnetic susceptibility

$$Q_{\alpha\beta} - G\left(\chi_{\alpha\beta} - \frac{1}{3}\delta_{\alpha\beta}\sum_\gamma \chi_{\gamma\gamma}\right). \tag{2.15}$$

We call $Q_{\alpha\beta}$ the *tensor order parameter*. It is real, symmetric, and of zero trace. The normalization constant G may be chosen at will; it is often convenient to define G by setting $Q_{zz} = 1$ in a fully oriented system. At this stage we can make the following remarks.

(i) The choice of the magnetic response as a starting point is a pure matter of convention; we might as well have used another static response function, such as the electric polarizability or the dielectric constant $\epsilon_{\alpha\beta}$.

Another possibility would be to define an order parameter through the dynamic dielectric tensor $\epsilon_{\alpha\beta}(K)$ at some standard frequency ω such as the yellow D-line of sodium. This has the advantage of being directly related to the refractive indices, which are easily obtained. A particularly accurate determination of order versus temperature, using refractive indices measured by interferometric techniques, has been carried out recently on MBBA [14].

However, we prefer to define Q through the magnetic susceptibility χ, because the relation between χ and molecular properties is well under-stood, while the relation between $\epsilon_{\alpha\beta}$ and molecular properties is much more obscure: we come back to this point later.

(ii) Our definition of an order parameter $Q_{\alpha\beta}$ covers a wider class of liquid crystals than simple nematics. When the axes α, β are chosen properly to diagonalize the symmetric matrix Q, the most general structure is

$$Q_{\alpha\beta} = \begin{vmatrix} Q_1 & 0 & 0 \\ 0 & Q_2 & 0 \\ 0 & 0 & -(Q_1+Q_2) \end{vmatrix} \tag{2.16}$$

This would correspond to a 'biaxial nematic'—a mesomorphic phase the existence of which is still unproven at the present time.† In usual (uniaxial) nematics the diagonal form simplifies to

$$Q_{\alpha\beta} = G \begin{vmatrix} \frac{1}{3}(\chi_\perp-\chi_\parallel) & 0 & 0 \\ 0 & \frac{1}{3}(\chi_\perp-\chi_\parallel) & 0 \\ 0 & 0 & \frac{2}{3}(\chi_\parallel-\chi_\perp) \end{vmatrix} \tag{2.17}$$

as is easily seen from eqn (2.14). It is the structure of the thermodynamic free energy as a function of $Q_{\alpha\beta}$ which decides whether the optimum $Q_{\alpha\beta}$ is of the form (2.16) or (2.17), i.e., if we have a uniaxial or a biaxial nematic.

2.1.2.3.   Relation between microscopic and macroscopic approaches. When the molecules may be approximately taken as rigid, we might hope to find a simple connection between the macroscopic tensors $\chi_{\alpha\beta}$ or $\epsilon_{\alpha\beta}$, and the microscopic quantities $S_{ij}^{\alpha\beta}$ introduced in eqn (2.7). In fact, our level of knowledge is not at all the same for χ as for ϵ.

Magnetic susceptibilities: since the magnetic couplings between neighbouring molecules are very small, $\chi_{\alpha\beta}$ is, to a reasonable ap-proximation, simply a sum of individual molecular responses. Let us call A_{ij} the magnetic polarizability tensor of one molecule, referred to the molecular frame ($\mathbf{i}$, $\mathbf{j} = \mathbf{a}$, $\mathbf{b}$, $\mathbf{c}$). Then, from the transformation rule of tensors

$$\chi_{\alpha\beta} = c \sum_{ij} A_{ij} \langle i_\alpha j_\beta \rangle,$$

where $c =$ number of molecules cm^{-3}. Using the definition (2.7) the anisotropic part of χ is seen to be

$$\chi_{\alpha\beta} - \tfrac{1}{3}\delta_{\alpha\beta}\chi_{\gamma\gamma} = cA_{ij}S_{ij}^{\alpha\beta}. \tag{2.18a}$$

† Biaxial nematics have been discussed theoretically by M. J. Freiser, *Mol. Cryst.* **14,** 165 (1971) and by R. Alben, *Phys. Rev. Lett.* **30,** 778 (1973).

The temperature dependence of the coefficient A_{ij} is expected to be weak. Thus, in principle, the A_{ij}'s may be derived from a study of the diamagnetism in the crystalline phase (assuming that the molecular orientations in this phase are known). The $S_{ij}^{\alpha\beta}$ can be simplified according to the rules of eqns (2.9) and (2.10). It is interesting to compare the experimental data on $\chi_{\alpha\beta}$ with the predicted values (2.18a) using the order parameters S_{ij} derived from n.m.r. measurements. This has been carried out for PAA. by Saupe and Maier [9]. They use a simplified model with uniaxial rods, in which case eqn (2.18a) reduces to

$$\chi_\parallel - \chi_\perp = c(A_\parallel - A_\perp)S(T). \qquad (2.18b)$$

The resulting values for $S(T)$ do show a temperature dependence very similar to those which are derived from n.m.r. data. The absolute magnitudes of S (as derived from χ) are smaller (by $\sim15\%$) than the values from n.m.r.: this difference may be due to uncertainties in $A_\parallel - A_\perp$ (i.e. depending on the properties of the crystal phase).

Dielectric constants, refraction indices, etc: the theory linking this type of quantity to the order parameters S_{ij} or S is in a much less satisfactory state, since the effective electric field seen by one molecule is a superposition of the field due to external sources, plus the field due to all other dipoles. The latter contribution is large; in the standard theory of dielectric response in isotropic dense media, it is included approximately in the form of a Lorentz field [10].

But in liquid crystals there is no clear-cut way of extrapolating this procedure; even in the simplest cases the dielectric response does not depend only on the angular distribution function f_a, but involves a correlation function $g(oa \mid ra')$ for *two* molecules, as a function of their relative distance $\mathbf{r}$ and orientations ($\mathbf{a}$ and $\mathbf{a}'$). All the approximations which have been proposed in the literature to relate S and $\epsilon_\parallel - \epsilon_\perp$ involve arbitrary assumptions about the correlation function g. Thus, the values of $S(T)$ derived from refractive index data may be slightly incorrect.

2.1.2.4. More general studies

Raman effect. Another probe of the alignment, which is interesting and gives in principle more information, has been used recently [15]. It involves the Raman scattering due to a $C\equiv N$ group attached at the end of the molecule (roughly parallel to the rod axis). The scattered light originates from a modulation of the electric polarizability of the C≡N bond; the light amplitude is thus proportional to

$$a = \alpha + \beta_{ij} S_{ij},$$

where S_{ij} is the instantaneous (i.e. non-averaged) order, while α and β depends on the polarizations of the light beams. The Raman intensity is proportional to the average

$$\langle |a|^2 \rangle = \alpha^2 + 2\alpha\beta\langle S \rangle + \beta^2 \langle S^2 \rangle,$$

(where we omit all indices).

It may be studied as a function of the polarizations, thus giving a measure of $\langle S \rangle = S$, and also of $\langle S^2 \rangle$; this second quantity was not obtained in the more conventional methods described above. Unfortunately the interpretation of these results is also complicated in practice by local field or multiple scattering corrections.

Quasi-elastic scattering of X-rays, neutrons, etc. A large body of data on X-ray scattering by nematics is available [16]. Some neutron scattering measurements have also been obtained recently [17]. In general these data cannot be interpreted very quantitatively; too many unknown correlation functions are involved.

However, there is one limiting case where, in principle, the situation becomes tractable [18]; this is the case of large-angle scattering, where all interference effects between different molecules become negligible. The measured intensities are then characteristic of one single molecule, and could allow reconstruction of the complete distribution function for orientations (for instance, with simple rods, the function $f(\theta)$ as it has been defined earlier). The large-angle data have not yet been analysed in this sense.

2.2. Statistical theories of the nematic order

The statistical mechanics of liquids is difficult; the statistical mechanics of nematics is still worse. Even for the simplest physical models, no exact solution has been worked out. We shall review but briefly here the approximate descriptions which are commonly employed.

2.2.1. Mean field calculations for hard rods

2.2.1.1.   *The Onsager approach.*   As discussed in Chapter 1, we can prepare solutions of hard-rod macromolecules with well-defined length L and diameter D (e.g. tobacco mosaic virus). Onsager in 1949 [19] discussed the statistics of such systems, with the following assumptions:

(1) The only forces of importance correspond to steric repulsion; the rods cannot interpenetrate each other[†]

(2) The volume fraction $\Phi = c \cdot \frac{1}{4}\pi L D^2$ (c = concentration of rods) is much smaller than unity.

(3) The rods are very long ($L \gg D$). It will turn out in practice that the values of Φ of interest near the isotropic nematic transition are such that $\Phi L/D \sim 4$. Thus the requirements (2) and (3) are in fact linked.

To sketch Onsager's derivation, let us first start from the more familiar case of a dilute gas of hard spheres (concentration c, radius r, $cr^3 \ll 1$). Here the free energy (per sphere) has the form [20]:

$$F = F_0 + k_B T \{\log c + \tfrac{1}{2} c \beta_1 + C(c^2)\} \qquad (2.19)$$

where F_0 is an additive constant, β_1 is the *excluded volume* (i.e. the volume which is not allowed for the centre of sphere '1' when sphere '2' is fixed at the origin), and $O(c^2)$ means terms of the order of c^2. Since the centres cannot come closer than a distance $2r$

$$\beta_1 = \tfrac{4}{3}\pi (2r)^3.$$

Let us now go to our system of hard rods. Here, we must specify not only the overall concentration c, but also the angular distribution of the rods; let us call ($c f_a \, d\Omega$) the number of rods per unit volume pointing in a small solid angle $d\Omega$ around a direction labelled by the unit vector a. Note incidentally that the sum of this over all solid angles must give the total concentration c, i.e.

$$\int f_a \, d\Omega = 1. \qquad (2.20)$$

The free energy is now given by a natural extension of eqn (2.19), namely:

$$\begin{aligned} F = F_0 + k_B T \int f_a \log(4\pi f_a c) \, d\Omega \\ + \tfrac{1}{2} c \iint f_a f_{a'} \beta_1(aa') \, d\Omega \, d\Omega' + O(c^2). \end{aligned} \qquad (2.21)$$

The second term in eqn (2.21) describes the drop in entropy associated with molecular alignment (i.e. a non-constant f). The third term describes the excluded volume effects; $\beta_1(aa')$ is the volume excluded by one rod in direction a as seen by one rod in direction a'. The calculation of β_1 is simple for long rods, where end effects are ignored, and is

[†] The effect of Coulomb repulsion between charged rods could also be included if desired; as shown by Onsager, it essentially amounts to an increase in the effective diameter of the rods.

explained in Fig. 2.2. The result is:

$$\beta_1 = 2L^2D \, |\sin \gamma| \qquad (L \gg D) \qquad (2.22)$$

where γ is the angle between **a** and **a**′.

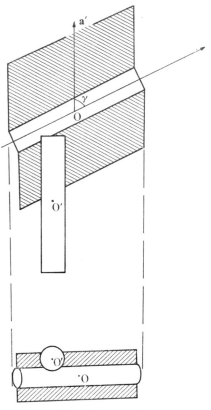

FIG. 2.2.

It must be emphasized that eqn (2.21), limited to order c, represents an approximation of the 'mean field' type: correlations between different rods are not taken into account.

We can obtain a self-consistent equation for the distribution function f_a by specifying that the free energy (2.21) is a minimum for all variations of f_a which satisfy the constraint (2.20). This amounts to writing

$$\delta F = \lambda \int \delta f_a \, d\Omega \qquad (2.23)$$

(where λ is an unknown Lagrange multiplier) and gives the self-consistent equation:

$$\log (4\pi f_a) = \lambda - 1 - c \int \beta_1(\mathbf{aa}')f_{a'} \, d\Omega'; \qquad (2.24)$$

λ is then determined by the normalization condition (2.20).

Equations (2.23) and (2.24) show that the concentration c enters in the problem only through the combination $cL^2D = \text{const} \times \Phi L/D$.

Equation (2.24) always has an 'isotropic' solution ($f_a = 1/4\pi$, independent of a) but if $\Phi L/D$ is large enough it may also have anisotropic solutions, describing a nematic phase. It is difficult to solve the non-linear integral equation (2.24) exactly. Onsager used a variational method, based on a trial function of the form

$$f_a = (\text{const}) \times \cosh(\alpha \cos\theta) \qquad (2.25)$$

where α is a variational parameter, and θ is the angle between a and the nematic axis. (The constant factor is chosen to normalize f according to eqn (2.20)). In the region of interest, α turns out to be large (~ 20) and the function f is strongly peaked around $\theta = 0$ and $\theta = \pi$. The order parameter is

$$S = \tfrac{1}{2} \int f_a \tfrac{1}{2}(3\cos^2\theta - 1)\sin\theta \, d\theta$$

$$\approx 1 - 3/\alpha \qquad (\alpha \gg 1). \qquad (2.26)$$

Minimizing the energy F (eqn 2.21) with respect to α, one obtains a function $F(c)$ showing a first-order phase-transition from isotropic ($\alpha = 0$) to nematic ($\alpha \geqslant 18\cdot6$) (we shall come back to the fundamental symmetry reasons which make the transition first-order later in this chapter). The volume fraction Φ occupied by the rods, in the nematic phase, just at the transition point, is

$$\Phi^c_{\text{nema}} = 4\cdot5 D/L \qquad (2.27)$$

At the same point, the value of Φ for the isotropic phase (in equilibrium with the nematic phase) is significantly smaller

$$\Phi^c_i = 3\cdot3 D/L$$

Note that Φ^c_n and Φ^c_i are independent of T in this model: hard rods are an 'athermal' system.

Of particular interest is the value of the order parameter S_c in the nematic phase just at the transition. This turns out to be quite high ($S_c \approx 0\cdot84$). Thus the Onsager solution leads to a rather abrupt transition between a strongly-ordered nematic, and a completely-disordered isotropic, phase.

Problem: Discuss the existence of a nematic phase for a system of *thin hard discs* (radius of the disc b much larger than its thickness e), in the Onsager approximation.

4

Solution: Derive first the excluded volume $\beta_1(\gamma)$ for two discs with their plane making an angle γ. Define axes xyz such that the first disc is centred at the origin, and is in the xy plane (Fig. 2.3).

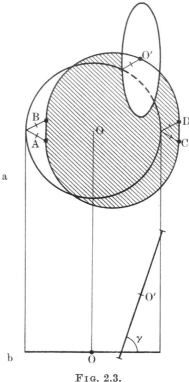

F I G. 2.3.

The direction of intersection of the planes of the two discs is parallel to Oy. The centre O' of the second disc is at an elevation z above the xy plane. The vector linking the point of contact c to O' has the following components:

$$\xi = z/\tan \gamma$$

$$\eta = \sqrt{(b^2 - z^2/\sin^2\gamma)}$$

$$z$$

The intersection of the excluded volume for O' by a plane $z = $ constant is then made of two half-circles (derived by the translations (ξ, η) and $(\xi, -\eta)$ from the disc O) plus a rectangular region of dimensions $2b$ and 2η (see Fig. 2.6). The area inside this curve is $\pi b^2 + 4\eta b$ and the excluded volume is

$$-\beta_1(\gamma) = 2 \int_0^{b \sin \gamma} \{\pi b^2 + 4\eta b\} \, dz$$

$$= 4\pi b^3 \sin \gamma$$

Thus, the dependence of β_1 on γ is identical in form to eqn (2.23) and the Onsager results can be transposed immediately, by performing the substitution:

$$L^2 D \rightarrow 2\pi b^3$$

Thus, we expect to have a transition from isotropic to nematic, with the concentrations of the two phases at equilibrium given by

$$b^3 c_n = (2/\pi^2) \times 4\cdot 5 = 0\cdot 9$$
$$b^3 c_i = (2/\pi^2) \times 3\cdot 3 = 0\cdot 67$$

There are in fact a few observations suggesting the existence of 'stacked phases' for certain flat dye molecules [21]. Extended calculations for ellipsoids can be found in A. Isihara, *J. chem. Phys.* **19**, 1142 (1951). The equation of state for 'hard plates' of rectangular shape (restricted to a finite set of orientations) has been discussed by Shih and Alben, *J. Chem. Phys.* **57**, 3057 (1972).

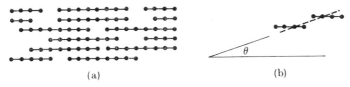

(a)   (b)

FIG. 2.4. Flory's lattice model for long rods: (a) conformation where all rods are parallel (the partition function for this case can be calculated exactly); (b) approximate method to describe tilted molecules.

2.2.1.2. The Flory calculation.

A slightly different mean-field calculation for the hard-rod problem has been used by Flory [22]. He describes a rod as a set of points inscribed on a lattice (see Fig. 2.4a). The number x of points on each rod plays the role of the parameter L/D in the preceding discussion. To describe a 'tilted' rod Flory uses the picture of Fig. 2.4b where the rod is replaced by a family of smaller units, each unit being still aligned along one same lattice direction. One attractive feature of the model is that, when all rods are parallel ($S = 1$), the partition function can be calculated exactly. The approximate form of the free energy derived for $S < 1$ becomes the exact result for $S = 1$. Thus the Flory approach is useful for a dense, highly-ordered, phase. On the other hand, his treatment of the angular distribution function f_a is rather crude, and this limits the accuracy of the result in the isotropic phase. Thus the Onsager and Flory calculation supplement each other but neither of them can be entirely reliable on the whole range of concentration c.

Numerically, Flory obtains higher value for the volume fractions at the transition

$$\Phi_n^c \sim 12\cdot 5 D/L \qquad \Phi_i^c \sim 8 D/L \qquad (L/D \gtrsim 10).$$

The critical value of the order parameter S^* is not quite meaningful, in view of the approximations in $f(\theta)$, but it turns out to be even larger than in the Onsager solution.

In Fig. (2.5) we show some experimental data on $\Phi^c(L/D)$ for poly-benzyl-L-glutamate, in typical solvents. There is a decrease of Φ^c with L/D corresponding roughly to an inverse law. The coefficient is not too meaningful, because various complications are involved with the physical system:

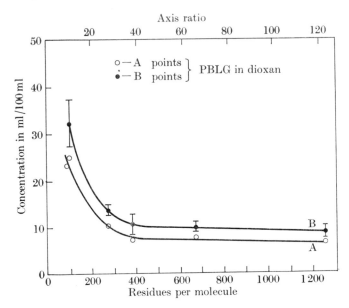

FIG. 2.5. Critical volume fraction Φ_c for a suspension of long rods: poly(benzyl-L-glutamate) in dioxan. N is the number of peptide units per rod. The length L is estimated to be $N \times 1\cdot 5$ Å. The filled circles give $\Phi_{c(n)}$ and open circles give $\Phi_{c(1)}$ (after Robinson, Ward, and Bevers, *Discuss. Faraday Soc.*, **25**, 29 (1958)).

(1) van der Waals attractions between rods, and other possible effects on contact [23].

(2) Polydispersity: if the rods are not uniform in length, the shorter ones are much more disordered than the long ones. This might have some effect on the equilibrium curves.

2.2.1.3. Strict lattice models. In the models of Onsager and Flory, the molecules have a continuous distribution of orientations. This corresponds to the actual physical situation. However, for purposes of calculation, it is tempting to work with a system where the number of allowed orientations is finite.

This is again obtained with a lattice where each molecule occupies a row of x consecutive points. The difference from the Flory model is that the molecules can point only in the directions defined by the nearest-neighbour vector (three directions for a simple cubic lattice, five directions for a face-centred cubic lattice): Fig. 2.4a is allowed, but Fig. 2.4b is now eliminated.

In the mean-field approximation, the statistics of such systems are not too hard to work out. At low densities, the system is isotropic. At higher densities, one direction is preferred (nematic order). A recent review of the results, and of some extensions of the method, has been

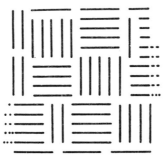

Fig. 2.6. One possible arrangement for close-packed molecules with purely repulsive forces, on a lattice; in this arrangement there is no long-range nematic order. The molecules are grouped in small 'disordered domains'.

given by M. Cotter [26]. The qualitative features of the transition do not differ very much from what we have described above. But the strict lattice models suffer from at least two conceptual shortcomings.

(1) The reduction of the continuous rotation group to a discrete point group destroys some important features of the fluctuations; this point will become clearer later, when we discuss light scattering by nematics (Chapter 3).

(2) It is not quite certain that such a lattice model, even for a large p (long molecule), will actually lead to a transition from isotropic to nematic at high enough densities. The two-dimensional problem with dimers ($p = 2$) is known to show no ordering, even at the maximum (close packed) density [27]. The trouble may in fact be more general; for any p, at the close-packed density, the nematic conformation of Fig. 2.4a is not the only one allowed in a strict lattice model. This is explained in Fig. 2.6, which shows a typical 'disordered domain' configuration.†

† I am indebted to J. Viellard-Baron for pointing out this effect.

A detailed examination of these disordered domains is also useful on more general grounds; with purely repulsive, hard core interactions the domains have a good chance to be present, independently of the details of the model. This 'hesitation' between nematic and disordered configurations may in fact explain why higher virial expansions converge poorly for hard-rod systems.

2.2.2. Mean field theory with S^2 interaction (Maier–Saupe)

With long, hard, rods in three dimensions we have a model which appears to lead to a nematic–isotropic transition. But this transition differs from the observations on actual thermotropic systems in many respects: the transition density is too low, and the jump in density at the transition is too high; the order parameter S_c at threshold is too high; Like all models involving only infinite repulsive forces, the system is 'athermal,' so that the transition density, for instance, is independent of temperature.

Clearly, in the present situation we are still in need of a very simple phenomenological theory, applicable independently of the detailed form of the interactions. This is the analogue, for nematics, of the 'molecular field approximation' introduced by P. Weiss for ferromagnets. Such a theory was worked out in detail by Maier and Saupe [32].

In general, to specify the orientation of the molecules, we would require a distribution function involving three Euler angles. However we decide to ignore the details of the molecular structure. We assume that each molecule has a well-defined long axis **a** (with polar angles θ and ϕ) and we discuss only the distribution function $f(\theta, \phi)$ for the long axis.

The next step is to introduce a convenient thermodynamic potential, which will be a minimum in the equilibrium state. Here, since we work in practice at fixed pressure rather than at fixed volume (and since our liquids do show some change in volume with temperature), we find it convenient to use the free enthalpy per molecule (or chemical potential) G. G depends on the angular distribution function f_a:

$$G(p, T) = G_i(p, T) + k_B T \int f_a \log(4\pi f_a) \, d\Omega + G_1(p, T, S) \quad (2.28)$$

where G_i is the free enthalpy of the isotropic phase. The second term, as in eqn (2.22) reflects the decrease in entropy due to an anisotropic angular distribution. The last term G_1 describes the effects of inter-molecular interactions. We *assume* that G_1 is quadratic in S:

$$G_1 = -\tfrac{1}{2}U(p,\,T)S^2 \tag{2.29}$$

The value of G_1 is decreased when S increases; thus U is positive. In the original presentation of Maier and Saupe [32] it was assumed that U is due entirely to van der Waals forces, and is temperature independent. In actual fact contributions from the steric repulsions (as in the Onsager calculation) may be non-negligible, and they do depend on T; the overall temperature dependence may be rather complex.

We can now minimize G with respect to all variations of f which satisfy the constraint (2.20). The variation equation is

$$\delta G = \lambda \int \delta f(\theta\phi)\, \mathrm{d}\Omega \tag{2.30}$$

From eqn (2.29) the variation of the term G_1 has the form

$$\delta G_1 = -US\, \delta S = -US \int \tfrac{1}{2}(3\cos^2\theta - 1)\, \delta f(\theta\phi)\, \mathrm{d}\Omega.$$

Thus

$$\delta G = \int \delta f[+k_{\mathrm B}T\{\log(4\pi f)+1\} - US(3\cos^2\theta - 1)/2]\, \mathrm{d}\Omega$$

and eqn (2.30) gives:

$$\log(4\pi f) = \lambda - 1 + (US/k_{\mathrm B}T)(3\cos^2\theta - 1)/2 \tag{2.31}$$

The (correctly normalized) resulting form of the distribution function is:

$$f(\theta) = \exp(m\cos^2\theta)/4\pi Z \tag{2.32}$$

with

$$m = \tfrac{3}{2}(US/k_{\mathrm B}T) \tag{2.33}$$

The normalization constant Z is defined by:

$$Z = \int_0^1 \mathrm{e}^{mx^2}\, \mathrm{d}x \tag{2.34}$$

(and may be expressed in terms of error functions).

We must now write a self-consistency condition for S, using eqn (2.2):

$$S = -\tfrac{1}{2} + \tfrac{3}{2}\langle\cos^2\theta\rangle$$

$$= -\frac{1}{2} + \frac{3}{2Z}\int_0^1 x^2\mathrm{e}^{mx^2}\, \mathrm{d}x$$

$$S = -\frac{1}{2} + \frac{3}{2}\frac{\partial Z}{Z\,\partial m}. \tag{2.35}$$

Eqns (2.33) and (2.35) may be solved graphically, as shown on Fig. 2.7 to obtain the values $S(T)$ and $m(T)$ relative to a certain temperature T. When $k_B T/U$ is small, there are two solutions for (2.33) and (2.35) which correspond to local minima in $G(S)$. One is associated with $S = 0$ and would correspond to an isotropic fluid. The other one corresponds to point M on the Figure, and describes a nematic phase.† To decide which solution is realized physically, we must compare the values of G

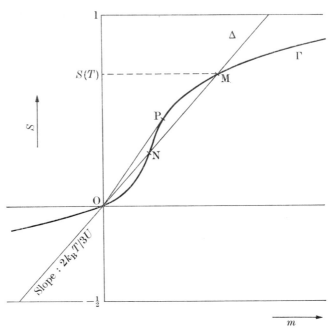

FIG. 2.7. Graphical solution for the self-consistency equation giving the order parameter $S(T)$ in the Maier–Saupe approximation. The curve Γ is defined by eqns (2.35) and (2.34). The straight line Δ is derived from eqn (2.33). When $T < T_c$, Δ intersects Γ, at the origin, at the point M (and also eventually at a third point N, of lower S value than M). The point M gives the physical state of minimum free energy G. The point N would correspond to an unstable state (local maximum in G).

associated with both minima. When T is below a temperature T_c defined by:

$$\frac{k_B T_c}{U(T_c)} = 4 \cdot 55 \tag{2.36}$$

the nematic phase is the stable one. For higher temperatures, the isotropic fluid is stable; there is a first-order transition at $T = T_c$. The

† There may exist, in certain temperature ranges, other solutions to eqns (33) and (35) than those associated with the points O and M on Fig. 2.7 (e.g. point N). But they correspond to a maximum of G, not to a minimum.

order parameter for T just below T_c is

$$S_c \equiv S(T_c) = 0 \cdot 44.$$

Thus, with the expression (2.28) for the chemical potential, the value of S_c must be the same for all nematic–isotropic transitions. In particular, for a given compound, if we shift the transition point T_c by applying pressure, we should retain the same value of S_c; this appears to be verified by n.m.r. experiments under pressure [33, 34]. It must also be emphasized that the amount of order at T_c is much smaller in the Maier–Saupe theory than in the Onsager model ($S_c \sim 0 \cdot 44$ instead of $0 \cdot 84$).

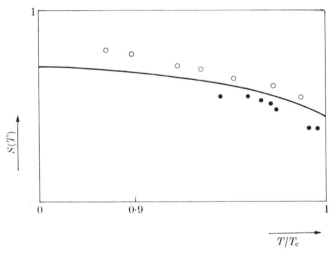

FIG. 2.8. Plot of the order parameter versus reduced temperature T/T_c. Continuous line: Maier–Saupe approximation. Open circles: n.m.r. data on 4,4′-bisethoxyazoxy-benzene. Filled circles: n.m.r. date on 4,4′-bismethoxyazoxybenzene (PAA). (After A. Saupe, *Angewandte Chemie*, **7**, 97 (1968)).

To reach a definite prediction on the temperature variation of S below T_c we must make an assumption concerning the temperature dependence of $U(T)$. As explained below in eqn (2.29), this dependence is non-trivial. If we assume with Maier and Saupe that U is essentially *independent* of T, the order parameter becomes a universal function of T/T_c. This function is plotted in Fig. 2.8 together with some experimental points.† There is good qualitative agreement.

† Many other studies derive $S(T)$ from refractive index measurements; as pointed out earlier in this chapter, the relationship between birefringence and S involves delicate questions on dipole–dipole interactions in a dense medium: thus the 'experimental' $S(T)$ curves already involve some approximations.

However, difficulties begin to appear if we use an independent experiment to determine the temperature variation of U. Consider for instance the heat of transition ΔH. This can be computed from the equation linking the entropy $\sum$ to the chemical potential G given in eqn (2.28) $\sum = -(\partial G/\partial T)_p$.

Writing that the G values for both phases are equal at the transition point, one arrives at:

$$\Delta H = T_c\,\Delta\sum = \tfrac{1}{2}S_c^2\left\{U - T_c\left(\frac{\partial U}{\partial T}\right)_p\right\}. \tag{2.37}$$

At $T = T_c$, U is given by eqn (2.36) and we have

$$T_c\left(\frac{\partial U}{\partial T}\right)_p\bigg/U = 1 - \frac{2\Delta H}{4\cdot55S_c^2 k_B T_c}$$

This is approximately $\tfrac{2}{3}$ for PAA at atmospheric pressure; thus the temperature dependence of U near T_c is strong, and the agreement obtained when neglecting it may be in fact somewhat accidental.†

2.3. Short-range order effects

2.3.1. Computer calculations

All the calculations which have been presented up to now are of the 'mean field' type; they neglect angular correlations between neighbouring rods, which are certainly important at the actual physical densities, and in particular near the transition point.

In the near future, direct computer calculations (sampling statistically the allowed configurations for a certain number of hard objects) will give quasi-exact correlations and equations of state. At the moment, this approach is still laborious and expensive; it has been carried out only in two dimensions, for 170 hard ellipses [28], but the results in this case are very interesting.

Hard 'circles' are known to have a solid–fluid transition.† If we go to ellipses, we find *two* transitions; one positional (solid $\rightleftharpoons$ fluid) and one orientational.‡ When the ellipses are not too elongated, the orientational transition takes place in the solid state, and is reminiscent of the rotational transitions observed in certain physical solids (such as ammonium halides). On the other hand, when the ellipses are strongly

---

† A detailed discussion of thermodynamic derivatives near T_c has been given by R. ALBEN *Mol. Cryst. liquid Cryst.* **10**, 21 (1970).

† The solid phase has certain pathological fluctuations, and should really be called quasi-solid. But these refined features are not observable in the Monte-Carlo calculations.

‡ The two successive transitions have also been discussed in 3 dimensions, for a certain lattice model (in the mean field approximation) by Chandrasekhar, Shashidar, and Tara *Mol. Cryst. liquid Cryst.* **10**, 337 (1970).

elongated, the orientational transition occurs inside the liquid phase, and represent a change from nematic to isotropic.

This numerical evaluation of the equation of state also provides us with a 'touchstone', which can be used to test various approximate theories attempting to include short-range effects. The most useful theory of this type (still retaining a tractable mathematical apparatus) is the so-called 'scaled particle theory' which essentially represents a local version of the mean field, and produces some correlations between particles. For hard ellipses, the scaled particle theory gives a critical density in good agreement with the Monte Carlo results [29].

In three dimensions, computer calculations on hard rods or similar objects with blunted ends, are currently under way, but the results are

(a)   (b)

Fig. 2.9. (a) Nematic phase. (b) Isotropic phase just above the clearing point T_c. In case (b), short-range order persists over distances $\xi(T)$ much larger than the molecular size.

not available yet. At present, the only available tools are phenomenological theories, which can be constructed independently of the detailed nature of the interactions. Two different domains must be distinguished here:

(1) Below T_c the fluctuations of the magnitude of S are weak, and the main effect is related to fluctuations in the *orientation* of the optical axis. This will be discussed in Chapter 3, after the introduction of continuum elasticity.

(2) Just above T_c we have sizable fluctuations both in magnitude and in orientation; they are described rather well by a simple (Landau-type) theory, which will be described qualitatively below.

2.3.2. Landau free energy above T_c

In the isotropic phase, there is no long-range order in the direction of alignment of the molecules; the tensor order parameter $Q_{\alpha\beta}$ vanishes on the average. However, if we were to look closely enough at the molecules (Fig. 2.9b) we would see that *locally* they are still parallel to each other.

This local order persists over a certain characteristic distance $\xi(T)$ which is called the *coherence length*.

In a rough qualitative sense, in the isotropic phase (and in this phase only) one may speak of nematic 'swarms'; i.e. small nematic droplets of size $\xi(T)$, the orientations of successive droplets being incorrelated. This picture is qualitatively suggestive, but often leads to incorrect exponents in various physical laws (because it implies abrupt changes in orientation at the border between two swarms). To reach a more accurate description, it is useful to start from a continuum theory, where the free energy is expanded as a power series of the order parameter $Q_{\alpha\beta}$ and of its spatial derivatives; this approach ensures that the variations of Q are smooth. It is analysed in ref. [35]; a critical examination of experimental data on short range order effects, comparing them to the continuum theory, is given in ref. [36].

The main features which emerge are listed below. We shall discuss first the order of the transition.

For purely geometrical reasons, the nematic $\rightleftharpoons$ isotropic transition *must* be of first order:† this was already recognized by Landau [37]. The argument may be summarized as follows; the free energy F may be expanded in powers of the order parameter, and the following terms occur (in the absence of external aligning fields H)

$$F = F_0 + \tfrac{1}{2}A(T)Q_{\alpha\beta}Q_{\beta\alpha} + \tfrac{1}{3}B(T)Q_{\alpha\beta}Q_{\beta\gamma}Q_{\gamma\alpha} + O(Q^4)$$

(Summation over repeated indices is implied). All these terms are invariant by any rotation of the axes (x, y, z) as they should be. There is no term linear in Q; this ensures that the state of minimum F is a state of zero Q that is to say, isotropic. But it is very important to realise that there is a *non-vanishing term* of order Q^3: the reason for this is that there is no symmetry relation between a state $(Q_{\alpha\beta})$ and the state $(-Q_{\alpha\beta})$. For

$$Q = \begin{vmatrix} -\xi & 0 & 0 \\ 0 & -\xi & 0 \\ 0 & 0 & 2\xi \end{vmatrix}$$

(with $\xi > 0$) may describe a small alignment of the molecules along the z axis while the state

$$Q = \begin{vmatrix} \xi & 0 & 0 \\ 0 & \xi & 0 \\ 0 & 0 & -2\xi \end{vmatrix}$$

† If we vary the presence p, we have a certain transition line in the (p, T) diagram. All along this line the transition will be of first order, except possibly at one point.

corresponds to molecules which tend to lie in the x, y plane: there is no reason for these two states to have the same free energy.

As soon as there is a non-vanishing term of order Q^3 in the expansion, the phase transition must be of *first order;* a qualitative explanation of this point is given in Fig. 2.10. Indeed, in all cases which have been studied up to now, the nematic–isotropic transition has been found to be of first order; discontinuities have been observed in the density, in the heat content, etc. However it must be emphasized that these discontinuities are small: the transition is 'weakly first order.' This means that the coefficient B is relatively small.

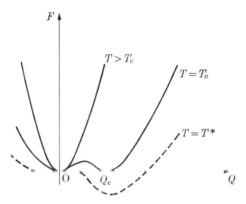

FIG. 2.10 Free energy as a function of order parameter for various temperatures. For $T > T_c$ the absolute minimum of F corresponds to $Q = 0$ (isotropic phase). For $T < T_c$ the minimum corresponds to $Q \neq 0$ (nematic phase). For $T = T_c$ we can have co-existence of an isotropic phase and a nematic phase with a finite amount of order ($Q = Q_c$): the transition at T_c is of first order. The (lower) temperature T^* corresponds to the vanishing of the Q^2 term in F. Below T^* the isotropic phase is completely unstable with respect to nematic ordering.

2.3.3. Static pretransitional effects

Since the transition is only weakly of first order, we expect that short-range order effects will be important at temperature just above the transition point T_c; in particular the coherence length $\xi(T)$ will be rather large—typically a few hundred angstroms (ten times the molecular length). This reacts on many physical properties. Here we shall outline them only very qualitatively; for a quantitative analysis, see ref. [35].

2.3.2.1. Magnetic birefringence.

The swarms are rather easily aligned by a magnetic field, if they are large enough. There is a magnetic birefringence

$$\Delta n = n_{\parallel} - n_{\perp} = CH^2 \qquad (2.39)$$

(where $n_{\parallel}$ and $n_{\perp}$ are the indices measured respectively along H and normal to H). The coefficient C is large just above T_c (typically a hundred times larger than in simple organic liquids such as nitrobenzene).

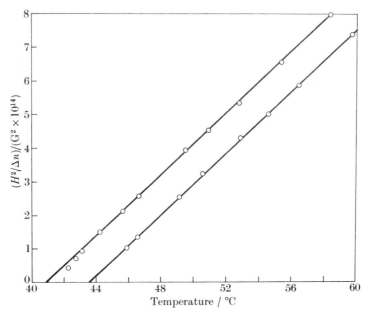

Fig. 2.11. Magnetic birefringence (Cotton–Mouton effect) in MBBA. The points give the *inverse* of the magnetic birefringence coefficient, as a function of temperature, for two different samples (with slightly different transition temperatures). In both samples the plots are nearly linear, in agreement with the Landau approximation (after ref. 36. Used with permission from McGraw Hill Co.)

A strict swarm picture would predict that C should be proportional to the volume (ξ^3) of the swarms. But the continuum theory shows that in fact

$$C \sim \xi^2. \qquad (2.40)$$

2.3.2.2. Electric birefringence (Kerr effect).   Here there is also an alignment effect but it is much more complicated [38a]. The temperature dependence of the Kerr constant $C' = (n_{\parallel} - n_{\perp})/E^2$ is not well described by eqn (2.40). A number of parasitic effects probably come into play; in impure samples, Helfrich has noticed that the anisotropic conduction currents induced by E inside a swarm will lead to charge accumulation at the surface of the swarm [38b]; this is related to the Carr–Helfrich effect described in Chapter 5 and gives rise to local forces, flows, and flow birefringence. Even at ultrahigh (optical) frequencies, when conduction is absent, the constant C' is anomalous

[38c]; it may be that the polarization charges present even in this case also lead to forces and flows.

2.3.2.3. Light scattering. The scattering of light by the isotropic phase is much smaller than in the nematic phase (for this reason T_c is often called the *clearing point*). However, just above T_c, there will remain a sizeable scattering intensity I, due to the birefringent swarms. The polarization dependence of I is indeed in agreement with what is expected for the scattering by a random distribution of anisotropic domains [36] [37].

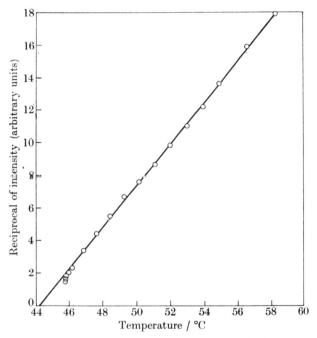

FIG. 2.12. Reciprocal of the light-scattering intensity in the isotropic phase of MBBA (after ref. [36]. Used with permission from McGraw-Hill Co.)

Since the swarms are much smaller than the optical wavelength, the intensity I is essentially independent of scattering angle. It has been verified by the MIT group [36] that I and the magnetic birefringence coefficient C have the same temperature dependence near T_c; to a first approximation this is simply:

$$C \sim I \sim 1(T - T^*),\qquad(2.41)$$

where T^* is a temperature slightly smaller than T_c. (Typically $T_c - T^* \sim 1\mathrm{K}$). Physically T^* is the temperature at which the size of the swarms would become infinite $\{\xi(T^*) = \infty\}$; it represents the lowest temperature down to which one could supercool the isotropic phase. It must be emphasized that in a strict swarm model, one would expect $I \sim \xi^3$, while the more correct 'Landau type,' theory gives $I \sim \xi^2$. We may understand the latter result by the following qualitative argument.

The light-scattering amplitude at one point is proportional to the fluctuations of the dielectric tensor, i.e. to Q (we purposely omit all component indices).

The intensity is proportional to

$$I = \int \langle Q(\mathbf{R}_1)Q(\mathbf{R}_2) \rangle \exp(i\mathbf{k} \cdot \mathbf{R}_{1,2}) \, d\mathbf{R}_{1,2},$$

where $\mathbf{k}$ is the scattering-wave vector. In the Landau approximation the $\langle QQ \rangle$ correlations have the Ornstein–Zernike form

$$\langle Q(0)Q(\mathbf{R}) \rangle \approx (e^{-R/\xi})/R$$

As pointed out above, the swarms are small when compared to the optical wavelength ($k\xi \sim 0$). Then

$$I \approx (e^{-R/\xi})/R \int 4\pi R^2 \, dR = 4\pi \xi^2$$

In the Landau approximation, the temperature dependence of ξ is of the form

$$\xi = \xi_0 \left(\frac{T^*}{T - T^*} \right)^{\frac{1}{2}}$$

where ξ_0 is a molecular length. This is compatible with the experimental law for I.

More direct measurements of ξ have been attempted; a small $\mathbf{k}$ dependence has been detected on I, and allows in principle for a determination of ξ [39a, b]. This experiment is difficult, and the first data [39a] were not very accurate. More recent data from the MIT group [39b] appear to agree with the above temperature dependence.

On the whole, all these experiments show that, in the isotropic phase of nematics, short-range order effects are rather well explained in terms of a simple 'mean field' or 'Landau' approximation. This leaves two questions open:

(1) For most known second-order transitions, the Landau approximation fails (in magnetic systems, in superfluid helium etc.). For instance the response function—measured in our case by the magnetic birefringence coefficient C—diverges not according to $(T - T^*)^{-1}$

but rather as $(T - T^*)^{-1}$ where $\gamma \sim 1\cdot25$ to $1\cdot40$. Why are these complications absent at the nematic–isotropic transition?

(2) Why is T^* so close to T_c? Experimentally

$$\frac{T_c - T^*}{T_c} \sim 2 \times 10^{-3}$$

Such a small number cannot be understood in terms of the Maier–Saupe theory, or in fact of any theory where the interactions are described in terms of a single coupling constant. An adequate picture must involve a 'small parameter' of some sort.

One possible answer to these questions is based on the following argument.[†] The Landau approximation does work in certain systems, such as the superconducting metals. In these metals, the correlation length, far from T_c, is $\xi_0 \sim 1000$ Å, and is much larger than the interelectron distance $a \sim 2$ Å $(a/\xi_0 \sim 2 \times 10^{-3})$. When $a \ll \xi_0$, the Landau approximation becomes adequate.[‡] With our nematogens, we may have something similar, where ξ_0 would be related to the *length* of the molecules, while a would be related to their *width* $(a\xi/_0 \sim 1/5)$.

Finally, apart from the static effects of the swarms which we have discussed here, there are also some very interesting *dynamic* effects, which will be discussed at the end of Chapter 5.

2.4. Mixtures

2.4.1. *Importance of mixed systems*

By mixing *two nematogens* X and Y one can often obtain a material with a lower melting point, while the 'clearing point' T_c is not much depressed. Many of the commercial room-temperature nematics 'Merk IV,' the tolanes, etc) are mixtures of this type, with the composition of the solid–liquid eutectic of minimal melting point (Fig. 2.13).

Miscibility studies are also important from a more fundamental point of view, to identify new phases. The rule which is used is the following:[†] if two phases (I) and (II) are continuously miscible without crossing any (first- or second-order) transition line, they have the same symmetry. Of course, a nematic phase can often be assigned simply from its optical appearance, especially from the defects which are present (see Chapter 4). However, these texture studies are not sufficient: for instance, a

† I am strongly indebted to Prof. M.H. Cohen for a discussion on this point.
‡ See for instance: P.G. de Gennes (1966). *Superconductivity of metals and alloys*, p. 174. Benjamin, New York.
      † The rule is discussed in more detail in Chapter 7.

smectic C may have a nematic-like texture. To distinguish between the two we can take an X-ray picture. This method is fundamental but slow. A simpler method (among others) is to perform miscibility studies. This can often be carried out under the microscope (in a concentration gradient) and is faster.

Mixtures of nematogens X with molecules Y of different shapes are also useful. For instance, the internuclear distances of molecules Y can be measured accurately by n.m.r. if the solvent X is nematic [see ref. 6]. Also, if Y is a chiral molecule, it twists the nematic into a cholesteric:

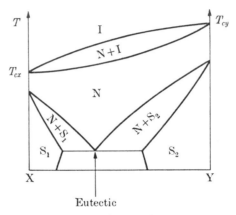

FIG. 2.13. Typical phase-diagram for a mixture of two nematogens X, Y which are well miscible in the nematic phase N, but not entirely miscible in the solid phase S. (This is often obtained when the terminal chains of X and Y differ widely in this length). I represents the isotropic phase.

this leads to a new form of polarimetry, where the chiral solute Y is not probed by the optical properties of its own molecule, but rather by the long-range distortions which it creates in X [40]. In many cases it may be of interest to *enhance* a certain property (such as the magnetic anisotropy, or the conductance) in a nematic phase X by adding a suitable solute Y.

The selective solution of certain isomers by a nematic solvent is also of interest for gas chromatography. For instance, if we compare the following species

(meta xylene)   (para xylene)

we find that the para-derivative accommodates better in nematic solvents (because of its elongated shape). The applications to chromatography have been discussed by H. Kelker.

For all these reasons, data on the phase diagrams of mixtures are very important. Some useful examples are given by Gray (ref. [1] of Chapter 1) and a general discussion has been given by Billard [41]. We shall present here only a few major facts.

2.4.2. General trends

Two nematogens X and Y usually show a continuous miscibility in the nematic phase, and, if X and Y are not too difficult chemically, this solution is even *nearly ideal*. This property is convenient; it allows us to predict the nematic–isotropic (N–I) transition-curves of the mixtures in terms of data taken on the *pure* compounds (X and Y) only; namely the clearing points T_{cX}, T_{cY}, and the transition enthalpies ΔH_X, ΔH_Y.† The relevant formulae were established long ago by Schroder and Van Laar [42]. Note that this situation of good miscibility holds not only between usual (biaromatic) nematogens, but also between them and the cholesterol esters.

In mixtures X–Y where X is a nematogen and Y is not we must distinguish two cases:

(1) If Y is very different in shape from X, we usually find a nematic phase N only for very small concentrations of Y (typically below five per cent).

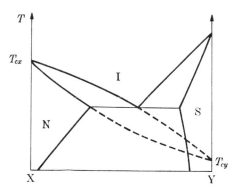

FIG. 2.14. Phase diagram for a mixture XY where X is a nematogen (i.e.: has a nematic phase) while Y is not, but Y is still close to having a nematic phase. By extrapolation of the equilibrium curves for the nematic mixtures, one can ascertain the hidden transition point T_{cY} of Y. The phase S which masks the transition may be solid or smectic.

† In principle, certain specific-heat data are also required, but in practice they give only weak corrections.

(2) If Y is not too different from X, the concentration domain of the N phase widens. The mixed N phase is again nearly ideal and follows the Schroder–Van Laar formulae. In this case, using the formulae backwards, and knowing T_{cX}, ΔH_X, one can *assign* a set of values T_{cY}, ΔH_Y which would represent the transition N–I for the pure Y compound [43]. This transition is hidden in reality by another phase of Y (see Fig. 2.14). But knowledge of the fictitious T_{cY} can be of great interest in discussions of nematic order inside a chemical series Y_1, Y_2, ... where certain members have a nematic phase, and others do not.

In a few favourable cases, one can in principle start with two compounds X and Y which are not nematogens, and find a mixture $X + Y$ which is nematic in a certain domain of concentrations and temperatures. In this case, both T_{cX} and T_{cY} are hidden by other low temperature phases.

REFERENCES

CHAPTER 2

[1] See, for instance, ABRAGAM, A. *The principles of nuclear magnetism*, Chapter 8. Oxford University Press. (1961).

[2] SPENCE, R. D., GUTOWSKY,H. S., and NOLM, C. H. *J. chem. Phys.* **21**, 1891 (1953).

[3] SAUPE, A. and ENGLERT, G. *Phys. Rev. Lett.* **11**, 462 (1963).

[4] SAUPE, A. *Angew. Chem. (Int. Edn.)* **7**, 97 (1968); *Mol. Cryst.* **16**, 87 (1972).

[5] PINCUS, P. *J. Phys. (Fr.)* **30**, (Suppl C4) 8 (1969).

[6] DIEHL, P. and KHETRAPAL, C. L., in *NMR, basic principles and progress*, vol. 1. Suringer, Berlin (1969).

[7] SAUPE, A. *Z. Naturf.* A19, 161 (1964).

[8] SNYDER, L. C. *J. chem. Phys.* **43**, 4041 (1965).

[9] SAUPE, A. and ENGLERT, G. *Mol. Cryst.* **1**, 503 (1966); NEHRING, J. and SAUPE, A. *Mol. Cryst. liquid Cryst.* **8**, 403 (1969).

[10] CABANE, B. and CLARKE, W. G. *Phys. Rev. Lett.* **25**, 91 (1970).

[11] ROWELL, J. C., et al., *J. chem. Phys.* **43**, 3442 (1965).

[12] A general review of this field has been given by LUCKHURST, G. R. *Mol. Cryst.* (to be published).

[13] GASPAROUX, H., REGAYA, B., and PROST, J. *C.r. hebd. Séanc. Acad. Sci., Paris, Ser. B* **272**, 1168 (1971); *J. Phys. (Fr.)* **32**, 953 (1971).

[14] BALZARINI, D. *Phys. Rev. Lett.* **25**, 914 (1970).

[15] PRIESTLEY, E. B. and PERSHAN, P. S. *Mol. Cryst. liquid Cryst.* (to be published).

[16] For a discussion of results up to 1967 see the review by CHISTIAKOV, *Sov. Phys. Usp.* **9**, 551 (1967). A more recent study on single-domain samples is described in DELORD, P. *J. Phys. (Fr.)* **30**, (Suppl. C4) 14 (1969). Specific effects of smectic short-range order in the nematic phase have also been found by DE VRIES and others: they will be disscused separately in Chapter 7.

[17] RISTE, T. *Mol. Cryst. liquid Cryst.* (to be published).

[18] DE GENNES, P. G. *C.r. Acad. Sci., Paris, B* **247**, 62 (1972).

[19] ONSAGER, L. *Ann. N.Y. Acad. Sci.* **51**, 627 (1949); see also WADATI, M. and ISIHARA, A. *Mol. Cryst. liquid cryst.* **17**, 95 (1972).

[20] See LANDAU, L. D. and LIFCHITZ, E. M. *Statistical physics*, § 71. Pergamon London (1958).

[21] DREYER, J. F. *J. Phys. (Fr.)* **30**, (Suppl. C4) 114 (1969).

[22] FLORY, P. J. *Proc. R. Soc.* A234, 73 (1956).

[23] Some of these effects, related to the existence of flexible side chains in materials such as PBLG, have been analysed in detail by FLORY, P. J. and LEONARD, M. *J. Am. chem. Soc.* **87**, 2102 (1965).

[24] DI MARZIO, E. *J. chem. Phys.* **35**, 658 (1969).

[25] COTTER, M. and MARTIRE, D. *Mol. Cryst.* **7**, 295 (1964).

[26] COTTER, M. *Mol. Cryst. liquid Cryst.* (to be published).

[27] KASTELEJN, P. W. *Physica* **27**, 1209 (1962); LIEB, E. H. *J. Math. Phys.* **8**, 2339 (1967).

[28] VIELLARD BARON, J. *J. Phys.* (*Fr.*) **30**, (Suppl. C4) 22 (1969); *J. chem. Phys.* (to be published).

[29] TIMLING, K. *Philips Res. Rep.* **25**, 223 (1970).

[30] COTTER, M. and MARTIRE, D. *J. chem. Phys.* **52**, 1902, 1909 (1970); **53**, 4500 (1970).

[31] ALBEN, R. *Mol. Cryst. liquid Cryst.* **13**, 193 (1971).

[32] MAIER, W. and SAUPE, A. *Z. Naturf.* **A13**, 564 (1958); **A14**, 882 (1959); **A15**, 287 (1960).

[33] DELOCHE, B., CABANE, B., and JEROME, D. *Mol. Cryst. liquid Cryst.* **15**, 197 (1971). The values of dT_c/dp quoted in this reference are probably incorrect, because the pressure was transmitted by helium gas, which is somewhat soluble in the nematic.

[34] McCOLL, J. and SHIH, C. *Phys. Rev. Lett.* **29**, 85 (1972); McCOLL, *Phys. Lett.* **A38**, 55 (1972).

[35] DE GENNES, P. G. *Phys. Lett.* **A30**, 454 (1969); *Mol. Cryst. liquid Cryst.* **12**, 193 (1971); see also STEPHEN, M. J. and FAN, C. P. *Phys. Rev. Lett.* **25**, 500 (1970).

[36] LITSTER, J. D. in *Critical phenomena* (R. E. Mills, ed.) p. 393. McGraw-Hill, New York (1971).

[37] LANDAU, L. D. *Collected papers* (D. Ter Haar, ed.) p. 193. Gordon and Breach, New York (1965).

[38] HELFRICH, W. and SCHADT, M. (a) *Mol. Cryst. liquid Cryst.* **17**, 355 (1972); (b) *Phys. Rev. Lett.* **27**, 561 (1971). A theory of this effect can be constructed along the lines of ref. [35], but does not give very good numerical results. (c) PROST, J. and LALANNE B. (to be published in *Phys. Rev.* A (1973)).

[39] (a) CHU, B., BAK, C. S., and LIN, F. L. *Phys. Rev. Lett.* **28**, 1111 (1972); (b) LITSTER, D. and STINSON, T. *Phys. Rev. Lett.* **30**, 688 (1973).

[40] See BILLARD, J. *C.r. Acad. Sci., Paris, B* **274**, 333 (1972).

[41] BILLARD, J. *Bull. Soc. fr. Minér. Cristallogr.* **95**, 206 (1972).

[42] SCHRODER, I. *Z. phys. Chem.* **II**, 449 (1893); VAN LAAR, J. J. *Z. phys. Chem.* **63**, 216 (1908); **64**, 257 (1908).

[43] BILLARD, J. and DOMON, M. (to be published).

STATIC DISTORTIONS IN A NEMATIC SINGLE CRYSTAL

Avec des courbes pareilles à celle des osiers quand on prépare la corbeille

J. GIONO

3.1. Principles of the continuum theory

3.1.1. Long-range distortions

In an ideal, nematic, single crystal, the molecules are (on the average) aligned along one common direction $\pm\mathbf{n}$. The system is uniaxial, and the tensor order parameter has the form

$$Q_{\alpha\beta} = Q(T)(n_\alpha n_\beta - \tfrac{1}{3}\delta_{\alpha\beta}) \tag{3.1}$$

However, in most practical circumstances, this ideal conformation will not be compatible with the constraints which are imposed by the limiting surfaces of the sample (e.g. the walls of a container) and by external fields (magnetic, electric, etc) acting on the molecules. There will be some deformation of the alignment; the order parameter $Q_{\alpha\beta}$ will vary from point to point. Three typical examples are shown in Fig. 3.1.

For most of the situations of interest, the distances l over which significant variations of $Q_{\alpha\beta}$ occur are much larger than the molecular dimensions a (typically $l \gtrsim 1~\mu$m, while $a \sim 20$ Å).

Thus the deformations may be described by a *continuum theory* disregarding the details of the structure on the molecular scale. To construct such a theory, one possible starting point would be the free energy density F, expressed as a function of $Q_{\alpha\beta}$ as in eqn (2.38). When $Q_{\alpha\beta}$ becomes a function of $\mathbf{r}$, we must add in F new terms involving the gradients of $Q_{\alpha\beta}$. This approach is indeed useful to study space-dependent properties above the nematic → isotropic transition, because in this region $Q_{\alpha\beta}$ is small, and the structure of the gradient terms is simple. Below T_c this approach would become rather clumsy, because, for large $Q_{\alpha\beta}$, there are many phenomenological coefficients involved. It is better then to start from the following observation: in a weakly distorted system $(a/l \ll 1)$, at any point, the local optical properties are still those of a uniaxial crystal; the magnitude of the anisotropy is unchanged; it is only the orientation of the optical axis (**n**) which has been rotated. In terms of an order parameter $Q_{\alpha\beta}$ this means that:

$$Q_{\alpha\beta}(\mathbf{r}) = Q(T)\{n_\alpha(\mathbf{r})n_\beta(\mathbf{r}) - \tfrac{1}{3}\delta_{\alpha\beta}\} \qquad (3.2)$$

$+$terms of higher order in (a/l)

(For similar reasons, the changes in density of the liquid, induced by a long-range distortion, are very small.)

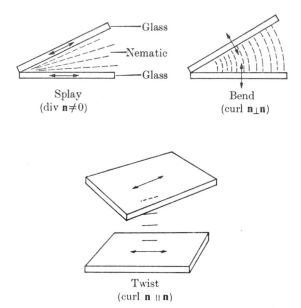

FIG. 3.1. The three types of deformation occurring in nematics. The Figure shows how each type may be obtained separately by suitable glass walls. The twisted geometry between parallel walls, commonly called '*plage tordue*', was used as early as 1911 by C. Mauguin.

The distorted state may then be described entirely in terms of a vector field **n**(**r**).† The 'director' **n** is of unit length but of variable orientation. It is assumed that **n** varies slowly and smoothly with **r** (except possibly on a few singular points or singular lines).

This type of description was initiated by Oseen [1] and Zocher [2]. More recently, it has been examined critically by Frank [3], and its relations to the hydrostatic properties of nematic liquids have been considered by Ericksen [4]. The mathematical aspects of the theory are

† This is similar to the conventional description of Bloch walls in ferromagnets where the changes of the *length* of the magnetization in the wall are neglected: the description is correct provided that the wall thickness is much larger than the interatomic distance.

reviewed in a forthcoming book by F. M. Leslie. In the present chapter, we shall insist mainly on the physical properties of the distorted conformations.

3.1.2. The distortion free-energy

Let us impose to our nematic a certain state of distortion, described by a variable director $\mathbf{n}(\mathbf{r})$. We make the following assumptions on this distorted system.

The variations of $\mathbf{n}$ are slow on the molecular scale:

$$a \, \nabla \mathbf{n} \ll 1.$$

(This does *not* imply, however, that $\mathbf{n}$ remains nearly parallel to one fixed direction; a simple counter-example is the helical deformation shown in Fig. (3.7)).

The only forces of importance between molecules are *short range*: this omits some possible long-range electric effects, which will be discussed separately.

Let us then call F_{d} the free energy (per cm³ of nematic material) due to the distortion of $\mathbf{n}$. F_{d} will vanish if $\nabla \mathbf{n} = 0$, and, with our assumptions, it may be expanded in powers of $\nabla \mathbf{n}$. The following conditions must be imposed on F_{d}:

(1) F_{d} must be even in $\mathbf{n}$; as explained in Chapter 1, the states $(\mathbf{n})$ and $(-\mathbf{n})$ are undistinguishable.

(2) There are no terms linear in $\nabla \mathbf{n}$. The only terms of this form which are invariant by rotations are

$$\begin{cases} \operatorname{div} \mathbf{n}; \text{ this is ruled out by (1)} \\ \mathbf{n}. \operatorname{curl} \mathbf{n}; \text{ this changes sign by the transformation } x \rightarrow -x, \; y \rightarrow -y, \\ z \rightarrow -z, \end{cases}$$

and is thus forbidden in a centrosymmetric material. (In cholesterics, which are not centrosymmetric, this term will be present).

(3) Terms in F_{d} which are of the form $\operatorname{div} \mathbf{u}$, where $\mathbf{u}(\mathbf{r})$ is an arbitrary vector field, may be discarded. This is a consequence of the identity:

$$\int \operatorname{div} \mathbf{u} \, d\mathbf{r} \equiv \int d\boldsymbol{\sigma}.\mathbf{u},$$

where $\int d\boldsymbol{\sigma}$ represents a surface integral and $d\boldsymbol{\sigma}$ is normal to the surface at each point. The integral is taken on the limiting surface of the nematic. The identity shows that such terms describe only certain contributions to the surface energies, and not to the volume energies: for the present discussion of bulk properties, we may omit them.†

---

† The question of surface energies will be discussed later in this section.

We shall now list all possible terms in F_d of order $(\nabla \mathbf{n})^2$. This represents a rather tedious operation; readers which are mainly interested in the physical results should skip this part and proceed directly to eqn 3.15.

To construct F_d, let us now consider explicitly the spatial derivatives of $\mathbf{n}$: they form a tensor of rank two $\partial_\alpha n_\beta$ (where we put $\partial_\alpha \equiv \partial/\partial x_\alpha$). As usual it is convenient to separate this tensor into a symmetric part

$$e_{\alpha\beta} = \tfrac{1}{2}(\partial_\alpha n_\beta + \partial_\beta n_\alpha) \tag{3.3}$$

and an antisymmetric part, related to the curl of $\mathbf{n}$:

$$(\text{curl } \mathbf{n})_z = \frac{\partial}{\partial x} n_y - \frac{\partial}{\partial y} n_x \quad \text{etc.} \tag{3.4}$$

In general a symmetric tensor $e_{\alpha\beta}$ has six independent components, but here these components are further restricted by the fact that $\mathbf{n}$ is a unit vector. This is most conveniently expressed by going to a local (orthogonal) frame of reference with the z axis parallel to local direction of $\mathbf{n}$. Then all gradients of n_z vanish, because of the identity

$$0 = \nabla(n_z^2 + n_x^2 + n_y^2) = 2n_z \nabla n_z + 0 = 2\nabla n_z. \tag{3.5}$$

Thus

$$\left. \begin{aligned} e_{zz} &= 0 \\ e_{zx} &= \tfrac{1}{2}(\text{curl } \mathbf{n})_y \\ e_{zy} &= -\tfrac{1}{2}(\text{curl } \mathbf{n})_x \end{aligned} \right\} \tag{3.6}$$

It will also be useful to recall that:

$$e_{xx} + e_{yy} + e_{zz} = e_{xx} + e_{yy} = \text{div } \mathbf{n}. \tag{3.7}$$

From requirement (1), F_d will be a quadratic function of the components $e_{\alpha\beta}$ and of curl $\mathbf{n}$. It is convenient to separate the contributions as follows:

$$F_d = F_e + F_c + F_{ec}$$

where F_e comes from terms quadratic in $e_{\alpha\beta}$, F_c from terms quadratic in curl $\mathbf{n}$ and F_{ec} represents cross terms. Let us start with F_e; finding the number of independent terms in F_e is formally equivalent to finding the number of elastic constants in a medium of symmetry C_∞ around z. The most general form for F_e can thus be found in text books on elasticity [5] and involves five arbitrary constants:

$$F_e = \lambda_1 e_{zz}^2 + \lambda_2(e_{xx} + e_{yy})^2 + \lambda_3 e_{\alpha\beta} e_{\beta\alpha} + \lambda_4 e_{zz}(e_{xx} + e_{yy}) + \lambda_5(e_{xz}^2 + e_{yz}^2). \tag{3.8}$$

(In the λ_3 term we have used the convention of summing over repeated indices). Taking into account the properties of eqn (3.6), this reduces to three terms

$$F_e = \lambda_2(\text{div } \mathbf{n})^2 + \lambda_3 e_{\alpha\beta} e_{\beta\alpha} + \tfrac{1}{4}\lambda_5(\mathbf{n} \times \text{curl } \mathbf{n})^2 \tag{3.9}$$

We now use the identity†

$$e_{\alpha\beta} e_{\beta\alpha} = (\text{div } \mathbf{n})^2 + \partial_\alpha(n_\beta\,\partial_\beta n_\alpha) - \partial_\beta(n_\beta\,\partial_\alpha n_\alpha) + \tfrac{1}{2}(\text{curl } \mathbf{n})^2 \tag{3.10}$$

The second and third term in eqn (3.10) must be dropped in agreement with requirement (3). Recalling that

$$(\text{curl } \mathbf{n})^2 \equiv (\mathbf{n}.\text{curl } \mathbf{n})^2 + (\mathbf{n} \times \text{curl } \mathbf{n})^2$$

we see that F_e is finally a sum of three contributions of the form:

$$(\text{div } \mathbf{n})^2, \quad (\mathbf{n}.\text{curl } \mathbf{n})^2, \quad \text{and} \quad (\mathbf{n} \times \text{curl } \mathbf{n})^2 \tag{3.11}$$

Turning now to the terms F_c, which are quadratic in the curl, in a medium of C_∞ symmetry, they must have the structure

$$F_c = \mu_1(\text{curl } \mathbf{n})_z^2 + \mu^2\{(\text{curl } \mathbf{n})_x^2 + (\text{curl } \mathbf{n})_y^2\}$$
$$= \mu_1(\mathbf{n}.\text{curl } \mathbf{n})^2 + \mu_2(\mathbf{n} \times \text{curl } \mathbf{n})^2 \tag{3.12}$$

Finally we discuss the cross terms F_{ec}. The only term linear in $(\text{curl } \mathbf{n})_z$ compatible with C_∞ symmetry is

$$(\text{curl } \mathbf{n})_z(e_{xx} + e_{yy}) = (\mathbf{n}.\text{curl } \mathbf{n})\text{div } \mathbf{n} \tag{3.13}$$

This is odd in $\mathbf{n}$ and thus ruled out by requirement (2). The only terms linear in $(\text{curl } \mathbf{n})_x$ and $(\text{curl } \mathbf{n})_y$ compatible with C_∞ symmetry are

$$(\text{curl } \mathbf{n})_x e_{xz} + (\text{curl } \mathbf{n})_y e_{yz} \equiv 0 \quad (\text{see eqn (3.6)})$$
$$(\text{curl } \mathbf{n})_y e_{zx} - (\text{curl } \mathbf{n})_x e_{zy} = \tfrac{1}{2}(\mathbf{n} \times \text{curl } \mathbf{n})^2$$

Thus the most general form for F_{ec} is

$$F_{ec} = \nu(\mathbf{n} \times \text{curl } \mathbf{n})^2 \tag{3.14}$$

Regrouping eqns (3.11), (3.12), and (3.14) we may write the distortion energy in the form

$$F_d = \tfrac{1}{2}K_1(\text{div } \mathbf{n})^2 + \tfrac{1}{2}K_2(\mathbf{n}.\text{curl } \mathbf{n})^2 + \tfrac{1}{2}K_3(\mathbf{n} \times \text{curl } \mathbf{n})^2, \tag{3.15}$$

Eqn (3.15) is the fundamental formula of the continuum theory for nematics.

† I am indebted to Dr. L. Brun for correcting a mistake in eqn (3.10).

3.1.3. Discussion of the distortion-energy formula

3.1.3.1.   *Three elastic constants.* The constants K_i $(i = 1, 2, 3)$ introduced in eqn (3.15) are respectively associated with the three basic types of deformation displayed in Fig. 3.1:

$$K_1: \text{conformations with div } \mathbf{n} \neq 0 \qquad \text{(splay)};$$

$$K_2: \text{conformations with } \mathbf{n} \cdot \text{curl } \mathbf{n} \neq 0 \quad \text{(twist)};$$

$$K_3: \text{conformations with } \mathbf{n} \times \text{curl } \mathbf{n} \neq 0 \quad \text{(bend)}.$$

It is possible to generate deformations which are pure splay, pure twist, or pure bend. Thus each constant K_i must be *positive;* if not, the undistorted nematic conformation would not correspond to a minimum of the free energy F_d.

A remark on dimensions and orders of magnitude: F_d is an energy per cm³, $\mathbf{n}$ is dimensionless; thus from eqn (3.15) the elastic constants K_i have the dimension of energy/cm (or dynes). By a purely dimensional argument, we expect the K's to be of order U/a where U is a typical interaction energy between molecules, while a is a molecular dimension. Taking $U \sim 2$ kcal/mole (0·1 eV or 10^3 K) and $a \simeq 14$ Å we expect $K_i \sim 1\cdot4\ 10^{-13}$ erg/1·4 10^{-7} cm $= 10^{-6}$ dynes. This is indeed the correct order of magnitude; for PAA at 120°C the measured elastic constants [17] are

$$K_1 = 0\cdot7 \times 10^{-6} \text{ dynes}$$

$$K_2 = 0\cdot43 \times 10^{-6} \text{ dynes}$$

$$K_3 = 1\cdot7 \times 10^{-6} \text{ dynes}$$

Note that the bending constant is much larger than the others, while the twist constant is small (this difference can be understood qualitatively by looking at models with hard rods). It must also be emphasized that the K_i values decrease rather strongly when T increases (they behave roughly like the square of the order parameter†) but their ratio is nearly independent of T (except for some special cases to be discussed in Chapter 7).

It is also useful to estimate the magnitude of the distortion energy, per molecule, for a typical distortion taking place in a distance l: this will be roughly $F_d a^3 \sim (K/l^2)a^3 \sim U(a/l)^2$. Thus, in the continuum limit $(a \ll l)$ it represents only a very small fraction of the total energy.

---

† The detailed behaviour of the elastic constants versus temperature is reviewed (and compared with certain theoretical proposals) in a forthcoming paper by H. Gruler, *Z. Naturforsch.,* to be published.

The number of independent elastic constants has been a matter of wide discussion in the past; from a microscopic analysis. Oseen [1] produced eqn (3.15) plus a term of the form (3.13), because, at those early times (1933) the equivalence of $\mathbf{n}$ and $-\mathbf{n}$ was not fully recognised. Frank [3] showed that the term (3.13) should be discarded, but then added to eqn (3.15) an extra term, closely related to the coefficient λ_3 of eqn (3.9). Ericksen [6] first recognised that this extra term could be integrated by parts and lumped into the other ones, as shown here through eqn (3.10). Thus it has taken about 30 years to define the

TABLE 3.1 *Elastic constants of PAA*

$t/°C$	K_1/dynes	K_2/dynes	K_3/dynes	Reference
120	$5 \cdot 0 \times 10^{-7}$	$3 \cdot 8 \times 10^{-7}$	$10 \cdot 1 \times 10^{-7}$	[1]
125	$4 \cdot 5 \times 10^{-7}$	$2 \cdot 9 \times 10^{-7}$	$9 \cdot 5 \times 10^{-7}$	[1]
129	$3 \cdot 85 \times 10^{-7}$	$2 \cdot 4 \times 10^{-7}$	$7 \cdot 7 \times 10^{-7}$	[1]
120	$7 \cdot 01 \times 10^{-7}$	$4 \cdot 26 \times 10^{-7}$	—	[2]
124·9	$6 \cdot 06 \times 10^{-7}$	$3 \cdot 7 \times 10^{-7}$	—	[2]
130	$4 \cdot 84 \times 10^{-7}$	$2 \cdot 89 \times 10^{-7}$	—	[2]
129	—	$(3 \cdot 1 \pm 0 \cdot 6)$ $\times 10^{-7}$	—	[3]
—		$\times 10^{-7}$	—	

[1] V. Zvetkov, *Acta Phys.-chim. URSS* **6**, 866/(1937). (Method: Fredericks.) The χ values used here are those of: V. Zvetkov and A. Sosnovsky, *Acta Phys.-chim. URSS*, **18**, 358 (1943).
[2] A. Saupe, *Z. Naturforsch.* **15a**, 815 (1960). (Method: Fredericks.)
[3] G. Durand, L. Leger, F. Rondelez, and M. Veyssie, *Phys. Rev. Lett.*, **22**, 227 (1969); and *Orsay Liquid Crystal Group in liquid crystals and ordered fluids* (ed. by J. F. Johnson and R. S. Porter), Plenum Press, p. 447 (1970). (Method: transition, cholesteric ⇌ nematic induced by a magnetic field).

distortion free-energy unambiguously! A compilation of values of the elastic constants in PAA and MBBA, as obtained by different methods, is listed in Tables 3.1 and 3.2.

3.1.3.2. The one-constant approximation. Experimental applications of eqn (3.15), and the possible methods of determination for the three elastic constants K_i will be reviewed in the following Sections. However, as we shall see, in many cases the full form of eqn (3.15) is still too complex to be of practical use; either because the relative values of the three elastic constants K_i are unknown, or because the equilibrium equations derived from (3.15) are prohibitively difficult to solve. In such cases, a further approximation is often useful; this amounts to assuming all three elastic constants equal

$$K_1 = K_2 = K_3 = K. \tag{3.16}$$

TABLE 3.2 *Elastic constants of MBBA†*

Temperature t (°C)	K_1 (10^{-7} dyne)	K_2 (10^{-7} dyne)	K_3 (10^{-7} dyne)	Ratios	Reference
~ 22 $\left(\tau = \dfrac{t-t_{\mathrm{NI}}}{t_{\mathrm{NI}}+273\cdot 2} = 0\cdot 055\right)$	5·8		7	$\dfrac{K_3}{K_1} \simeq 1\cdot 25 \pm 0\cdot 05$	[1]
~ 24 $(\tau = 0\cdot 064)$		$K_{2,\mathrm{max}} = 3\cdot 8$			[1]
22	6·2±0·6		8·6±0·4		[2]
22			7·3±1·5		[3]
$(\tau = -0\cdot 04)$		3·4±0·3	8±0·8		[4]
22		3·34±0·04			[5]
22	5·3±0·5	2·2±0·7	7·45±1·1	$\dfrac{K_3}{K_1} = 1\cdot 4 \pm 0\cdot 2$	[6]
25	3·5				[7]
22		$3\cdot 35 \times 10^{-7}$			[8]
23			8·1	$\dfrac{K_3}{K_1} = 1\cdot 38$	[19]
				$\dfrac{K_3}{K_2} = 2\cdot 89$	
24			7·2±1		[10]
26	3·2±1		6·1±1	$0\cdot 85 < \dfrac{K_3}{K_1} < 1\cdot 4$	[11]
22				$\dfrac{K_3}{K_1} = 1\cdot 16 \pm 0\cdot 05$	[12]
23·5±0·5 $(\tau = -0\cdot 029)$	3·88		4·66		[13]

† In many of the methods used, what is measured is the ratio K/χ_a of an elastic constant K to a diamagnetic anisotropy χ_a. To make meaningful comparisons, we have systematically used the χ_a values of H. GASPAROUX *et al.*, *C.r. Acad. Sci.* (*Paris*), **272B**, 1168 (1971).

[1] HALLER, I. *J. chem. Phys.* **57**, 1400 (1972).
[2] ROBERT, J. and LABRUNIE, G. Proceeedings 4*th* international liquid crystal conference, Kent, 1972, to be published.
[3] GALERNE, Y., DURAND, G., VEYSSIE, M. and PONTIKIS, V. *Phys. Letts.* **38A**, 449, (1972).
[4] WILLIAMS, C. and CLADIS, P. E. *Solid State Commun.* **10**, 357, (1972).
[5] CLADIS, P. E. *Phys. Rev. Letts.* **28**, 1629, (1972).
[6] RONDELEZ, F. and HULIN, J. P. *Solid State Commun.*, **10**, 1009, (1972).
[7] REGAYA, B., GASPAROUX, H. and PROST, J. *Revue Phys. appl.* (*France*), **7**, 83, (1972).
[8] LEGER, L. *Solid State Commun.*, **10**, 697, (1972).
[9] LEGER, L. *Solid State Commun.*, and *Mol. Cryst. Liquid Cryst.*, to be published.
[10] MARTINAND, J. L. and DURAND, G., *Solid State Commun.*, **10**, 815, (1972).
[11] PIERANSKI, P., BROCHARD, F. and GUYON, E. *J. Phys.* (*France*), **33**, 681 (1972).
[12] WAHL, J. and FISCHER, F., *Proceedings 4th International liquid crystal conference, Kent* 1972 to be published.
[13] GRULER, H., unpublished results.

The free energy then takes the form

$$F_d = \tfrac{1}{2}K\{(\text{div } \mathbf{n})^2 + (\text{curl } \mathbf{n})^2\}$$
$$= \tfrac{1}{2}K\, \partial_\alpha n_\beta\, \partial_\alpha n_\beta \tag{3.17}$$

(the latter form of F_d differing from the preceding one only by surface terms).

The numerical values quoted above for PAA show that eqn (3.17) cannot be quantitatively correct. Nevertheless the simpler form of eqn (3.17) makes it a valuable tool to reach a qualitative insight on distortions in nematics.

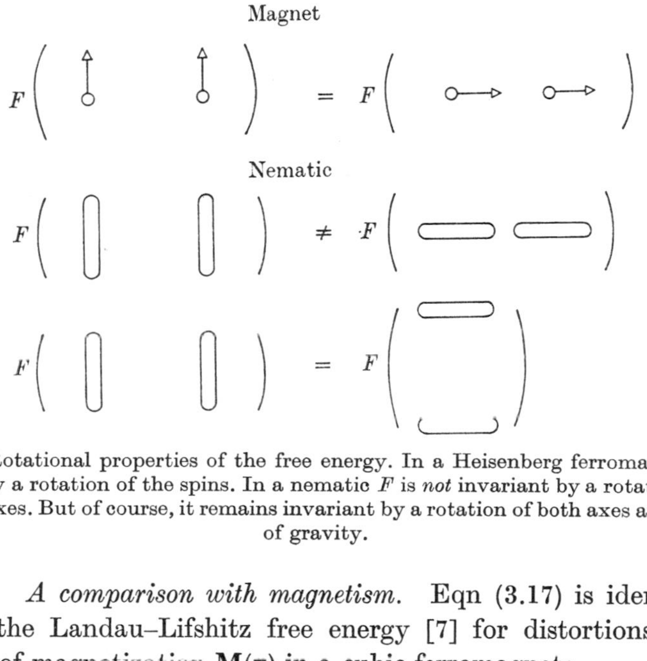

Magnet

Nematic

FIG. 3.2. Rotational properties of the free energy. In a Heisenberg ferromagnet, F is invariant by a rotation of the spins. In a nematic F is *not* invariant by a rotation of the molecular axes. But of course, it remains invariant by a rotation of both axes and centres of gravity.

3.1.3.3. *A comparison with magnetism.* Eqn (3.17) is identical in form to the Landau–Lifshitz free energy [7] for distortions in the direction of magnetization $\mathbf{M}(\mathbf{r})$ in a cubic ferromagnet:

$$F_m = \tfrac{1}{2}A\, \partial_\alpha M_\beta\, \partial_\alpha M_\beta. \tag{3.18}$$

But it must be realised that F_m is rigorous for a Heisenberg ferromagnet, while eqn (3.17) represents only an arbitrary simplification of the correct eqn (3.15) for a nematic. The source of this difference is as follows. In a Heisenberg ferromagnet, rotation of the spins (keeping the atoms fixed) does not change the energy; as shown in ref. [7] this immediately leads to eqn (3.18). On the other hand, in a nematic phase, a rotation of the molecules (keeping their centres of gravity fixed) changes the energy (see Fig. 3.2). Thus the free energy (3.15) is invariant only by a *simultaneous* rotation of both molecular axes and centres of gravity.

3.1.3.4. Equilibrium conditions (in the absence of external fields). To obtain the conditions for equilibrium in the bulk we write that the total distortion energy $\mathscr{F}_\mathrm{d} = \int F_\mathrm{d}\,\mathrm{d}\mathbf{r}$ is a minimum with respect to all variations of the director $\mathbf{n}(\mathbf{r})$ which keep $\mathbf{n}^2 = 1$. According to the rule of Lagrange, the latter condition may be taken into account by writing

$$\delta\mathscr{F}_\mathrm{d} = \int \mathrm{d}\mathbf{r}\tfrac{1}{2}\lambda(\mathbf{r})\,\delta(\mathbf{n}^2) = \int \mathrm{d}\mathbf{r}\lambda(\mathbf{r})\mathbf{n}\cdot\delta\mathbf{n}(\mathbf{r}), \qquad (3.19)$$

where $\lambda(\mathbf{r})$ is an arbitrary function of $\mathbf{r}$, and $\delta\mathbf{n}$ is an arbitrary variation of the vector $\mathbf{n}$.

In our case F_d, as given by eqn (3.15), is a quadratic function of the gradients $g_{\alpha\beta} = \partial_\alpha n_\beta$, and is also a function of $\mathbf{n}$ itself. Imposing a small variation $\delta\mathbf{n}(\mathbf{r})$ at all points we have

$$\delta\mathscr{F}_\mathrm{d} = \int \left\{\frac{\partial F_\mathrm{d}}{\partial n_\beta}\,\delta n_\beta + \frac{\partial F_\mathrm{d}}{\partial g_{\alpha\beta}}\,\partial_\alpha(\delta n_\beta)\right\}\mathrm{d}\mathbf{r} \qquad (3.20)$$

We integrate the second term by parts, and neglect the surface term†

$$\delta\mathscr{F}_\mathrm{d} = \int \left\{\frac{\partial F_\mathrm{d}}{\partial n_\beta} - \partial_\alpha\left(\frac{\partial F_\mathrm{d}}{\partial g_{\alpha\beta}}\right)\right\}\delta n_\beta\,\mathrm{d}\mathbf{r}.$$

Inserting this in eqn (3.19) and writing that the Lagrange condition must be satisfied for any function $\delta\mathbf{n}(\mathbf{r})$ we reach the equilibrium equation

$$h_\beta \equiv -\frac{\partial F_\mathrm{d}}{\partial n_\beta} + \partial_\alpha\left(\frac{\partial F_\mathrm{d}}{\partial g_{\alpha\beta}}\right) = -\lambda(\mathbf{r})n_\beta \qquad (3.21)$$

We call the vector $\mathbf{h}$ the *molecular field* (a notation which is derived from magnetism). Equation (3.21) then states that, in equilibrium, *the director must be at each point parallel to the molecular field*. (In this form, we dispose of the arbitrary function $\lambda(\mathbf{r})$).

Inserting eqn (3.15) for F_d into eqn (3.21), one arrives to a rather complicated form for the molecular field.

$$\mathbf{h} = \mathbf{h}_\mathrm{S} + \mathbf{h}_\mathrm{T} + \mathbf{h}_\mathrm{B},$$

where the three parts refer to splay, twist, and bending respectively, and are given by

$$\left.\begin{aligned} \mathbf{h}_\mathrm{S} &= K_1\,\nabla(\mathrm{div}\ \mathbf{n}) \\ \mathbf{h}_\mathrm{T} &= -K_2\{A\ \mathrm{curl}\ \mathbf{n} + \mathrm{curl}(A\mathbf{n})\} \\ \mathbf{h}_\mathrm{B} &= K_3\{\mathbf{B}\times\mathrm{curl}\ \mathbf{n} + \mathrm{curl}(\mathbf{n}\times\mathbf{B})\} \end{aligned}\right\} \qquad (3.22)$$

where $A = \mathbf{n}\cdot\mathrm{curl}\ \mathbf{n}$ and $\mathbf{B} = \mathbf{n}\times\mathrm{curl}\ \mathbf{n}$.

† The surface term is discussed in § 3.5.

In the one constant approximation $\mathbf{h}$ becomes simpler:† from eqn (3.17) and (3.21)

$$\mathbf{h} = K \, \nabla^2 \mathbf{n} \qquad (3.23)$$

A general discussion of eqn (3.21) has been given by Ericksen [8]. However, in practice, the equilibrium equations are rarely used in this general form; it is often more convenient to express the unit vector $\mathbf{n}$ in terms of suitably chosen polar angles, and to write that $\mathscr{F}_\mathrm{d}$ is a minimum with respect to all variations in these angles.

Let us show this by an example: the pure twist deformation shown in Fig. 3.1 and in more detail on Fig. 3.7. The director $\mathbf{n}$ is a function of y only, with the following components

$$n_z = \cos \theta(y) \qquad n_x = \sin \theta(y) \qquad (3.24)$$

Inserting this form into eqn (3.15) shows that

$$F_\mathrm{d} = \tfrac{1}{2} K_2 \left(\frac{\partial \theta}{\partial y} \right)^2 \qquad (3.25)$$

The energy $\mathscr{F}_\mathrm{d}$ (calculated per unit area at the $y0z$ plane) is

$$\mathscr{F}_\mathrm{d} = \tfrac{1}{2} K_2 \int \left(\frac{\partial \theta}{\partial y} \right)^2 \mathrm{d}y$$

We write that $\delta \mathscr{F}_\mathrm{d} = 0$ for an arbitrary variation in form of the twist $\delta \theta(y)$. This gives

$$0 = \delta \mathscr{F}_\mathrm{d} = K_2 \int \frac{\partial \theta}{\partial y} \frac{\partial}{\partial y} \, \partial \theta \, \mathrm{d}x$$

$$= -K_2 \int \frac{\partial^2 \theta}{\partial y^2} \, \delta \theta \, \mathrm{d}x + \text{surface terms}$$

Thus the condition for local equilibrium is

$$K_2 \frac{\partial^2 \theta}{\partial y^2} = 0 \qquad (3.26)$$

and this may readily be integrated to give

$$\frac{\partial \theta}{\partial y} = \text{constant.} \qquad (3.27)$$

---

† There is an apparent discrepancy between (3.23) and the form obtained from (3.22) by putting the three elastic constants equal. In fact the two expressions are equivalent, for the following reason: one can always add to F a term of the form $\tfrac{1}{2}\mu(\mathbf{r})\mathbf{n}^2 = \tfrac{1}{2}\mu$ without changing the distortion energy. This changes $\mathbf{h}$ into $\mathbf{h} + \mu\mathbf{n}$: two fields differing only by $\mu(\mathbf{r})\mathbf{n}$ are equivalent.

6

Distortions of this type have been produced and were studied very early on by Mauguin [9]—they are called *'plages tordues'* (twisted areas) in the French literature. To understand in detail how these distortions are produced, however, we must discuss how one can impose the orientation of **n** at the boundaries of the sample; this will be done in 3.1.2.4. below.

A final word of caution: when we start with a postulated form of the distortion, such as eqn (3.24) above, we must always check at the end of the calculation that the molecular field **h** is collinear with **n**, in agreement with (3.21). For the present case this is true; div **n** and **B** are equal to zero, **h** reduces to $\mathbf{h}_T$, $A = \partial\theta/\partial y = $ constant, and $h_T = -2K_2A^2\mathbf{n}$.

3.1.4. Boundary effects

Equation (3.15) defining the distortion energies in the bulk of the nematic phase must be, in principle, supplemented by a description of the energies associated with the *surface* of the sample. We shall now show, however, that in most practical conditions, the surface forces are strong enough to impose a well-defined direction to the director **n** at the surface; this is what we call 'strong anchoring.' Then, instead of minimizing the sum of bulk+surface energies, it is sufficient to minimize only the bulk terms, with fixed boundary conditions for **n**.

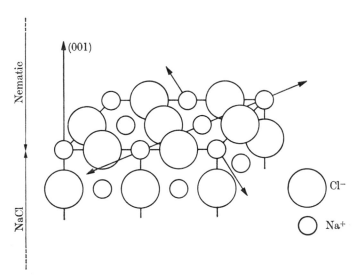

FIG. 3.3. Two easy directions for PAA molecules at the surface of a sodium chloride crystal (surface plane 001).

3.1.4.1. Easy directions at the surface.   Let us restrict our attention to a plane surface, separating the nematic from an external medium (solid, liquid, or gas). We assume first that the bulk of the nematic is undistorted, with a certain (constant) director **n**; thus there is no bulk contribution from eqn (3.15). What are the 'easy' directions i.e. the directions of **n** which minimize the energy of the surface region? The answer depends on the nature of the external medium.

(1) If the external medium is a *single crystal*, and the interface corresponds to a well-defined crystallographic plane, the easy directions form a discrete manifold; this has been shown experimentally in a certain number of cases by Grandjean [10]. For

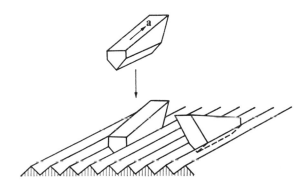

F ɪ ɢ. 3.4. A mechanical model showing how certain molecules may have an easy axis at a finite angle from the crystallographic axes of the surface plane.

instance, with *p*-azoxyanisole on a (001) face of sodium chloride, there are two preferred directions lying in the plane of the surface, namely (110) and (1$\bar{1}$0) (see Fig. 3.3). It often happens, as in this example, that the easy directions are parallel to some simple crystallographic axes; but from a purely geometrical point of view there is no rule which imposes this; a counter-example is shown in Fig. 3.4. When there is more than one easy direction we say that the problem is *degenerate*.

(2) Chatelain has shown that when a glass surface is carefully *rubbed in one direction*, the nematic molecules of materials such as PAA tend to line up along this direction [11*a*]. A clear cut microscopic

explanation of this effect is still lacking,† but in any case the rubbing has created one easy direction in the plane of the glass.‡ Another very powerful technique giving this type of anchoring has been invented very recently [11b]: certain thin films (nominal thickness 3–30 nm) evaporated on glass at an *oblique* incidence, give an excellent anchoring, with the easy axis in the plane of incidence. Depending on

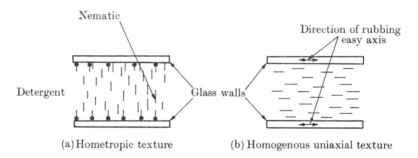

(a) Hometropic texture   (b) Homogenous uniaxial texture

Fig. 3.5. Two types of nematic single crystals obtained between parallel glass walls.

the angle of incidence, the size of the constituent grains etc. the easy axis may be either in the plane of the glass, or tilted with respect to this plane. (E. Guyon, P. Pieranski, M. Boix, *Lett. App. Eng. Sci.* **1**, 19, 1973).

(3) If the external medium is *isotropic* (a fluid, a clear glass surface, etc …) we can have the following three situations:

(*a*) The normal to the surface is the easy direction. In the early days, this condition was difficult to achieve; it could sometimes be obtained on glass surfaces etched with chromic acid. More recently it has been found that suitable detergents, when incorporated by small amounts in the nematic, tend to attach their polar head on a glass surface, as shown in Fig. 3.5. This imposes a normal orientation. With such a treatment, between two parallel glass plates, it is thus easy to prepare a 'homotropic texture,' i.e a single domain with optical axis

† Fatty acids brought on the surface by the rubbing process may play an important rôle. Another explanation is discussed in the problem on page 77.

‡ It has not always been proved, however, that the easy direction is exactly in the plane of the glass.

normal to the walls. Similar results are obtained by treating the
surfaces with polyamides or with certain lipids, such as lecithin [12].

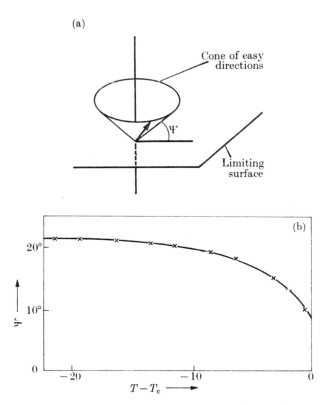

FIG. 3.6. (a) 'Conical' boundary conditions at the interface between a nematic and an
isotropic medium. (b) Variation of the angle ψ with temperature for the MBBA–air
interface (after ref. [13]).

(b) All directions in the plane of the surface are easy directions;
this is probably found for the free surface of PAA.

(c) The easy directions make an angle ψ with the surface, and are
thus distributed on a cone (Fig. 3.6). It has been shown (by a study of
optical reflectance) that this is the situation for the free surface of
MBBA [13]. The angle ψ is somewhat dependent of temperature, and
probably also very sensitive to surface contaminants.

In cases (b) and (c) there is a continuous manifold of easy directions; we call these cases *continuously degenerate*.

3.1.4.2.   Reaction of bulk distortions on the surface state.   We now
turn our attention to cases where the nematic arrangement is distorted
in the bulk, and investigate the effects of the distortion on the surface
properties. Let us start with our simple situation, corresponding to pure
twist (Fig. 3.7). We assume that the surface (the $x0z$ plane) has one easy

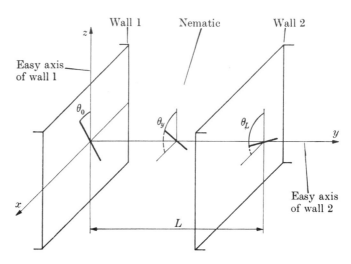

FIG. 3.7. '*Plage tordue*' between two polished glass plates.

direction ($0z$). At a distance y from the surface the angle between **n** and
$0z$ is $\theta(y)$. The bulk distortion energy (per unit area in the $y0z$ plane) is,
as we have seen:

$$\mathcal{F}_{\text{bulk}} = \int_0^L \tfrac{1}{2}K_2\left(\frac{\partial\theta}{\partial y}\right)^2 \mathrm{d}y$$

and writing that $\mathcal{F}$ is a minimum leads to the condition $\partial\theta/\partial y =$
constant (eqn. 3.27).

To impose a non-zero value of the bulk twist $\partial\theta/\partial y$ we fix the value
of θ at a large distance L from the surface which we call $\theta(L)$. On the
other hand, the value of θ near the surface will turn out to be small
(since $\theta = 0$ is the easy direction). Thus the bulk twist will be nearly
equal to $\theta(L)/L$.

Near the surface, eqn (3.27) no longer holds: a possible curve for $\theta(y)$
in this region is shown in Fig. 3.8. One important property of this
profile is the following: provided that in all the surface region θ is small,

the equations ruling θ will be linear; if we change the bulk torsion $\theta(L)$, the profile $\theta(y)$ will simply scale up or down. The 'extrapolation length' b defined in Fig. 3.8 is independent of the magnitude of the torsion. All the information of interest for the continuum theory is contained in b.

To derive the profile $\theta(y)$ near the surface, and the length b, from microscopic calculations is far beyond our present means. However, we can obtain a qualitative estimate of b through a simple argument; let us

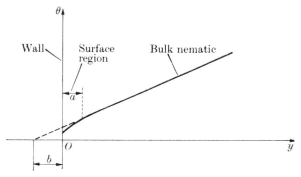

FIG. 3.8. Shape of the twisted conformation near the first wall of Fig. (3.7). In a region of molecular thickness a near the surface the twist $\partial\theta/\partial y$ is dependent on detailed molecular properties. Out of this region the twist is essentially constant. b is the 'extrapolation length': depending on the strength of the alignment along the easy axis at the wall, b may be $\sim a$ or $\gg a$.

assume that eqn (3.27) holds down to the surface, i.e.

$$\frac{\partial\theta}{\partial y} = \frac{\theta(L)-\theta(0)}{L} \tag{3.28}$$

but that there is a surface term in the energy, favouring $\theta(0) = 0$.

$$\mathscr{F}_{\text{surf}} = \tfrac{1}{2}A\theta^2(0) \qquad (\theta(0) \ll 1) \tag{3.29}$$

where A is a positive constant with the dimensions of surface tension. Then the total energy (bulk+surface) per cm² of wall is:

$$\mathscr{F} = \tfrac{1}{2}A\theta^2(0)+\tfrac{1}{2}K_2 L\left\{\frac{\theta(L)-\theta(0)}{L}\right\}^2. \tag{3.30}$$

Minimizing this with respect to $\theta(0)$ we find

$$\theta(0) = \frac{K_2}{A}\frac{\partial\theta}{\partial y} \tag{3.31}$$

$$\mathscr{F} = \mathscr{F}_{\text{bulk}}+\frac{1}{2}\frac{K_2^2}{A}\left(\frac{\partial\theta}{\partial y}\right)^2. \tag{3.32}$$

Comparing eqn (3.31) with the definition of the extrapolation length b on Fig. 3.8 we see that, in our approximation

$$b = K_2/A. \tag{3.33}$$

The constant A is of order U_{WN}/a^2 where U_{WN} is the anisotropic part of the interaction between the wall and one nematic molecule lying against the wall, and a represents an average molecular dimension. We have already seen that $K_{22} \sim U/a$, where U is the nematic–nematic interaction. This leads to:

$$b \sim a\frac{U}{U_{WN}}. \tag{3.34}$$

Equation (3.34) is the fundamental formula for boundary effects. In practice there are two possibilities:

(1) *Strong anchoring:* if U_{WN} is comparable to (or larger than) U, the extrapolation length b is comparable to the molecular dimensions. Also from eqn (3.32) we see that the ratio

$$\frac{\mathscr{F}_{\text{surf}}}{\mathscr{F}_{\text{bulk}}} = \frac{K_2}{AL} = \frac{b}{L} \sim \frac{a}{L} \tag{3.35}$$

Thus, in the continuum limit ($a/L \ll 1$) we may neglect the surface energy, put $b = 0$, and use the effective boundary condition $\theta(0) = 0$. We call this situation 'strong anchoring.' We expect it to hold usually for a crystalline wall.

(b) *Weak anchoring:* if $U_{WN} \ll U$, the extrapolation length b may become much larger than the molecular dimensions a. The angle of rotation of the molecules at the surface is $\theta(0) \sim b\theta(L)/L$. If the torsion is strong enough ($\theta(L) \sim 1/b$), the angle $\theta(0)$ becomes large;† an external constraint can disrupt the surface alignment. We call this 'weak anchoring': it might be found with certain glass surfaces polished by the method of Chatelain [11].

The above results have been obtained for a particularly simple geometry (twist deformation, plane surface, etc.) and for a non-degenerate problem (one single easy direction on the surface). A complete formal discussion, covering all cases (and in particular the continuously degenerate cases), has not yet been carried out. However, it appears plausible that the results will remain meaningful; we are thus led to the following formulation:

In all cases where $U_{WN} \sim U$ (strong anchoring) the **n**-dependent terms in the surface energies may be omitted. The effect of the surface is simply to impose certain easy directions to **n**.

---

† Of course if $\theta(0) \sim 1$ the surface energy term (29) must be amended.

Problem: study the contact of a nematic with a rubbed solid surface, assuming that rubbing has created an undulating surface, departing from the plane $x0y$ by a small amount (see Fig. 3.9):

$$\zeta(x, z) = u \cos(qx) \qquad (qu \ll 1)$$

Assume that at all points of the surface, the director must be *tangential*. Derive the energy of the conformation where the director $\mathbf{n}$, at large distances from the surface, takes a fixed orientation ($n_y = 0$, $n_x = \sin \theta_0$, $n_z = \cos \theta_0$).

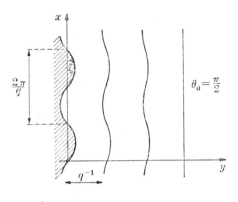

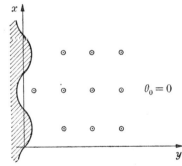

FIG. 3.9. Contact between a nematic and an undulating wall; the molecules are supposed to be tangential to the local wall surface. For $\theta_0 = 0$ this can be obtained without distortion. For $\theta_0 = \pi/2$ there must be some distortions extending up to a thickness q^{-1} in the nematic.

Solution: for $\theta_0 = 0$ there is no distortion energy. For $\theta_0 = \pi/2$ the arrangement is distorted; at the surface, we must have (to first order in $\nabla \xi \sim qu$)

$$n_y = -\frac{\partial \zeta}{\partial x} = uq \sin(qx)$$

Inside the nematic n_y is small (of first order in qu), $n_x = 1$ (to order q^2u^2) and $n_z = 0$. Making for simplicity the one constant approximation (eqn 3.17), we

obtain for the free energy
$$\mathscr{F} = \int \tfrac{1}{2} K (\nabla n_y)^2 \, \mathbf{dr},$$

and the condition $\mathscr{F}$ = minimum leads to $\nabla^2 n_y = 0$. The solution compatible with the value of n_y at the boundary is

$$n_y = uq \sin qx \, e^{-qy}.$$

The energy per unit area of wall is

$$\mathscr{F}_{\text{surf}} = \tfrac{1}{2} K \int u^2 q^4 \, e^{-2qy} \{ \sin^2 qx + \cos^2 qx \} \, dy = \tfrac{1}{4} K u^2 q^3$$

A similar calculation for arbitrary values of the angle θ_0 gives

$$\mathscr{F}_{\text{surf}} = \tfrac{1}{4} K u^2 q^3 \sin^2 \theta_0$$

In particular, comparing this with eqn (3.29), we find

$$A = \tfrac{1}{2} K u^2 q^3$$

The corresponding extrapolation length b would be of order

$$\frac{K}{A} = \frac{2}{u^2 q^3} = \frac{1}{4 \pi^3} \cdot \frac{\lambda^3}{u^2} \qquad \left(\lambda = \frac{2\pi}{q} \right)$$

Taking $\lambda = 10^{-4}$ cm, $u = 3 \times 10^{-5}$ cm, would give $b = 10^{-4}$ cm $= 1 \, \mu$m. This effect has been noticed independently, and is discussed in much more detail, by D. Berreman, *Phys. Rev. Lett.*, **28**, 1683, (1972).

3.1.5. Nematics transmit torques

Conventional liquids cannot transmit static torques, but nematics can. To understand this property, we shall start once more from the problem of pure one-dimensional twist between two parallel plates (Fig. 3.7). The first plate ($y = 0$) imposes one preferred direction (θ_0) and the other one ($y = L$) imposes another preferred direction (θ_L). Strong anchoring is assumed on both plates. In this situation, as already shown, the optimum state for the nematic is a uniform twist

$$\frac{\partial \theta}{\partial y} = \frac{\theta_L - \theta_0}{L} = \text{constant}.$$

The distortion energy $\mathscr{F}_d$ (per unit area of the plates) is

$$\mathscr{F}_d = \frac{K_2}{2L} (\theta_L - \theta_0)^2.$$

$\mathscr{F}_d$ *depends on the relative orientation of the two plates.* Thus the second plate (at $y = L$) is submitted to a torque, due to the nematic near it, of magnitude (per cm² of wall)

$$-\frac{\partial \mathscr{F}_d}{\partial \theta_L} = \frac{K_2}{L} (\theta_0 - \theta_L) = C. \tag{3.36}$$

Similarly the first plate is submitted to a torque from the nematic

$$-\frac{\partial \mathscr{F}_d}{\partial \theta_0} = \frac{K_2}{L}(\theta_L - \theta_0) = -C. \tag{3.37}$$

Eqn (3.37) may also be interpreted by saying that the first plate exerts the torque C on the nematic, and eqn (3.36) means that the same torque C is transmitted from the nematic to the second plate; thus nematics transmit torques. Note that $C = -K_2(\partial\theta/\partial y)$. Now consider the inside of the nematic and imagine it cut along a plane parallel to the plates $(y = y_0)$; the left half exerts on the right half a torque $-K_2(\partial\theta/\partial y)_{y_0}$. Finally let us consider a slab $(y_0 < y < y_1)$ in an arbitrary state of twist $\theta(y)$; the slab suffers from the left side a torque $-K_2(\partial\theta/\partial y)_{y_0}$ and from the right side a torque $+K_2(\partial\theta/\partial x)_{y_1}$. If we want the slab to be in equilibrium (and there are no other forces acting on it) these two torques must balance each other

$$K_2\left\{\left(\frac{\partial\theta}{\partial y}\right)_{y_1} - \left(\frac{\partial\theta}{\partial y}\right)_{y_0}\right\} = 0, \qquad \text{hence} \qquad \frac{\partial\theta}{\partial y} = \text{constant}.$$

Thus the equilibrium equation (3.27) may be interpreted as expressing the balance of torques.

We shall present later (in §3.5) a more general discussion of stresses and torques in nematics, after the introduction of magnetic and electric field effects.

3.2. Magnetic field effects

3.2.1. Molecular diamagnetism

Most organic molecules are diamagnetic, the diamagnetism being particularly strong when the molecule is aromatic. A benzene ring, for instance, when it experiences a magnetic field **H** *normal* to its plane, builds up a current inside the ring, which tends to reduce the flux going through it; thus the lines of force tend to be expelled, as shown in Fig. 3.10(*a*), and this raises the energy. On the other hand, if the field **H** is applied *parallel* to the ring, no current is induced, the lines of force are nearly undistorted and the energy is not raised [Fig. 3.10(*b*)]. Thus a benzene molecule tends to choose an orientation such that **H** is in the plane of the ring.

Typical nematogenic molecules such as MBBA or PAA have two aromatic rings (see the formulas in Chapter 1). If we have a nematic monodomain with a magnetic field **H** parallel to its optical axis **n**, **H** will

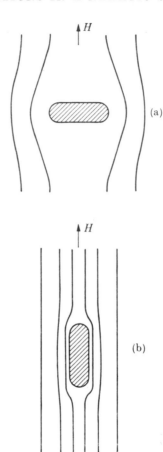

FIG. 3.10. Anisotropic diamagnetism of an aromatic ring. (The ring is normal to the sheet). In case (a) the lines of force are more distorted and the energy is higher.

be in the plane of the rings. But, if **H** is normal to **n**, for most molecules in the sample **H** will be at a finite angle with the rings; thus the lowest energy is obtained when the optical axis is *parallel to the field*. This is indeed what is observed for both PAA and MBBA.

Of course, the coupling which we have just discussed is very small. In fact, quantum mechanical calculations show that the coupling energy per molecule is of order $(\mu H)^2/E$ where μ is a Bohr magneton and E an electronic excitation energy.

Typically for $H \sim 10^4$ oersteds, $\mu H \sim 1$ K, $E \sim 10$ eV $\sim 10^5$ K and $(\mu H)^2/E \sim 10^{-5}$ K is very small when compared to $k_B T$. Thus a single molecule (e.g. in the vapour phase) would not be aligned in practice by any achievable field **H**, because of thermal agitation. However if instead of considering an isolated molecule, we go to a large nematic sample, we have now ($N \sim 10^{22}$) molecules which can only rotate in unison. Then the coupling energy is of order $N(\mu H)^2/E_{el} \gg k_B T$ and the sample will indeed put its optical axis parallel to $\mathbf{H}$. We shall see in this section that a sample of linear dimensions $\geqslant 0.1$ mm can be aligned in fields $H \sim 10^3$ oersteds. This provides a method of preparing nematic single crystals which supplements very usefully the technique of Chatelain (using rubbed glass walls).

In smaller samples there is often a competition between the alignment favoured by the wall and that favoured by the field, giving rise to *curved conformations*. We shall see that, from an experimental study of these conformations, one may extract useful informations on the Frank elastic constants.

To investigate these effects we need first to write down quantitatively the effect of the field $\mathbf{H}$ on a nematic of director $\mathbf{n}$; the magnetization $\mathbf{M}$ induced by $\mathbf{H}$ is of the form:

$$\left. \begin{array}{ll} \mathbf{M} = \chi_{\parallel}\mathbf{H} & \text{if } \mathbf{H} \text{ is parallel to } \mathbf{n} \\ \mathbf{M} = \chi_{\perp}\mathbf{H} & \text{if } \mathbf{H} \text{ is perpendicular to } \mathbf{n} \end{array} \right\} \qquad (3.44)$$

Here both $\chi_{\parallel}$ and $\chi_{\perp}$ are negative (diamagnetism) and small ($\sim 10^{-7}$ to 10^{-6} in c.g.s. electromagnetic units). If $\mathbf{H}$ makes an arbitrary angle with $\mathbf{n}$, the formula for $\mathbf{M}$ becomes

$$\mathbf{M} = \chi_{\perp}\mathbf{H} + (\chi_{\parallel} - \chi_{\perp})(\mathbf{H}.\mathbf{n})\mathbf{n}. \qquad (3.45)$$

In usual nematics the difference

$$\chi_a = \chi_{\parallel} - \chi_{\perp} \qquad (3.46)$$

is *positive*.† From the equation (3.45) giving the magnetization we can derive the free energy (per cm³):

$$F = F_d - \int_0^H \mathbf{M}.d\mathbf{H} = F_d - \tfrac{1}{2}\chi_{\perp}H^2 - \tfrac{1}{2}\chi_a(\mathbf{n}.\mathbf{H})^2. \qquad (3.47)$$

The term $\tfrac{1}{2}\chi_{\perp}H^2$ is independent of the molecular orientation (i.e. independent of $\mathbf{n}$) and may be omitted in all the cases to be discussed below. The last term is the interesting one; note that for $\chi_a > 0$ this term is minimized when $\mathbf{n}$ is collinear with $\mathbf{H}$, as stated before.

As already seen in §3.1.5, it is sometimes convenient to discuss a distorted state in terms of torques, rather than in terms of energies.

† $\chi_a = 1.21 \ 10^{-7}$ for PAA at 122°C.
  $1.23 \ 10^{-7}$ for MBBA at 19°C
More detailed references on χ_a are listed under (15).

According to a classic formula of high school physics, the magnetic torque $\mathbf{\Gamma}_M$ (per cm³) acting on the magnetization $\mathbf{M}$ is

$$\mathbf{\Gamma}_M = \mathbf{M} \times \mathbf{H} = \chi_a(\mathbf{n}.\mathbf{H})\mathbf{n} \times \mathbf{H}. \tag{3.48}$$

To write the equilibrium conditions under a field $\mathbf{H}$ we may operate in two (equivalent) ways:

(1) We minimize $\int F \, d\mathbf{r}$, where F is given by eqn (3.47). The result is still that $\mathbf{n}$ must, at each point, be parallel to a certain molecular field $\mathbf{h}$, defined by an equation similar to (3.21), but F substituted for F_d. This amounts to insert in $\mathbf{h}$, in addition to the terms in eqn (3.22) a magnetic contribution

$$\mathbf{h}_M = \chi_a(\mathbf{n}.\mathbf{H})\mathbf{H}. \tag{3.49}$$

(2) We may also write that for any volume element in the nematic the bulk torque (3.48) is balanced at equilibrium by the surface torques (3.36, 3.37) due to neighbouring regions.

We shall see examples of both approaches in the following discussions. Historically, equations equivalent to (3.47) and (3.48) seem to have been recorded first by Zocher [2].

3.2.2. Definition of a magnetic coherence length

3.2.2.1a. Simple twist. Let us consider the competing effects of a wall and of a magnetic field on the alignment of a nematic sample. Let us take the plane of the wall as the $(x0z)$ plane, the nematic lying in the region $y > 0$. We assume that, at the wall, there is one easy direction for the molecules ($\pm 0z$) and that "strong anchoring' (in the sense of §3.1.5) prevails.

We shall consider first the case where the magnetic field is along the x direction; i.e. normal to the wall's easy axis, but in the plane of the wall (Fig. 3.11). Certainly, if we go far enough from the wall (y large and positive) we will find the nematic aligned along $\mathbf{H}$, as explained in Section 3.2.1. Closer to the wall, there will be a transition layer, where the nematic molecules stay parallel to the $x0z$ plane, but make a variable angle $\theta(y)$ with the z direction. This is a situation of pure twist, which is particularly simple to analyse.

The magnetic torque $\mathbf{\Gamma}_M$ (eqn 3.48) is parallel to the y axis, and of magnitude

$$\Gamma_M = \chi_a H^2 \sin\theta \cos\theta. \tag{3.50}$$

Consider a slab (of area 1 cm² in the $x0z$ plane) extending from y to $y + dy$ in the nematic. As seen §3.1.5 this slab experiences a surface

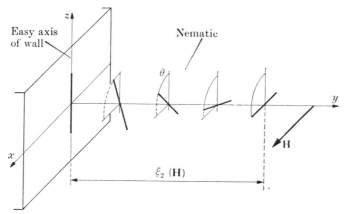

Fɪɢ. 3.11. Competition between wall-alignment and field-alignment. Case of pure twist.

torque $-K_2(d\theta/dy)|_y$ at (y) and a torque $+K_2(d\theta/dy)|_{y+dy}$ at $(y+dy)$. Furthermore, it is submitted to the bulk torque $\Gamma_M\,dy$. Writing that the sum of these three contributions vanishes, we arrive at the equilibrium equation:

$$K_2\frac{d^2\theta}{dy^2}+\chi_a H^2\sin\theta\cos\theta = 0. \qquad (3.51)$$

Let us define a length $\xi_2(H)$ through the equation

$$\xi_2(H) = (K_2/\chi_a)^{\frac{1}{2}}/H \qquad (3.52)$$

In terms of ξ_2, eqn (3.51) takes the form

$$\xi_2^2\frac{d^2\theta}{dy^2}+\sin\theta\cos\theta = 0.$$

It can be solved by the following trick; multiplying by $d\theta/dy$ we get

$$\xi_2^2\frac{d}{dy}\left\{\frac{1}{2}\left(\frac{d\theta}{dy}\right)^2\right\}+\frac{d}{dy}\left(-\frac{1}{2}\cos^2\theta\right) = 0.$$

This may be integrated to give

$$\xi_2^2\left(\frac{d\theta}{dy}\right)^2 = \cos^2\theta + \text{const.} \qquad (3.53)$$

Far from the wall $(y \to +\infty)$ we expect $\theta = \pi/2$ and $d\theta/dy = 0$. Thus the integration constant in eqn (3.53) must vanish.

$$\xi_2\frac{d\theta}{dy} = \pm\cos\theta. \qquad (3.54)$$

Both choices of sign are permissible, corresponding to a 'right-handed' or a 'left-handed' transition region. (These conformations are related to each other by a mirror reflection in the $y0z$ plane). Choosing for instance the $+$ve determination in eqn (3.54) we have

$$\frac{dy}{\xi_2} = \frac{d\theta}{\cos\theta} = -\frac{du}{\sin u},\qquad(3.55)$$

where we have found convenient to introduce $u = (\pi/2)-\theta$. Eqn (3.55) may be integrated in terms of

$$t = \tan(u/2)$$

$$\sin u = \frac{2t}{1+t^2}\qquad du = \frac{2\,dt}{1+t^2}$$

giving:

$$\frac{dy}{\xi_2} = -\frac{dt}{t}$$

$$t = \exp(-y/\xi_2).\qquad(3.56)$$

The integration constant in eqn (3.56) has been chosen to ensure that for $y = 0$ (at the wall) $u/2 = \pi/4$, $u = \pi/2$, and $\theta = 0$ as required.

Eqn (3.56) shows that the thickness of the transition layer is essentially $\xi_2(H)$, as defined in eqn (3.52). Taking $K_2 = 10^{-6}$, $\chi_a = 10^{-7}$, and $H = 10^4$ oerstedts, we get from (52) $\xi_2(H) \sim 3\ \mu m$. We meet here for the first time one of the fascinating properties of liquid crystals: with a rather weak external perturbation we can induce distortions on a scale comparable to an optical wavelength.

3.2.2.2.   Generalization.   The competition between the effects of a wall and the effects of a field may occur in a number of different geometries. One example is shown on Fig. 3.12 where the wall again

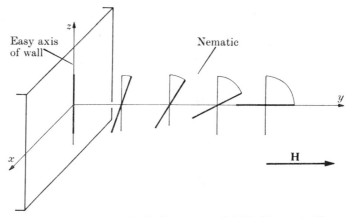

Fᴵɢ. 3.12. Competition between wall-alignment and field-alignment. The case shown involves a mixture of bending and splay.

imposes one easy direction Oz, but the field **H** is now normal to the wall (along y). In this case the distortion is a combination of bend and splay.

An interesting experiment, probing optically the transition layer, has been carried out recently on MBBA (J. Prost and H. Gasparoux, (1972) *C.r. hebd. Séan. Acad. Sci. Ser. C.* **273**, 355).

The algebra is slightly more involved, and will not be discussed here in detail. But the same general features are found; the effects of the wall decrease exponentially at large distances, and the thickness of the transition layer is a certain weighted mean of the lengths

$$\xi_1 = \left(\frac{K_1}{\chi_a}\right)^{\frac{1}{2}} \frac{1}{H} \qquad \xi_3 = \left(\frac{K_3}{\chi_a}\right)^{\frac{1}{2}} \frac{1}{H}, \tag{3.57}$$

related to the elastic constants K_1 and K_3 for splay and bend. The three lengths ξ_i ($i = 1, 2, 3$) are usually of comparable magnitude. In the one constant approximation ($K_i = K$) they become equal:

$$\xi(H) = \left(\frac{K}{\chi_a}\right)^{\frac{1}{2}} \frac{1}{H}. \tag{3.58}$$

We call $\xi(H)$ the *magnetic coherence length* of the nematic. The general significance of ξ may be explained with the following '*gedanke experiment*';[†] a large nematic sample is first aligned under a magnetic field **H**. Then, in a small region around one point 0 in the nematic, we act on the molecules with some other perturbing force of arbitrary strength. This creates a long range distortion in the surrounding nematic. It may be [16] shown that the angle between **n** and **H** decreases, at large distances r from 0, according to:

$$\frac{1}{r} \exp -\{r/\xi(\mathbf{H})\}.$$

Thus the nematic alignment is perturbed only in a region of linear dimensions ξ.

3.2.3. *The Frederiks transition*

3.2.3.1. *Experimental set-up.* Let us now consider a nematic single crystal, of thickness d (~ 20 μm), oriented between two solid plates. Various geometrical possibilities are shown on Fig. 3.13. At the surface of the plates, we assume strong anchoring. The easy direction imposed by both surfaces may be in the plane of the surfaces (cases 1 and 2) or

† Gedanke experiment = thought experiment, i.e. one carried out in the imagination, not in reality.

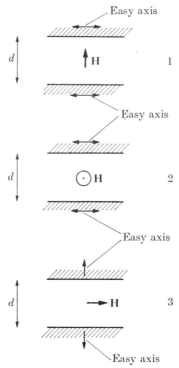

FIG. 3.13. The Frederiks transition for a nematic slab under a magnetic field **H**. At low H the molecules are parallel to the easy axis of the wall. For $H > H_c$ the molecules near the centre of the slab rotate towards **H**.

normal to it (case 3). A magnetic field **H** is applied *normal to the easy axis*. This choice implies that the magnetic torque

$$\Gamma_M = \chi_a(\mathbf{n}.\mathbf{H})\mathbf{n}\times\mathbf{H} \qquad \text{(see eqn 3.28)}$$

vanishes in the unperturbed configuration ($\mathbf{n}.\mathbf{H} = 0$). Thus the unperturbed configuration does satisfy the conditions for local equilibrium, even in the presence of **H**. However, it is clear that for large **H** (i.e., when $\xi(\mathbf{H}) \ll d$) the optimum state will correspond to a different conformation, with molecules aligned along H in most of the slab, except for two thin transition regions (of thickness ξ) near each wall. There will be a phase transition, at a certain critical value H_c of the field, between the unperturbed conformation and the distorted conformation (see Fig. 3.14).

A transition of this type was detected optically and studied by Frederiks in 1927 [17]. He was mainly concerned with case 3; he showed that the critical field H_c was inversely proportional to the sample thickness d.

$$H_c d = \text{const.} \tag{3.59}$$

(typical values being in the range of 10^4 oersteds for $d = 10 \ \mu\text{m}$). Soon afterwards H. Zocher [2] set up a first version of the continuum theory and showed that Frederiks law (3.59) is a natural consequence of it.

Let us first review briefly the methods which can be used to monitor the transition. They fall essentially into two classes: macroscopic measurements and optical measurements.

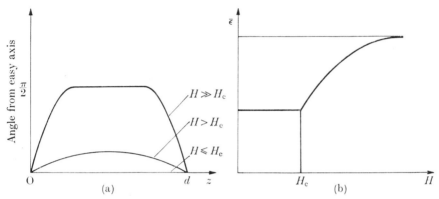

FIG. 3.14. (a) Progressive stages of the Frederiks transition as observed at different points in the slab. (b) Typical plot of apparent dielectric constant $\bar{\varepsilon}$ versus field in a Frederiks transition of type 1 or 3.

(1) Macroscopic measurements. Any anisotropic property, such as the dielectric constant [18], (or the thermal conductivity [19]) may be used as a probe of the average state of alignment. Usually the probing field (electric field, or thermal gradient) is applied *normal* to the slab. This turns out to be convenient in cases 1 and 3. The average dielectric constant $\bar{\varepsilon}(H)$ of the slab behaves, as a function of H, as shown on Fig. 3.14b. For $H < H_c$ the nematic is undistorted and $\bar{\varepsilon}(H) = \bar{\varepsilon}(0)$. For $H = H_c$ there is a discontinuity in the slope $d\bar{\varepsilon}/dH$, which allows for a rather precise determination of H_c. (This, as we shall see later, gives one elastic constant). For $H \gg H_c$, $\bar{\varepsilon}(H)$ saturates, since nearly all the sample is aligned along H. From a study of the saturation law one can derive further information about the elastic constants.

The method does not work for case 2, where the molecules stay constantly normal to the probing field. In fact, it could still be used, if the probing field was now made *parallel* to the slab; this is feasible if one measures certain transport properties such as electric inductance or diffusion of a dye.

(2) Optical observations. The distortions induced by H are usually very weak on the scale of an optical wave-length; this makes them rather hard to detect. To understand this more precisely, consider for instance the following experiment, corresponding to case 2: a light beam is admitted normal to the slab (along Z), and is linearly polarized parallel to the rubbing direction of the plates (we call this direction X). We explore the state of polarization of the transmitted beam, under a field H applied along Y.

The answer is a consequence of a certain 'adiabatic theorem' which was already recognized by Mauguin [9]: even when $H > H_c$, and when a good fraction of the sample has become aligned along Y, the outgoing beam is still linearly polarized along X (inside the sample, the direction of polarization remains everywhere parallel to the local optical axis. Since this axis is fixed along X at both plates, the light beam must thus come out polarized along X). Thus, in this experiment, it is *not* possible to detect the transition. In fact, changes in the outgoing polarization start to occur only at much higher fields, namely when $\xi(H)$ becomes comparable to the optical wave-length λ; then the adiabatic theorem breaks down, and for instance if we operated between crossed polarizers the dark field observed at low H would become progressively brighter.

Thus, to follow the transition optically, more advanced techniques are required, such as *conoscopy* (Fig. 3.15). Convergent light illuminates the slab. The interference between the two allowed polarizations for one beam, depends on the beam direction, and gives rise to a characteristic interference pattern. The method is particularly well suited to case 3; here, for $H < H_c$, the configuration is homeotropic, the interference pattern has full rotational symmetry around the normal to the slab, and is made of dark and bright circles. For $H > H_c$ the symmetry is reduced and the circles become distorted. Finally, for $H \gg H_c$ one obtains the characteristic pattern of a birefringent slab, which corresponds to hyperbolas with one axis parallel to $\mathbf{H}$.

The conoscopic method has been applied first by Meyer [20]. It is used not only in case 3, but also in some other geometries; in case 2 for instance, we start at $H = 0$ with a set of hyperbolas with one axis parallel to the rubbing direction. For $H > H_c$ the main effect is a rotation of the pattern, by an angle

$$\delta = \tfrac{1}{2} \tan^{-1}\left(\frac{\overline{\sin 2\phi}}{\overline{\cos 2\phi}}\right)$$

where ϕ is the local tilt of the optical axis induced by $\mathbf{H}$, and the bars

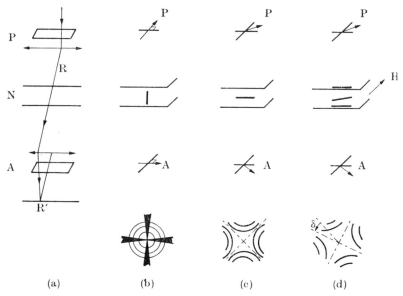

(a)   (b)   (c)   (d)

FIG. 3.15. Principle of conoscopy measurements. (a) Set up with polarizer P, analyser A, and light converging on the nematic sample N. To each direction for the incident light ray R, corresponds one image point R' in the observation plane; (b) homeotropic texture → circles; (c) planar texture → hyperbolas; (d) planar texture by a field N → hyperbolas expanded and rotated by an angle δ.

represent an average over the thickness of the slab [21]. The aspect of $\delta(H)$ is shown on Fig. 3.16.

3.2.3.2. Calculation of the critical fields. It may be shown rigorously that the Frederiks transition is of second order, i.e. that the distortions found just above the threshold field are small. Accepting this result, for the three situations of Fig. 3.13 the critical field H_c may be derived by a

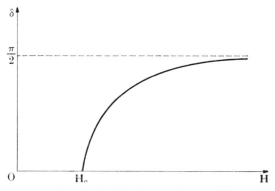

FIG. 3.16. Rotation angle δ versus field in the experiment of Fig. 3.4d (qualitative plot).

simple argument. Starting from the unperturbed state ($\mathbf{n} = \mathbf{n}_0$) we consider a slight deflection

$$\mathbf{n} = \mathbf{n}_0 + \delta\mathbf{n}(\mathbf{r})$$

where $\delta\mathbf{n}$ is normal to $\mathbf{n}_0$ (since $\mathbf{n}^2 = 1$) and is parallel to $\mathbf{H}$ (since this is the direction in which the molecules are solicited). It is natural to assume that the distortion depends only on z (where z is the normal to the slab). The distortion energy (3.15) reduces to

$$F_{\mathrm{d}} = \tfrac{1}{2}K_i\left(\frac{\partial\delta\mathbf{n}}{\partial z}\right)^2 \qquad \text{(case } i) \tag{3.60}$$

where $K_i = K_1$ (for case 1), K_2 (for case 2), K_3 (for case 3). The magnetic energy (3.47) gives a contribution

$$F_{\mathrm{M}} = -\tfrac{1}{2}\chi_a H^2\,\delta\mathbf{n}^2 \tag{3.61}$$

Since we have assumed strong anchoring, at both boundaries ($z = 0$ and $z = d$) $\delta\mathbf{n}$ must vanish. It is then convenient to analyse $\delta\mathbf{n}$ in a Fourier series

$$\left.\begin{aligned} \delta\mathbf{n} &= \sum_q \delta n_q \sin qz \\[2mm] q &= \nu\,\frac{\pi}{d} \qquad (\nu = \text{positive integer}) \end{aligned}\right\} \tag{3.62}$$

Inserting this form of $\delta\mathbf{n}$ in the free energy, and integrating over the thickness one obtains (per cm^2 of slab)

$$\mathscr{F} = \mathrm{d}(\bar{F}_{\mathrm{d}} + \bar{F}_{\mathrm{M}}) = \frac{d}{4}\sum_q \delta n_q^2 (K_i q^2 - \chi_a H^2) \tag{3.63}$$

If we want the unperturbed state to be stable, the increase in free energy $\mathscr{F}$ must be positive for all values of the parameters δn_q.

$$\chi_a H^2 < K_i q^2$$

The smallest value of q is $q = \pi/d$ ($\nu = 1$) corresponding to a distortion of half-wavelength d. Thus the threshold field $H_{\mathrm{c},i}$ corresponds to

$$\begin{aligned} \chi_a H_{\mathrm{c},i}^2 &= K_i(\pi/d)^2 \\ H_{\mathrm{c},i} &= (\pi/d)(K_i/\chi_a)^{\frac{1}{2}} \end{aligned} \tag{3.64}$$

Equation (3.64) may also be stated in the following terms; at the critical field the coherence length $\xi_i(H_{ci})$ is equal to d/π. The result (3.64) does show the $1/d$ dependence found experimentally by Frederiks. It also gives the principle of a simple *determination of the three elastic constants;* for applications of the method see the references listed under [17].

However, it must be emphasized that such determinations are meaningful only if:

(1) strong anchoring prevails [12]; and

(2) the easy direction is normal or parallel to the slab. If, instead, we had conical boundary conditions, there would be a non-zero magnetic torque on the molecules (and distortions would occur) for arbitrarily weak fields.

Problem: Starting from case (1) of Fig. 3.13, in low fields H, one twists the nematic by rotating the upper plate. What is the change in the critical field H_C as a function of the rotation angle ψ_0? (F. M. Leslie, **12**, 57, 1970).

Solution: Let us call z the normal to the plates (H is parallel to z) and look for a configuration of the form.

$$n_X = \cos\theta(z) \cos \phi \ (z)$$
$$n_Y = \cos\theta(z) \sin \phi \ (z)$$
$$n_Z = \sin\theta(z)$$

The free-energy density derived from eqn (3.15) and (3.47) is

$$F = \tfrac{1}{2}(K_1 \cos^2\theta + K_3 \sin^2\theta)(d\theta/dz)^2 + \tfrac{1}{2}\cos^2\theta(K_2 \cos^2\theta + K_3 \sin^2\theta)(d\phi/dz)^2$$
$$-\tfrac{1}{2}\chi_a H^2 \sin^2\theta$$

The unperturbed solution corresponds to $\theta = 0$ and $\phi = \phi_0(z) = \psi_0 z/L$ ($L =$ plate thickness). Let us look for a small deviation from this state $\theta \to \theta_1, (z)$ $\phi = \phi_0 + \phi_1(z)$. To second order in θ_1 and ϕ_1 we have

$$F = F_0 + \tfrac{1}{2}K_1(d\theta_1/dz)^2 + \tfrac{1}{2}(K_3 - 2K_2)(\psi_0/L)^2 \ \theta_1^2 + \tfrac{1}{2}K^2(d\phi_1/dz)^2 - \tfrac{1}{2}\chi_a H^2\theta_1^2$$

The term involving ϕ_1 is not coupled to θ_1, and is positive; it leads to no instability. But the terms involving θ_1 do lead to an instability provided that H is larger than a certain threshold $H_c(\psi_0)$. Operating as in eqns (3.62)(3.63) one obtains:

$$H_c^2(\psi_0) = H_c^2(0)[1 + \{(K_3 - 2K_2)/K_1\}(\psi_0/\pi)^2].$$

In practice, one cannot achieve values of the twist angle ψ_0 which are larger than $\pi/2$, because larger twists are always relaxed by the nucleation of a 'disclination loop' on some defect in the structure (for a definition of the disclination loop, see Chapter 4). This implies that:

$$\frac{H_c^2(\psi_0) - H_c^2(0)}{H_c^2(0)} \leqslant \frac{K_3 - 2K_2}{4K_1}.$$

Taking the accepted values of the constants K for PAA one finds that the right hand side is of order $\tfrac{1}{3}$. The experiment has been performed by Gerritsma, de Jeu, and van Zanten (*Phys. Lett.* **36A**, p. 389, 1971); it agrees very well with the formulas above,† except for very thin slabs ($\sim$7 μm) where surface irregularities, etc, probably become important.

At fields $H > H_c$, the set-up can be used as a magneto-optic device; take for instance the incident light polarized parallel to the easy axis of the first plate X. Depending on the size of H, we have two regimes:

(1) When H is not too high, the lightwave sees a slowly twisted nematic, and the polarization follows the angle $\phi(Z)$;

(2) When H is large the slab behaves essentially as a homeotropic monodomain, and the emerging polarization is not rotated. A detailed calculation of the optical transmission as a function of field has been carried out by Van Doorn *Phys. Lett.* **42A**, p. 537 (1973).

The transition between the two regimes is progressive, but reasonably sharp. This application (using electric rather than magnetic fields as the orienting agent) was invented independently by Helfrich and by Fergason.

Problem: A nematic slab (of thickness d) contains a homogeneous suspension of elongated magnetic grains. Assume that the grains always have their moment parallel to the director **n** ('ferronematic'). In the absence of external fields there is one easy direction ($\mathbf{n}_0$) either parallel or normal to the slab. A field **H** is now applied antiparallel to the magnetization. Compute the critical field H_c at which the conformation begins to be distorted (F. Brochard, 1970).

Solution: The two possible geometrical situations (a) and (b) are shown on Fig. 3.17. Again we look at small distortions

$$\mathbf{n} = \mathbf{n}_0 + \delta\mathbf{n}(z)$$

The magnetization **M** is of the form $\mathbf{M} = M\mathbf{n}$

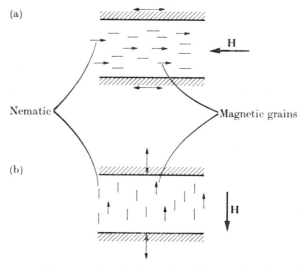

FIG. 3.17. Two possible geometries for the study of hysteresis cycles in ferronematics.

† The behaviour above the threshold was analysed by Shtrinkman *et al. Phys. Letters.* **37A**, 369, (1971).

where the magnitude M is fixed by the grain moments μ and concentration c_g ($M = \mu c_g$). We assume that the fields due to $\mathbf{M}$ itself are small in comparison with $\mathbf{H}$, and neglect them in what follows. The distortion free-energy is

$$F_d = \begin{cases} \tfrac{1}{2}K_1\left(\dfrac{\partial \delta n_z}{\partial z}\right)^2 + \tfrac{1}{2}K_2\left(\dfrac{\partial \delta n_y}{\partial z}\right)^2 & \text{(Case } a\text{)} \\[2ex] \tfrac{1}{2}K_3\left(\dfrac{\partial}{\partial z}\,\delta\mathbf{n}\right)^2 & \text{(Case } b\text{)} \end{cases}$$

The magnetic energy density (neglecting now all diamagnetic effects) is

$$-\mathbf{M}.\mathbf{H} \approx MH(1 - \tfrac{1}{2}(\delta n)^2).$$

As in the Fredericks transition, the instability occurs first for the Fourier component $\delta\mathbf{n}_q$ of wave vector $q = \pi/d$.

For this component the integrated free-energy is given by:

$$\frac{4\mathscr{F}}{d^2} = \begin{cases} (K_1\delta n_z^2 + K_2\delta n_y^2)\dfrac{\pi^2}{d^2} - MH(\delta n_z^2 + \delta n_y^2) & (a) \\[2ex] \left(K_3\dfrac{\pi^2}{d^2} - MH\right)\delta\mathbf{n}^2 & (b) \end{cases}$$

Case a: If $K_1 < K_2$ we have an instability with respect to splay and a critical field

$$H_{cs} = \frac{\pi^2 K_1}{Md^2}.$$

If $K_1 > K_2$ we have an instability with respect to twist, and a critical field

$$H_{ct} = \frac{\pi^2 K_2}{Md^2}.$$

Case b: Here the instability is associated with a bending distortion, and the critical field is

$$H_{cb} = \frac{\pi^2 K_3}{Md^2}.$$

Note the $1/d^2$ dependence of these critical fields. Typically for $M = 1$ gauss, $K = 10^{-6}$ dynes, and $d = 3$ μm, $H_c = 100$ gauss. The main experimental problem is to achieve a stable colloid with the required properties.

Problem: a horizontal field $\mathbf{H}$ is applied to the free surface of a nematic. Discuss the effect of $\mathbf{H}$ on the energy per unit area, for a small tilt of the free surface around a horizontal axis normal to $\mathbf{H}$ (Fig. 3.18).

Solution: we take the (xy) plane as representing the imperturbed surface ($\mathbf{H}$ being parallel to the x-axis). The perturbed surface has an altitude $\zeta = \epsilon x$ where $\epsilon(\ll 1)$ is the tilt angle. The director $\mathbf{n}$ is in the (x, y) plane, and makes a variable angle $\theta(z - \zeta)$ with the x-axis. Deep in the nematic $(-z + \zeta > \xi(\mathbf{H}))$ $\mathbf{n}$

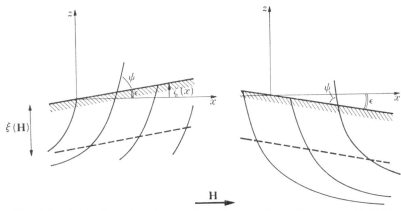

FIG. 3.18. Distortion of the free surface of a nematic under a magnetic field.

is parallel to $\mathbf{H}(\theta = 0)$. At the free surface, we assume strong anchoring $= \mathbf{n}$ must make a fixed angle ψ with the surface plane ($\psi = 0$ is the tangential case, $0 < \psi < \pi/2$ is a conical case, $\psi = \pi/2$ is the normal case). This means

$$\theta(z = \zeta) = \pm\psi + \varepsilon.$$

The two signs correspond to two types of distortion (see Fig. 3.18). Let us consider for instance the case $\varepsilon > 0$. Then the state of lowest distortion energy will correspond to $\theta(\zeta) = -\psi + \varepsilon$. The distortion energy is easily derived in the one constant approximation. The equations are similar in form to those of § 3.2.2, namely

$$\xi^2 \frac{d^2\theta}{dz^2} = \sin\theta\cos\theta$$

$$\xi^2 \left(\frac{d\theta}{dz}\right)^2 = \sin^2\theta$$

$$\xi \frac{d\theta}{dz} = -\sin\theta$$

($-$ determination, corresponding to the more favourable distortion)

The distortion energy per cm^2 is

$$\mathscr{F}_{\mathrm{d}} = \tfrac{1}{2}K \int_{-\infty}^{\zeta} \left\{ \frac{\sin^2\theta}{\xi^2} + \left(\frac{d\theta}{dz}\right)^2 \right\} dz = \frac{K}{\xi^2} \int_{-\infty}^{\zeta} \sin^2\theta \, dz$$

$$= \frac{K}{\xi} \int_{0}^{-\psi+\varepsilon} \sin\theta \, d\theta = \frac{K}{\xi}\{1 - \cos(\psi - \zeta)\}$$

Expanding to first order in ε we have

$$\mathscr{F}_{\mathrm{d}} - \mathscr{F}_{\mathrm{d}}^0 - \frac{K\sin\psi}{\xi}\varepsilon$$

Note that the other type of distortion, with θ going from 0 to $\psi + \epsilon$, would lead to a symmetrical formula

$$\mathscr{F}_{\mathrm{d}} = \mathscr{F}_{\mathrm{d}}^0 + \frac{K \sin \psi}{\xi} \epsilon.$$

To this energy must be added the conventional terms due to surface tension A and to gravity g;† calling ρ the density difference between the nematic and the over medium, this gives in general.

$$\mathscr{F}_{\mathrm{surface}} = \int\int \left\{ \tfrac{1}{2} A \left(\frac{\mathrm{d}\zeta}{\mathrm{d}x}\right)^2 + \tfrac{1}{2}\rho g \zeta^2(x) \pm \frac{K \sin \psi}{\xi} \frac{\mathrm{d}\zeta}{\mathrm{d}x} \right\} \mathrm{d}x \, \mathrm{d}y,$$

where we have generalized the result to arbitrary (slow) distortions $\zeta(x)$. For $\rho = 0$ the optimum slope would be

$$|\epsilon| = \left|\frac{\mathrm{d}\zeta}{\mathrm{d}x}\right| = \frac{K \sin \psi}{A\xi} \sim \frac{a}{\xi} \sin \psi \sim 10^{-3}$$

where we always call U a typical interaction energy, and put $A \sim U/2$. This tilt could be observed only if the densities are very nearly equal. If not, choosing a *definite sign* in $\mathscr{F}_{\mathrm{surface}}$ and minimizing gives the standard equation†

$$\frac{\mathrm{d}^2\zeta}{\mathrm{d}x^2} = \zeta/\lambda^2 \qquad \lambda = \sqrt{\left(\frac{A}{\rho g}\right)} \sim 1 \text{ mm}$$

The new term comes only in the boundary conditions at the walls of the container. Thus a surface of dimensions $\gg \lambda$ will remain horizontal, if the distortion is everywhere of the same type. But if we allow for alternating regions, using *both* types of distortion, then we can have a rugged surface; this will be discussed in Chapter 4.

3.3. Electric field effects in an insulating nematic

A static electric field E imposed on a nematic has many physical effects, some of which are quite complex. They have been reviewed recently by W. Helfrich: *Mol. Crystals* **21**, 187 (1973). In the present section we restrict our attention to the case of a *perfect nematic insulator*.‡

Even in this ideal situation, the coupling of an external electric field to a nematic medium involves at least two different processes. One is the anisotropy of the dielectric constant, and is similar in its consequences to the diamagnetic anisotropy described in §3.2. The second effect (invented by R. B. Meyer [22]) is far less trivial; in a deformed nematic, there should appear in many cases a spontaneous dielectric polarization. Conversely, an electric field may in some cases induce distortions in the bulk.

† See Landau and Lifshitz, *Fluid mechanics*, Pergamon, Oxford, ch. 7.

‡ Also, to avoid all possible injections of current carriers, the sample must be separated from all electrodes by suitable insulating layers.

3.3.1 Dielectric anisotropy

The static dielectric constants, measured along ($\epsilon_\parallel$) or normal ($\epsilon_\perp$) to the nematic axis are different. For a more general direction of the electric field **E**, the relation between electric displacement **D** and field has the form

$$\mathbf{D} = \epsilon_\perp \mathbf{E} + (\epsilon_\parallel - \epsilon_\perp)(\mathbf{n}.\mathbf{E})\mathbf{n}. \tag{3.65}$$

The difference

$$\epsilon_a = \epsilon_\parallel - \epsilon_\perp, \tag{3.66}$$

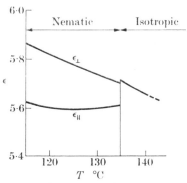

FIG. 3.19. Low-frequency dielectric constants for PAA. After W. Maier and G. Meier, *Z. Naturf.* **16a**, 470 (1961).

may be positive or negative, depending on the detailed chemical structure of the constituent molecules:

(1) if each molecule carries a permanent dipole moment parallel (or nearly parallel) to its long axis, the dipole can be oriented efficiently by a field **E** along the nematic axis (if the field is, say, along $+\mathbf{n}$, there will be more dipoles along $+\mathbf{n}$ than along $-\mathbf{n}$). But a field **E** normal to **n** has only weak effects. Thus in this case we have $\epsilon_\parallel > \epsilon_\perp$. In practice to realize large values of ϵ_a, it is efficient to attach at one end of the molecule a rather strongly polar group (assuming that it does not destroy the nematic order), pointing along the long axis, such as a —C≡N group.

(2) if there is a permanent dipole moment which is more or less normal to the long axis, the situation is reversed, and $\epsilon_\parallel < \epsilon_\perp$. This is the case in particular with PAA, where the moment associated with the N—O group is dominant. A plot of dielectric constants versus temperature for PAA is shown on Fig. (3.19).

The dielectric anisotropy leads, in principle, to a method of alignment by electric fields. The electric contribution to the thermodynamic potential is (per cm³)†

$$-\frac{1}{4\pi} \int \mathbf{D}.d\mathbf{E} = -\frac{\epsilon_\perp}{8\pi} E^2 - \frac{\epsilon_a}{8\pi} (\mathbf{n}.\mathbf{E})^2. \qquad (3.67)$$

The first term is independent of orientation. The second term is the interesting one; it favours parallel alignment ($\mathbf{n}$ collinear with $\mathbf{E}$) if $\epsilon_a > 0$, and perpendicular alignment if $\epsilon_a < 0$. For instance, pure PAA aligns perpendicular to E.‡

The electric analogue of the Frederiks transition has been known since the days of Frederiks and Zvetkov; recent experiments and detailed theoretical calculations have been worked out by Gruler and Maier [18a]. The formula for the threshold field is identical to eqn 3.64, if one makes the substitution

$$\tfrac{1}{2}\chi_a H^2 \rightarrow \frac{\epsilon_a E^2}{8\pi}.$$

It is of interest to compare the relative efficiencies of an $\mathbf{H}$-field and of an $\mathbf{E}$-field. With the above correspondence we get $\mathbf{E} = (4\pi\chi_a/\epsilon_a)^{\frac{1}{2}}\mathbf{H}$. Taking $\chi_a = 10^{-7}$, $\epsilon_a = 0{\cdot}1$, and $H = 1$ gauss, this gives $E \cong 3 \times 10^{-3}$ $ues \approx 1$ volt/cm. Thus roughly, one volt/cm is equivalent to one gauss. But in certain favourable cases ϵ_a can be made much larger. Above threshold ($E > E_c$) the detailed analysis differs for the electric case, because the distortions of the director field react on the field-distribution inside the slab. Studies of the alignment for large specimens (no wall effects) in *competing* H and E fields have been carried out by Carr [18b].

3.3.2. *Polarization induced by distortion (flexoelectric effect)*

In certain solids, a strain will induce a polarization $\mathbf{P}$. The source of the strain may be an external pressure; for this reason the effect is called *piezoelectric*. In liquid crystals, a splay or a bending distortion can create a polarization: the physical origin of the effect is exemplified on Fig. 3.20 which is extracted from the work of R. B. Meyer [22]. He has also used the word 'piezoelectric' to describe this effect. This terminology might create some misunderstandings, since pressure does not influence the director $\mathbf{n}$ in a nematic, and is thus unable to induce distortions and the associated polarizations: for this reason, we prefer the word '*flexoelectric*.'

† The potential used here is that one which must be minimized for fixed voltages applied on external conductors.
‡ In impure PAA conductivity effects come into play and the trend may be reversed: see Chapter 5.

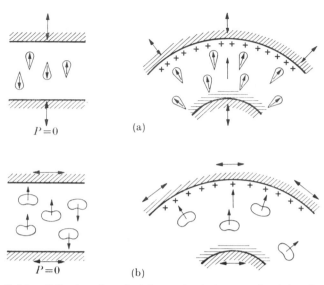

F IG. 3.20. Origin of the two flexoelectric constants in nematics (after R. B. Meyer). Case (*a*): wedge shaped molecules carrying permanent moments give $P \neq 0$ when a splay is imposed (for instance by suitable walls). Case (*b*): banana-shaped molecules with a moment normal to their long axis give $P \neq 0$ under bending.

To make the description of Fig. 2.20 quantitative, let us construct the most general form of the polarization $\mathbf{P}_d$ which is induced by weak distortions. This means that we want $\mathbf{P}_d$ to be proportional to the first-order space derivatives of the director $\mathbf{n}$ (higher derivatives being smaller by successive powers of a/l and negligible in the continuum limit). Also $\mathbf{P}_d$ must be an *even* function of $\mathbf{n}$, since, as repeatedly stated, the states $\mathbf{n}$ and $-\mathbf{n}$ are equivalent. Finally $\mathbf{P}_d$ must transform like a vector. The most general form of $\mathbf{P}_d$ satisfying these requirements is

$$\mathbf{P}_d = e_1\mathbf{n}(\mathrm{div}\ \mathbf{n}) + e_3\ (\mathrm{curl}\ \mathbf{n}) \times \mathbf{n}. \tag{3.68}$$

It involves two coefficients e_1 and e_3, with the dimensions of an electric potential, and of arbitrary sign; we call them flexoelectric coefficients. With molecules which are very asymmetric in shape and carry a strong electric dipole moment μ_e the flexoelectric coefficients might reach values of order μ_e/a^2 where a is a typical molecular dimension. In all other cases (and in particular if the molecules do not have a permanent moment) they will be smaller.

In principle e_1 and e_3 could be obtained from two types of experiment:

(1) Measuring the polarizations (or the surface changes) induced by an imposed distortion. This could be done with bent condensers, as

shown on Fig. 3.19. However, in practice, the charges on the condenser plates are easily screened out by impurity conduction; the experiment would require ultrapure materials.

(2) Using the inverse effect: when a field **E** is applied on a nematic single crystal, the alignment may become distorted, since a suitable distortion will imply a polarization $\mathbf{P}_d$ parallel to **E**. This inverse effect has apparently been observed by Schmidt, Schadt, and Helfrich on MBBA [23]. The principle is shown in Fig. 3.20. The sample is limited by

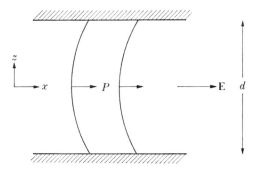

FIG. 3.21. Measurement of the piezoelectric coefficient e_3. The molecules are assumed to be weakly anchored on both plates. Under the field E, the molecular pattern distorts like a bow. The figure is drawn for e_3 positive.

two parallel glass surfaces, treated with lecithin to achieve a homeotropic texture. However, it is necessary for the proposed interpretation of the experiment that the boundary conditions do *not* correspond to strong normal anchoring; the angle of the molecules at the surface must not be fixed. A field E is applied in the plane of the slab (along X); since $\epsilon_\parallel < \epsilon_\perp$ in MBBA, if dielectric effects were present alone, they would stabilize the alignment along the normal (z) to the plates. But in practice a distortion is observed, as described in Fig. 3.21. Such a distortion is a natural consequence of the flexoelectric effect, if weak anchoring is assured at both boundaries. To see this, let us write down the polarization $P_{d(x)}$ corresponding to a distortion of the form:

$$n_z = \cos \phi(z) \approx 1 \qquad (\phi \ll 1)$$

$$n = \sin \phi(z) \approx \phi(z).$$

From eqn (3.68) we get:

$$P_{d(x)} = e_3 \frac{\partial \phi}{\partial z}. \tag{3.69}$$

For simplicity let us restrict ourselves to a case where the dielectric anisotropy ϵ_a is negligible. Then the free energy reduces to two contributions: the coupling of P_d with E, and the Frank elastic energy, which for the present case (with ϕ small) corresponds to a pure-bending type of deformation. The free energy per unit volume is thus:

$$F = -P_{d(x)}E + \tfrac{1}{2}K_3 \left(\frac{\partial \phi}{\partial z}\right)^2.$$

Inserting the form found for $P_{d(x)}$ and minimizing F with respect to $\partial \phi / \partial z$ we find:

$$\frac{\partial \phi}{\partial z} = \frac{e_3}{K_3} E$$

$$\phi = \frac{e_3}{K_3} Ez \tag{3.70}$$

(In this last equation we took the origin in the midplane of the slab). Clearly this form, leading to non-zero ϕ values on both plates, is valid only for weak anchoring. Typically, for fields E of order 300 volts cm^{-1} the curvature $\partial \psi / \partial z$ is in the range of 10 cm^{-1}.

In the experiment of ref. [23] the distortion was monitored by an interference method, measuring the difference Δl in optical path lengths for rays propagating (along z) through the slab, with respective polarizations along x and y. The difference is easily shown to be

$$\Delta l = d(n_e - n_0)\bar{\psi}^2 \qquad (\psi \ll 1)$$

where $n_e - n_0$ is the difference between extraordinary and ordinary indices, and $\bar{\psi}^2$ represents an average over the slab thickness d. Here we have:

$$\bar{\psi}^2 = \frac{d^2}{12}\left(\frac{\partial \varphi}{\partial z}\right)^2.$$

Thus, collecting our results, we find that:

$$\Delta l = \frac{1}{12}\left(\frac{e_3}{K_3}\right)^2 (n_e - n_0)E^2 d^3. \tag{3.71}$$

The proportionality to $E^2 d^3$ is well confirmed by the experimental data. From the measured Δl, one can thus extract $|e_3| = 4.5\ 10^{-5}$ cgs units.

The *sign* of e_3 must be obtained by a separate experiment: in ref. [23] the space-charge $\rho = -\text{div } P$ was probed through the resulting bulk force ρE; in suitable conditions (a sample with open ends) this force induces a hydrodynamic flow, the sign of which is easy to detect. From this it was concluded that $e_3 > 0$.

Another interesting consequence of the flexoelectric effect is that two distorted regions, remote from one another, in a nematic material, may interact significantly through the Coulomb forces associated with their polarization charge. These charges are small, but the Coulomb force is of such long range that the effect may be sizeable; in fact, if the flexoelectric coefficients were large, the resulting contribution to the free energy would become comparable to F_d: the description based on the Frank elasticity only (i.e. on F_d) would then be quite incorrect [24]. In practice, however, the Frank elasticity is found to provide an acceptable description for PAA; this may be due to the fact that the flexoelectric coefficients are not very large. Also, even if they are large, since PAA is usually not a very good insulator, the flexoelectric charges are screened out by impurity conduction.

3.4. Fluctuations in the alignment

3.4.1. Light-scattering experiments

Nematics are turbid in appearance. The scattering of visible light by nematics is higher, by a factor of the order of 10^6, than the scattering by conventional isotropic fluids. This was in fact one of the reasons which cast doubt on the very existence of liquid crystals in the early years; it was tempting to assume that they were made of a suspension of small crystallites in a fluid phase, with crystallite dimensions comparable to an optical wavelength. However it became progressively clear that the high scattering power was in fact an intrinsic property of well-defined nematic phases. The first detailed experimental studies in this field are due to P. Chatelain [25]. As we shall see, they give us a very direct probe of the *spontaneous fluctuations of the alignment* in a nematic medium.

Typical scattering geometries are shown on Fig. 3.22. In all cases the crucial parameters are the wave vectors $\mathbf{k}_i$, $\mathbf{k}_f$ of the ingoing and outgoing beam inside the sample, and also the corresponding polarizations, defined by the unit vectors $\mathbf{i}$ and $\mathbf{f}$. The difference

$$\mathbf{g} = \mathbf{k}_i - \mathbf{k}_f$$

is the 'scattering vector.' The plane ($\mathbf{k}_i$, $\mathbf{k}_f$) is the 'scattering plane,' and the angle between $\mathbf{k}_i$ and $\mathbf{k}_f$ is the 'internal scattering angle.'

In the most simple case (case I of Fig. 3.21) both beams ($\mathbf{k}_i$) and ($\mathbf{k}_f$) are normal to the nematic axis. Depending on whether the polarizations $\mathbf{i}$ and $\mathbf{f}$ are normal, or parallel, to the optical axis, the beams propagate with the 'ordinary' refractive index $n_\perp$ or with the 'extraordinary' index $n_\parallel$. (For PAA at 125°C with the sodium D-line, $n_\parallel \approx 1.83$ and

8

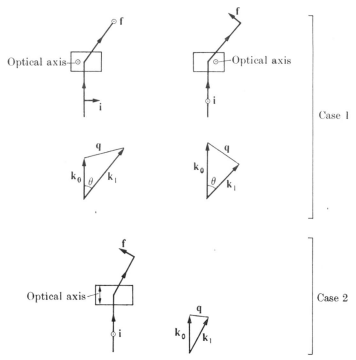

Case 1

Case 2

FIG. 3.22. A few typical geometries for the study of light scattering by a nematic single crystal; **i** and **f** are polarization vectors (for the three examples the choices shown for **i** and **f** ensure a large scattering intensity).

$n_\perp \approx 1.57$.) This allows us to compute the lengths of the wave vectors as a function of the optical frequency $\omega_0/2\pi$, using the relation[†]

$$k = n\omega_0/c \qquad (c = \text{speed of light}).$$

Finally one can then construct the scattering wave-vector **q** for a given scattering angle, as shown in Fig. 3.21. Similar (but somewhat more complicated) rules hold for case 2.

Having specified these geometrical parameters, we can now list the main results obtained by Chatelain:

(1) the scattering is intense only for crossed polarizations (i.e. when **i** and **f** are not parallel. The weak residual intensity observed for **i** parallel to **f** is probably due mainly to multiple scattering;

(2) the scattering is particularly strong at small **q**. The precise law of intensity versus **q** could not be derived, since, for technical reasons (collimation, multiple scattering, etc), Chatelain could not work at

[†] Note that, if the polarizations **i** and **f** are crossed, the lengths of $\mathbf{k_i}$ and $\mathbf{k_f}$ are different.

very small **q**s (which correspond to small scattering angles: see Fig. 3.22, case 2).

As already mentioned, a tempting model for the interpretation of these scattering data is to invoke small objects (of diameter D in the micron range) each of these being optically anisotropic, and sending out depolarized light with no phase coherence between different objects. This is the essence of the so called 'swarm theory' which has plagued the field of nematic liquid crystals for thirty years. In actual fact there is no reason for the director **n** to stay constant in a region of dimension D, and then to shift abruptly to another orientation; the energies associated with the abrupt shift would be prohibitively large. The actual situation is much more continuous, and can be described rigorously in terms of small fluctuations of **n(r)**: we shall now discuss this in detail.

3.4.2. Orientation fluctuations and correlations in a nematic single crystal [26]

We consider a nematic sample with optical axis z. The average director $\mathbf{n}_0$ is parallel to z. The fluctuations of the optical axis at any point **r** will be described by small, non-zero components $n_x(\mathbf{r})$, $n_y(\mathbf{r})$. To second order in n_x, n_y the distortion energy (3.15) reduces to:

$$\mathscr{F}_\mathrm{d} = \frac{1}{2} \int \left\{ K_1 \left(\frac{\partial n_x}{\partial x} + \frac{\partial n_y}{\partial y} \right)^2 + K_2 \left(\frac{\partial n_x}{\partial y} - \frac{\partial n_y}{\partial x} \right)^2 + K_3 \left[\left(\frac{\partial n_x}{\partial z} \right)^2 + \left(\frac{\partial n_y}{\partial z} \right)^2 \right] \right\} \, d\mathbf{r}. \tag{3.72}$$

We may also impose a magnetic field H along z: this will add a term (derived from eqn 47):

$$\mathscr{F}_\mathrm{mag} = \tfrac{1}{2} \int \chi_\mathrm{a} H^2 (n_x^2 + n_y^2) \, d\mathbf{r} + \text{const.} \tag{3.73}$$

It is convenient to analyse $n_x(\mathbf{r})$ and $n_y(\mathbf{r})$ in Fourier components, defined by

$$n_x(\mathbf{q}) = \int n_x(\mathbf{r}) \, e^{i\mathbf{q} \cdot \mathbf{r}} \, d\mathbf{r} \quad \text{etc}$$

In terms of these Fourier components the free energy becomes

$$\mathscr{F} = \mathscr{F}_0 + \tfrac{1}{2}\Omega^{-1} \sum_q \{ K_1 \, |n_x(\mathbf{q})q_x + n_y(\mathbf{q})q_y|^2 + K_2 \, |n_x(\mathbf{q})q_y - n_y(\mathbf{q})q_x|^2 +$$

$$+ (K_3 q_z^2 + \chi_\mathrm{a} H^2)\{ |n_x(\mathbf{q})|^2 + |n_y(\mathbf{q})|^2 \} \} \tag{3.74}$$

where Ω is the sample volume. For a given **q**, it is convenient to diagonalise the quadratic form in eqn (3.74) by a linear transformation $(n_x, n_y) \rightarrow (n_1, n_2)$. The meaning of the axes (1) and (2) is the following. We introduce for each **q** two unit vectors $\mathbf{e}_1$ and $\mathbf{e}_2$ in the (xy) plane:

$\mathbf{e}_2$ is normal to $\mathbf{q}$; $\mathbf{e}_1$ is normal to $\mathbf{e}_2$. The component of $\mathbf{n}(\mathbf{q})$ along $\mathbf{e}_\alpha$ is called $n_\alpha(\mathbf{q})$ ($\alpha = 1, 2$). $n_1(\mathbf{q})$ describes a periodic distortion which is a mixture of splay and bend. $n_2(\mathbf{q})$ describes a periodic distortion which is a mixture of twist and bend. In terms of n_1 and n_2 the free energy takes a very simple form

$$\mathscr{F} = \mathscr{F}_0 + \tfrac{1}{2}\Omega^{-1} \sum_q \sum_{\alpha=1.2} |n_\alpha(q)|^2 (K_3 q_\parallel^2 + K_\alpha q_\perp^2 + \chi_a H^2) \qquad (3.75)$$

where $q_\parallel = q_z$ is the component of the wave vector parallel to the optical axis, while $\mathbf{q}_\perp = \mathbf{q} \cdot \mathbf{e}_1$ is the normal component. In eqn (3.75) the various degrees of freedom are decoupled. We are now in a position to derive the thermal average of $|n_\alpha(q)|^2$. For this, we apply the *equipartition theorem*: for a classical system,[†] with free energy quadratic in the amplitudes $n_\alpha(\mathbf{q})$, the average $\mathscr{F}$, per degree of freedom, at thermal equilibrium, is equal to $\tfrac{1}{2}(k_\mathrm{B}T)$.

$$\langle \tfrac{1}{2}\Omega^{-1} |n_\alpha(\mathbf{q})|^2 (K_3 q_\parallel^2 + K_\alpha q_\perp^2 + \chi_a H^2) \rangle = \tfrac{1}{2}k_\mathrm{B}T$$
$$\langle |n_\alpha(q)|^2 \rangle = (\Omega k_\mathrm{B}T)/(K_3 q_\parallel^2 + K_\alpha q_\perp^2 + \chi_a H^2) \qquad (3.76)$$

where the brackets $\langle\ \rangle$ denote a thermal average. Eqn (3.76) is the central formula of fluctuation theory for nematics.

From eqn (3.76) we may derive the correlation between values of the director $\mathbf{n}$ taken at two different points $\mathbf{r}_1$ and $\mathbf{r}_2$ in the nematic. From the general theorem on Fourier transforms we have, for instance:

$$\langle n_x(\mathbf{r}_1) n_x(\mathbf{r}_2) \rangle = \Omega^{-2} \sum_{qq'} \langle n_x(\mathbf{q}) n_x(-\mathbf{q}') \rangle \exp\{\mathrm{i}(\mathbf{q}' \cdot \mathbf{r}_2 - \mathbf{q} \cdot \mathbf{r}_1)\}$$

The Fourier components $\mathbf{n}(\mathbf{q})$, $\mathbf{n}(\mathbf{q}')$ for different values of the wave vector are uncorrelated. Thus

$$\langle n_x(\mathbf{r}_1) n_x(\mathbf{r}_2) \rangle = \Omega^{-2} \sum_q \langle |n_x(\mathbf{q})|^2 \rangle \exp(-\mathrm{i}\mathbf{q} \cdot \mathbf{R})$$

$$\mathbf{R} = \mathbf{r}_2 - \mathbf{r}_1$$

We can now transform from $n_x(\mathbf{q})$ to $n_\alpha(\mathbf{q})$ and use the equipartition formula (3.76). The algebra is rather tedious, except in one case; if we make the one-constant approximation, the denominator of (3.76) becomes simply $K(q^2 + \xi^{-2})$ where ξ is the magnetic coherence length.

---

† Quantum corrections to the equipartition theorem come in only when $\hbar\omega_\alpha(\mathbf{q}) \gtrsim k_\mathrm{B}T$, where $\omega_\alpha(\mathbf{q})$ is a characteristic oscillation or relaxation frequency for the mode $(\alpha, \mathbf{q})$. These modes will be discussed in Chapter 5. For long wavelengths ($qa \ll 1$), their characteristic frequencies are indeed much lower than $k_\mathrm{B}T$.

Then:

$$\langle n_x(\mathbf{r}_1)n_x(\mathbf{r}_2)\rangle = \langle n_y(\mathbf{r}_1)n_y(\mathbf{r}_2)\rangle = (2\Omega)^{-2}\sum_q \langle|n_1(\mathbf{q})|^2+|n_2(\mathbf{q})|^2\rangle\exp(-i\mathbf{g}.\mathbf{R})$$

$$= (2\pi)^{-3}\int k_{\mathrm{B}}T/\{K(q^2+\xi^{-2})\}\exp(-i\mathbf{q}.\mathbf{R})$$

$$= (k_{\mathrm{B}}T/4\pi KR)\exp(-R/\xi) \tag{3.77}$$

Equation (3.77) is valid only in the continuum limit, i.e., for $R \gg a$. Let us discuss it first in the case of *zero* field. Then ξ is infinite (see eqn (3.58)) and the correlations $\langle n_x n_x\rangle$ decrease slowly with distance; the $1/R$ dependence is the direct analogue of that found for the distortion around a floating object in the Problem in §3.2.5.

Note that, in zero field, it is not possible to define a characteristic length D above which the correlations die out rapidly. This is in sharp contradiction with the swarm model, which assumes strong correlation at distances R smaller than the swarm diameter, and zero correlation at larger R.

When the differences between the three elastic constants K_i are allowed for, one still finds a slow decrease essentially like $1/R$. The exact formula being

$$\langle n_x(\mathbf{r}_1)n_x(\mathbf{r}_2)+n_y(\mathbf{r}_1)n_y(\mathbf{r}_2)\rangle$$

$$= (k_{\mathrm{D}}T/4\pi)\{(K_3K_1)^{-\frac{1}{2}}R_1^{-1}\}+\{(K_3K_2)^{-\frac{1}{2}}R_2^{-1}\} \quad (H-0)$$

where

$$R_\alpha = \sqrt{(x^2+y^2+z_\alpha^2)}$$

$$z_\alpha = z(K_\alpha/K_3)^{\frac{1}{2}}$$

This type of $1/R$ decrease is found for the transverse correlations of all physical systems where:

(a) the ordered state is characterized by a privileged axis ($\mathbf{n}_0$), but the direction of $\mathbf{n}_0$ is arbitrary

(b) the interactions are short range. For instance, similar correlations are found in the ordered phase of a Heisenberg ferromagnet in zero field.

In a finite field $\mathbf{H}$, eqn (3.77) shows that the range of the correlations of $\mathbf{n}$ is the magnetic coherence length ξ. We had defined ξ earlier by the range of the distortions induced in the nematic by a local perturbation (a wall, a floating object, etc). The equivalence of these two definitions for ξ is a consequence of general relations existing between response functions and correlation functions [23].

Problem: discuss the corrections to the fluctuation intensities $\langle|\delta n_\alpha(\mathbf{q})|^2\rangle$ when flexoelectric effects are included, for a pure (insulating) nematic, in zero external electric fields [24].

Solution: Let us call $\delta\mathbf{n}$ the vector $(n_x, n_y, 0)$. The flexoelectric polarization (eqn 3.68) has the components $P_{\|} = e_1$ div $\delta\mathbf{n}$ and $P_\perp = e_3 = (\partial/\partial z) = \delta\mathbf{n}$. Introducing the electric field $\mathbf{E} = -\nabla V$ and the displacement $\mathbf{D} = \boldsymbol{\varepsilon}.\mathbf{E} + 4\pi\mathbf{P}$, writing that div $\mathbf{D} = 0$ one arrives at the following equation for the Fourier component V_q of the potential

$$\mathbf{q}.\boldsymbol{\varepsilon}.\mathbf{q}\, V_q = 4\pi e q_{\|}(\mathbf{q}.\mathbf{n}_q) = 4\pi e q_{\|} q_\perp n_1(\mathbf{q})$$

$$e = e_1 + e_3$$

We can then compute the thermodynamic potential at constant $\mathbf{E}$, $G_\mathbf{E}$ (using eqn 3.75) or the thermodynamic potential at constant $\mathbf{D}$, $G_\mathbf{D} = G_\mathbf{E} + \mathbf{E}.\mathbf{D}/4\pi$; for the present problem they are in fact equal, because D_q is normal to $\mathbf{q}$, while $\mathbf{E}_q$ is parallel to $\mathbf{q}$. The result is

$$G = \mathscr{F}_\mathrm{d} + \Omega^{-1} \sum_q \frac{\mathbf{q}.\boldsymbol{\varepsilon}.\mathbf{q}}{8\pi} |V_q|^2$$

$$= \tfrac{1}{2}\Omega^{-1} \sum_q [|n_1(q)|^2\{K_1 q_\perp^2 + K_3 q_\|^2 + \chi_\mathrm{a} H^2 + (4\pi e^2 q_\|^2 q_\perp^2 / q_\|^2 \varepsilon_{\|} + q_\perp^2 \varepsilon_\perp)\}$$

$$+ |n_2(q)|^2 \{K_2 q_\perp^2 + K_3 q_\|^2 + \chi H^2\}]$$

Thus the contribution of the modes (1) to the thermodynamic potential is modified, and the correction brings in a novel angular dependence on $\mathbf{q}$. One can then derive the change in $\langle|n_1(\mathbf{q})|^2\rangle$ by the equipartition theorem. The shift in $|n_1^2|$ and the resulting effect on light scattering may become an efficient way of displaying flexoelectric effects in pure nematics.

3.4.3. Scattering of light by orientation fluctuations

The propagation of light is sensitive to fluctuations in the dielectric tensor

$$\epsilon_{\alpha\beta} = \epsilon_\perp \delta_{\alpha\beta} + (\epsilon_{\|} - \epsilon_\perp) n_\alpha n_\beta$$

(The reader may verify for himself that this form of $\boldsymbol{\varepsilon}$ gives $\epsilon = \epsilon_{\|}$ for an electric field parallel to $\mathbf{n}$, etc.).

The fluctuations of $\boldsymbol{\varepsilon}$ may come from two sources: (a) fluctuations in the magnitude of $\epsilon_{\|}$ and $\epsilon_\perp$, due to small, local, changes in the density, the temperature, etc; and (b) fluctuations in the orientation of $\mathbf{n}$; this is the dominant effect, specific to the nematic phase, which we shall now discuss.

In general, the analysis of light scattering in an *anisotropic* medium including the correct variations of refraction indices with beam orientation, and the correct definition of photometric intensities, is delicate.

For this reason we will restrict our attention to the limiting case where $\epsilon_a = \epsilon_{\parallel} - \epsilon_{\perp}$ is small, so that both the ingoing and the outgoing beams may be described as propagating in an isotropic medium. This approximation is not too good for a material like PAA (where $\epsilon_{\parallel} = n_{\parallel}^2 \approx 3\cdot3$ and $\epsilon_{\perp} = n_{\perp}^2 = 2\cdot3$ at 125°C), but it allows for a much more explicit display of the important physical features.

Let us first derive a formula for the *scattering cross section*. Our starting point is the formula giving the field radiated by a dipole $\mathbf{P}$, oscillating at the angular frequency ω, and located at point $\mathbf{r}$.

$$\mathbf{E}(\mathbf{r}') = (\omega^2/c^2 R)\exp(ikR)\mathbf{P}_v(r) \qquad (3.79)$$

where $k = \bar{n}\omega/c$, $\bar{n}$ being the average refraction index, and $\mathbf{P}_v$ is the component of $\mathbf{P}$ normal to the observation direction $\mathbf{R} = \mathbf{r}' - \mathbf{r}$ (c is the velocity of light).

Let us write that the dipoles P are induced by the ingoing radiation field $\mathbf{E}_{in}(\mathbf{r}) = E\mathbf{i} \exp(i\mathbf{k}_0.\mathbf{r})$ (E = amplitude, $\mathbf{i}$ = unit vector specifying the polarization):
$$\mathbf{P}(\mathbf{r}) = (4\pi)^{-1}(\varepsilon(\mathbf{r}) - \mathbf{1})\mathbf{E}_{in}(\mathbf{r}),$$

where $\mathbf{1}$ represents a unit tensor.

The outgoing field $E_{out}(\mathbf{r}')$ is obtained by summing all the contributions (3.79) over the volume of the sample. If $\mathbf{r}'$ is far enough from the scattering region the factor $1/R$ may be taken out of the integral. We may also write $kR = \mathbf{k}_1.\mathbf{R} = \mathbf{k}_1(r' - r)$ where $\mathbf{k}_1$ is the wave vector in the direction of the outgoing beam. Projecting E_{out} on the final polarization direction $\mathbf{f}$ ($\mathbf{f}$ normal to $\mathbf{k}_1$) we arrive at

$$\mathbf{f}.\mathbf{E}_{out}(\mathbf{r}') = (E/R)\exp(i\mathbf{k}.\mathbf{r}')\alpha$$
$$\alpha = \omega^2/4\pi c^2 \int_{(\Omega)} \{\mathbf{f}.(\varepsilon(\mathbf{r}) - 1).\mathbf{i}\}\exp(-i\mathbf{q}.\mathbf{r}) \, dr \qquad (3.80)$$
$$\mathbf{q} = \mathbf{k}_0 - \mathbf{k}$$

The length α is called the *scattering amplitude*. Note that for $q \neq 0$ the term $(-\mathbf{1})$ in $(\varepsilon(\mathbf{r}) - \mathbf{1})$ does not contribute to the integral $\int d\mathbf{r}$. In terms of the Fourier transform

$$\varepsilon(\mathbf{q}) = \int_{(\Omega)} \varepsilon(\mathbf{r})\exp(-i\mathbf{q}.\mathbf{r}) \, dr$$

we may write
$$\alpha = (\omega^2/4\pi c^2)\mathbf{i}.\varepsilon(\mathbf{q}).\mathbf{f} \qquad (3.81)$$

The differential cross section (per unit solid angle of the outgoing beam around the direction $\mathbf{k}_1$) is

$$\sigma = \langle|\alpha|^2\rangle \qquad (3.82)$$

where the brackets $\langle\ \rangle$ as always denote a thermal average.

Let us now restrict our attention to the contribution in $\boldsymbol{\varepsilon}(\mathbf{q})$ which are due to fluctuations in the optical axis $\mathbf{n}$, putting $\mathbf{n} = \mathbf{n}_0 + \delta\mathbf{n}$, with $\delta\mathbf{n} = (n_x n_y 0)$, and expanding $\boldsymbol{\varepsilon}$ to first order in $\delta\mathbf{n}$: using (3.78) we get

$$\mathbf{f}.\boldsymbol{\varepsilon}.\mathbf{i} = \mathbf{f}.\langle\boldsymbol{\varepsilon}\rangle.\mathbf{i} + \epsilon_a(\mathbf{f}.\delta\mathbf{n})(\mathbf{n}_0.\mathbf{i}) + \epsilon_a(\mathbf{f}.\mathbf{n}_0)(\delta\mathbf{n}.\mathbf{i})$$

The first term, being independent of $\mathbf{r}$, does not contribute to the Fourier transform at finite $\mathbf{q}$. The second and third terms are linear in $\delta\mathbf{n}(\mathbf{q})$. It is convenient to analyse $\delta\mathbf{n}(\mathbf{q})$ into the eigenmodes n_1, n_2, defined in the preceding paragraph

$$\delta\mathbf{n}(\mathbf{q}) = \mathbf{e}_1 n_1(\mathbf{q}) + \mathbf{e}_2 n_2(\mathbf{q})$$

When we square the scattering amplitude (3.81) and take the average, the cross terms involving $n_1 n_2^*$ disappear, the two modes being uncorrelated. Thus we are left with

$$\sigma = (\epsilon_a \omega^2/4\pi c^2)^2 \sum_{\alpha=1,2} \langle|n_\alpha(\mathbf{q})|^2\rangle(i_\alpha f_z + i_z f_\alpha)^2 \qquad (3.83)$$

where $i_\alpha = \mathbf{e}_\alpha.\mathbf{i}$, etc. The final formula for σ is obtained by inserting eqn (3.76) for the thermal averages.

$$\sigma = \Omega(\epsilon_a \omega/4\pi c^2)^2 \sum_{\alpha=1,2} k_B T/(K_3 q_\parallel^2 + K_a q_\perp^2 + \chi_a H^2)(i_\alpha f_z + i_z f_\alpha)^2 \quad (3.84)$$

Let us discuss first the order of magnitude of σ; it is proportional to the sample volume Ω, as it should be, and roughly given by

$$\sigma \sim \Omega(\epsilon_a \omega^2/4\pi c^2)^2(k_B T/K q^2) \qquad \text{(in zero magnetic field).} \quad (3.85)$$

To get a better feeling for this result let us compare it to the scattering by an isotropic liquid, of dielectric constant ϵ. Here the fluctuations of ϵ are mainly due to fluctuations in the density. Defining a local dilation $\theta(\mathbf{r})$ we may write $\epsilon = \bar{\epsilon} + \epsilon'\theta(\mathbf{r})$. The corresponding cross section, as derived from eqns (3.81) and (3.82) is

$$\sigma|_{\text{isotropic}} = \Omega(\epsilon'\omega^2/4\pi c^2)^2(\mathbf{f}.\mathbf{i})^2\langle|\theta(q)|^2\rangle \qquad (3.86)$$

To find $\langle|\theta(\mathbf{q})|^2\rangle$ we write down a compressional free energy

$$f_c = \tfrac{1}{2}\int W\theta^2(\mathbf{r})\,d\mathbf{r} = W/2\Omega \sum_q |\theta_q|^2, \qquad (3.87)$$

where W^{-1} is the isothermal compressibility. Applying the equipartition theorem we get $\langle|\theta_q|^2\rangle = \Omega k_B T/W$. Inserting this into (3.86) and comparing with σ we get

$$\sigma/\sigma|_{\text{iso}} \sim (\epsilon_a/\epsilon)^2(W/K q^2)$$

For an order of magnitude estimate we can put $\epsilon_a/\epsilon' \sim 1$, $K \sim U/a$ and $W \sim U/a^3$ where U is a typical binding energy. This gives:

$$\sigma/\sigma|_{\text{iso}} \sim 1/(qa)^2 \qquad (3.88)$$

$2\pi/q$ is comparable to an optical wavelength (or even larger for scattering at small angles). Thus $qa \sim 10^{-3}$ and the scattering in the nematic phase may be a million times larger than the scattering in the isotropic phase. Physically, we may say that in isotropic liquids, a long wavelength modulation of ϵ requires a uniform *dilation*, and this implies a finite elastic energy. On the other hand, in a nematic fluid, we may modify ϵ simply by a *rotation* of the optical axis. If the rotation is nearly uniform ($q \to 0$) this requires very little energy (see eqn (3.75), with $H = 0$). Thus eqn (3.84) does explain why the scattering is large, and particularly large at small q values. The polarization effects may also be accounted for; consider for instance the case I of Fig. 3.19, with an ingoing polarization $\mathbf{i}$ normal to the optical axis ($i_z = 0$), and 0 angle scattering ($\mathbf{k}_1$ nearly parallel to $\mathbf{k}_0$). Then, if $\mathbf{f}$ is parallel to $\mathbf{i}$ ($f_z = 0$) the polarization factor in (3.84) vanishes, while if $\mathbf{f}$ is normal to $\mathbf{i}$ ($f_z = 0$) we get a large cross section. Similar agreement is found for the other cases.

Let us end up this paragraph by a discussion of the information which could be obtained from more detailed scattering measurements, particularly in the small q region.† In most experiments, absolute values of cross sections are not measured. But, by a suitable choice of geometry, and using formulae such as (3.83) one could extract from the data the thermal averages $\langle |n_\alpha(\mathbf{q})|^2 \rangle$, except for a constant scale factor.

When $H = 0$, we expect from (3.76) that $\langle |n_\alpha(\mathbf{q})|^2 \rangle$ should be inversely proportional to $K_3 q_{\parallel}^2 + K_\alpha q_{\perp}^2$. Studying the dependence of the cross section on the angle between $\mathbf{q}$ and the optical axis, one should be able to extract the ratios K_α/K_3.

With a strong field $\mathbf{H}$, the fluctuations $\langle |n_\alpha(\mathbf{q})| \rangle^2$ are decreased; they become inversely proportional to $K_3 q_{\parallel}^2 + K_\alpha q_{\perp}^2 + \chi_a H^2$.‡ If χ_a is known, it becomes then possible to measure the *magnitudes* of the three elastic constants from relative intensity measurements at various $\mathbf{q}$'s. The experiment has been performed recently (with electric fields $\mathbf{E}$, instead of magnetic fields $\mathbf{H}$, as the orienting agent) [28].

---

† This limit is interesting for two reasons: (1) because the scattering is very large, (2) because (in cases I and II) the outgoing beam propagates in a direction for which the refraction indices are well known and only weakly dependent on the scattering angle.

‡ An increase in transmission under magnetic fields has been observed long ago by Moll and Ornstein, *Proc. Acad. Sci. Amsterdam*, **21**, 259 (1918).

In practice, there are some slight complications due to the anisotropy of the medium: for instance, as we have already mentioned, q does not tend to zero at zero scattering angle in case I with crossed polarizations, because the magnitudes of $\mathbf{k}_0$ and $\mathbf{k}_1$ are different. But, independently of these details, light scattering studies do allow for quantitative measurements of the elastic constants.

A final word of comment on the *temperature dependence* of the scattering. The main factors, displayed in eqn (3.85), are the dielectric anisotropy $\epsilon_a(T)$ and the elastic constant K. (The factor $k_B T$ gives but weak effects in the narrow range of existence of the nematic phase.) The intensity (at fixed $\mathbf{q}$) is essentially proportional to ϵ_a^2/K. Both ϵ_a and K tend to decrease rather strongly when $T \to T_c$ (the nematic–isotropic transition point). But the ratio ϵ_a^2/K is only weakly temperature dependent; in a mean field theory of the Maier–Saupe type, ϵ_a is linear in S while $K \sim S^2$; thus ϵ_a^2/K is independent of S. Indeed the temperature effects found by Chatelain [25] are small.

Problem: discuss the effect of a large magnetic field $\mathbf{H}$ on the birefringence of a nematic single crystal.

Approximate solution: the effect of $\mathbf{H}$ is to decrease the magnitude of the fluctuations in the nematic. Putting always $\mathbf{H}$ (and $\mathbf{n}_0$) parallel to the z-axis, we estimate the fluctuations of the director $\mathbf{n}$ at one point $\mathbf{r}$ by inverting the Fourier transform (3.76)

$$\langle n_x^2(\mathbf{r}) \rangle = (2\pi)^{-3} \int k_B T/\{K(q^2 + \xi^{-2})\} \, d\mathbf{q}.$$

(where for simplicity we make the one-constant approximation). The integral over $\mathbf{q}$ must be cut off at a certain $\mathbf{q}_{max} \sim 1/a$ (limit of validity of the continuum theory). Performing the integration gives

$$\langle n_x^2 \rangle = k_B T/2\pi^2 K (q_{max} - \pi/2\xi) = \langle n_y^2 \rangle$$

From this we get $\langle n_z^2 \rangle = 1 - \langle n_x^2 \rangle - \langle n_y^2 \rangle$ and the effective anisotropy of the dielectric constant

$$\langle \epsilon_\| - \epsilon_\perp \rangle = \epsilon_{ao} \langle n_z^2 \rangle$$

(where ϵ_{ao} is the dielectric anisotropy for full alignment). This result contains a term independent of H, and a term linear in $|H|$, in which we are primarily interested

$$\langle \epsilon_\| - \epsilon_\perp \rangle = \epsilon_a(T) + \epsilon_{ao} k_B T/2\pi K \xi = \epsilon_a(T) + \epsilon_{ao} k_B T \chi_a^{1/2}/2\pi K^{3/2}|H|$$

Taking $\epsilon_{ao} \approx 1$, $T = 400\text{K}$, $\chi_a = 10^{-7}$, $K = 10^{-6}$, and $H = 10^5$ oersteds, the correction term is of order 4×10^{-4}, corresponding to a change in refractive index anisotropy of order 10^{-4}. The anomalous exponent $|H|$ has been predicted independently by J. Alcantara† for a similar problem in magnetic systems. However

† To be published.

it must be emphasized that the solution is only approximate. But in any case it would be interesting to observe this effect, possibly with pulsed magnetic fields.

Problem: Discuss the correlations and the scattering of light in a nematic film floating at the surface of a fluid (two-dimensional nematic).

Solution: We assume that the film is of molecular thickness; such films might possibly be achieved with certain long molecules floating on a liquid surface. If the temperature is not too high, and if the surface density is suitable, we may postulate that locally, at any point $\rho(x, y)$ the nematic molecules are aligned along a certain direction $\mathbf{n}(x, y)$ in the film plane. However, as we shall see, there is no long-range order.

Let us start from the distortion free energy: there are two types of distortion (splay and bend). For simplicity, we shall make the one-constant approximation, and write

$$F_{\mathrm{d}} = \tfrac{1}{2}K(\nabla\theta)^2 = \tfrac{1}{2}K\left\{\left(\frac{\partial\theta}{\partial x}\right)^2 + \left(\frac{\partial\theta}{\partial y}\right)^2\right\}$$

Here F_{d} is an energy per cm^2 and K has the dimensions of energy (ML^2T^{-2}). θ is the angle between $\mathbf{n}$ and a fixed direction (x) in the plane of the film. Applying the equipartition theorem as in eqn (3.76) we get, for a two-dimensional Fourier component of wave vector $\mathbf{q}(q_x q_y, 0)$

$$\langle|\theta_q|^2\rangle = \frac{k_{\mathrm{B}}T}{Kq^2}$$

and by inversion

$$\sigma^2(\rho_{12}) = \langle\{\theta(\rho_1)-\theta(\rho_2)\}^2\rangle = \int\int (2\pi)^{-2}2\{1-\exp(i\mathbf{q}.\rho_{12})\langle|\theta_0|^2\rangle\,\mathrm{d}q_x\,\mathrm{d}q_y$$
$$= (k_{\mathrm{B}}T/\pi K)\mathrm{lu}(\rho_{12}u)$$

where $\rho_{12} = |\rho_2-\rho_1|$ and a is a cut-off distance, of the order of the molecular length. This formula shows that the mean-square deviation $\sigma^2(\rho_{12})$ diverges when the distance ρ_{12} between the two observation points becomes large; there is no long-range order. The distribution law for $\delta = \theta(\rho_2) - (\theta\rho_1)$ is a gaussian

$$p(\delta) = \sigma^{-1}(2\pi)^{-\frac{1}{2}}\exp(-\delta^2/2\sigma^2).$$

As we shall see, the correlation function which is directly of interest for light scattering experiments is

$$\langle\cos\{2\theta(\rho_1)-2\theta(\rho_2)\}\rangle = \int \cos(2\delta)p(\delta) = \exp(-2\sigma^2) = (a/\rho_{12})^x.$$

where $x(T) = 2k_{\mathrm{B}}T/\pi K$ is a temperature-dependent quantity, and is probably of the order of unity in the temperature range where local nematic order exists.

Let us now apply this result to the scattering of light by orientation fluctuations. We call $\mathbf{q}$ the projection of the scattering wave vector on the plane of the film. A dimensionless scattering amplitude at point ρ may be defined as

$$\alpha = \{2(\mathbf{i}.\mathbf{n})(\mathbf{f}.\mathbf{n})-(\mathbf{i}.\mathbf{f})\}e^{i\mathbf{q}.\mathbf{p}}$$

where $\mathbf{i}$ and $\mathbf{f}$ are the incident and emergent polarizations; for simplicity we assume that both light beams are nearly vertical, so that $\mathbf{i}$ and $\mathbf{f}$ are in the plane of the film. If θ_i and θ_f are the angles of $\mathbf{i}$ and $\mathbf{f}$ with the x-axis we may write

$$\alpha = \cos(2\theta - \theta_i - \theta_f)e^{i\mathbf{q}\cdot\boldsymbol{\rho}} = \cos 2(\theta - \bar{\theta})e^{i\mathbf{q}\cdot\boldsymbol{\rho}}$$

$$2\bar{\theta} = \theta_i + \theta_f$$

The scattered intensity is then proportional to

$$I(\mathbf{q}) = \int \langle \cos 2(\theta_1 - \bar{\theta})\cos 2(\theta_2 - \bar{\theta}) \rangle e^{i\mathbf{q}\cdot\boldsymbol{\rho}_{12}} \, d\boldsymbol{\rho}_{12}$$

$$= \tfrac{1}{2}\int \langle \cos 2(\theta_1 - \theta_2) \rangle e^{i\mathbf{q}\cdot\boldsymbol{\rho}_{12}} \, d\boldsymbol{\rho}_{12}$$

$$= \text{const } (a^x/q^{2-x})$$

Since $x > 0$, the divergence of $I(\mathbf{q})$ at small q is weaker than in a three-dimensional nematic.

This must be compared to the scattering caused by capillary waves at the liquid surface. The dimensionless scattering amplitude for the latter process is of the form

$$\alpha_c = \{\zeta(\boldsymbol{\rho})/b\}(\mathbf{i}.\mathbf{f})e^{i\mathbf{q}\cdot\boldsymbol{\rho}}$$

where ζ is the vertical displacement at the interface, and b is a molecular length. The corresponding intensity is

$$I_c(\mathbf{q}) = \langle |\zeta(\mathbf{q})|^2 \rangle (\mathbf{i}.\mathbf{f})^2/b^2 = (k_B T/Ab^2q^2)(\mathbf{i}.\mathbf{f})^2$$

where A is the surface tension

The ratio of the two intensities is thus

$$I/I_c = Ab^2/k_B T (qa)^x (\mathbf{i}.\mathbf{f})^{-2}$$

$Ab^2/k_B T$ will often be of order unity, but $qa \ll 1$. Thus the scattering due to orientation fluctuations can be dominant only if: (1) $x(T)$ in not too large, and (2) the polarization vectors $\mathbf{i}$ and $\mathbf{f}$ are orthogonal.

3.5. Hydrostatics of nematics

3.5.1. Free energy and molecular field

In §§3.1.5 and 3.2.2 we discussed the transmission of torques through nematics, for situations of simple twist. It is sometimes useful to define stresses and torques for more general cases, as was first done by Ericksen [29]. We shall describe this here, restricting our attention (for simplicity) to incompressible nematics, under a homogeneous magnetic field $\mathbf{H}$ (we postulate no electric field, and no flexoelectric effect). The free-energy density then includes three contributions:

$$F = F_d + F_m + F_g. \tag{3.88}$$

F_d is the distortion energy (eqn 3.15). F_m is the magnetic energy (defined in eqn 3.47) and $F_g = \rho\phi$ where ϕ is the gravitational potential, ρ the density.

One basic concept, introduced by eqns (3.21) and (3.49) is that of the *molecular field* **h(r)**

$$h_\alpha = \partial_\gamma \pi_{\gamma\alpha} - \frac{\partial F_d}{\partial n_\alpha} + \gamma_a(\mathbf{n}.\mathbf{H})H_\alpha \tag{3.89}$$

where we use the notation:

$$\pi_{\gamma\alpha} = \frac{\delta F_d}{\delta(\partial_\gamma n_\alpha)} = \frac{\delta F_d}{\delta g_{\gamma\alpha}}. \tag{3.90}$$

As explained in §3.1, the equilibrium condition for the director is that **h** be parallel to **n** at all points:

$$h_\alpha = \lambda(\mathbf{r})n_\alpha, \tag{3.91}$$

3.5.2. Stresses and forces

Equation (3.91) was obtained by considering the change in total free energy δf_{tot} due to small rotations of **n**. We shall now study δf_{tot} for another type of transformation, where the centers of gravity of the molecules are displaced in space, but *each molecule retains its orientation* There will be a non-zero δf_{tot} for such a case; to see this, we might start again from our usual simple case; a nematic slab of area S and thickness L, between two walls with tangential boundary condition. An angle α is imposed between the preferred axis of both walls, and the slab is under twist $\theta(0) - \theta(L) = \alpha$. The distortion energy is

$$f_d = \tfrac{1}{2}K_{22}(\alpha/L)^2 LS = \tfrac{1}{2}K_{22}\alpha^2 S/L \tag{3.92}$$

If we increase the thickness L by δL (keeping the same volume: $SL = $ constant) we get a non-zero δf_d:

$$\frac{\delta f_d}{f_d} = \delta S/S - \delta L/L = -2\delta L/L \tag{3.93}$$

(The nematic wishes to decrease f_d by increasing its thickness). The deformation just described may be obtained by displacing the molecules without changing their director: if **u(r)** is the displacement of a molecule initially located at **r**, we might have chosen

$$u_y = y(\delta L/L) \quad \text{(along the normal to the slab)}$$

$$\left.\begin{array}{l} u_x = -x(\delta L/L) \\ u_z = 0 \end{array}\right\} \text{ (in the plane of the slab)}$$

and obtained a non-zero δf_d, given by eqn 3.93.

Even a displacement **u(r)** corresponding to a pure rotation may change f_d if the molecules retain their initial orientation: this was already stressed on Fig. 3.2. Another example is the following

Pure bend

we start from a state of pure bending, then displace the centres of all molecules by a 90° rotation, and reach a state of pure splay: thus if the splay and the bend elastic constants differ, f_d will be changed.

Let us now compute δf_d in detail; we start with a molecule at point $\mathbf{r}$, with director $\mathbf{n(r)}$. Then we displace it from $\mathbf{r}$ to $\mathbf{r'} = \mathbf{r} + \mathbf{u(r)}$, keeping the director frozen: the final distribution of the director in space is given by a new function $\mathbf{n'(r')}$, and

$$\mathbf{n'(r')} = \mathbf{n'(r+u)} = \mathbf{n(r)}. \tag{3.94}$$

The new state of distortion involves the derivatives

$$\frac{\partial n'_\gamma}{\partial r'_\beta} = \frac{\partial n_\gamma}{\partial r_\alpha}\frac{\partial r_\alpha}{\partial r'_\beta} \tag{3.95}$$

Since $\mathbf{r} = \mathbf{r'} - \mathbf{u}$ we have

$$\frac{\partial r_\alpha}{\partial r'_\beta} = \delta_{\alpha\beta} - \frac{\partial u_\alpha}{\partial r'_\beta} \simeq \delta_{\alpha\beta} - \partial_\beta u_\alpha$$

$$\frac{\partial n'_\gamma}{\partial r'_\beta} - \frac{\partial n_\gamma}{\partial r_\beta} = -(\partial_\alpha n_\gamma)(\partial_\beta u_\alpha) \tag{3.96}$$

where we keep only the terms which are of first order in $\mathbf{u}$, and write as usual $\partial/\partial r_\beta \equiv \partial_\beta$.

Consider now a nematic element (d³r) around a point with position vector $\mathbf{r}$, which becomes $\mathbf{r'}$ (d³r′ = d³r because of our assumption of incompressibility). The charge in distortion energy of this element will be, from eqn (3.90) and (3.96):

$$\text{d}^3\mathbf{r}\,\pi_{\beta\gamma}(-\partial_\alpha n_\gamma)(\partial_\beta u_\alpha). \tag{3.97}$$

Thus we may write the total change δf_d as an integral over the initial sample volume:

$$\delta F_d = \int \sigma^d_{\beta\alpha}\,\partial_\beta u_\alpha\,\text{d}^3\mathbf{r}, \tag{3.98}$$

where we have introduced a second-rank tensor $\boldsymbol{\sigma}^d$ (the 'distortion stress tensor') by the equation

$$\sigma^d_{\beta\alpha} = -\pi_{\beta\gamma}\,\partial_\alpha n_\gamma. \tag{3.99}$$

Note that $\boldsymbol{\sigma}^d$ is *not symmetric* in general ($\sigma^d_{\beta\alpha} \neq \sigma^d_{\alpha\beta}$). In eqn (3.98) a displacement $\mathbf{u}$ corresponding to a pure rotation for the centres of gravity (and fixed **n**) will usually change the energy, as explained before eqn (3.94). The asymmetry in $\boldsymbol{\sigma}^d$ describes this effect. The only case where $\boldsymbol{\sigma}^d$ becomes symmetric is obtained when the three elastic constants are equal; then it can easily be verified that F^d becomes invariant separately by rotations in **u** space, or in **n** space.

Let us now return to eqn (3.98). To obtain a complete expression for the changes of the free energy, we must now make the following changes:

(1) since we restrict our attention to incompressible regimes

$$\text{div } \mathbf{u} = 0$$

we must use not the total free energy f_{tot} but a slightly different quantity

$$f = f_{\text{tot}} - \int p(\mathbf{r}) \text{ div } \mathbf{u} \text{ d}^3\mathbf{r}$$

where $p(\mathbf{r})$ is an unknown function of **r**, which we call the *pressure*. Here p plays the role of a Lagrange multiplier: the minimum of f for arbitrary displacements **u**, will coincide with the minimum of f_{tot} for displacements **u** which leave the density constant. The form of the function $p(\mathbf{r})$ will be found later from the conditions for equilibrium (see eqn 3.107). The introduction of the pressure leads to a modified form of stress; we shall put

$$\delta f = \int \sigma^e_{\beta\alpha} \, \partial_\beta u_\alpha \text{ d}^3\mathbf{r}$$

where

$$\sigma^e_{\beta\alpha} = \sigma^d_{\beta\alpha} - p\delta_{\alpha\beta}. \tag{3.100}$$

We call $\boldsymbol{\sigma}^e$ the '*Ericksen stress*.' Note that $\boldsymbol{\sigma}^e$, just like $\boldsymbol{\sigma}^d$, is not in general a symmetric tensor.

(2) We must include the effect of F_m and F_g. For simplicity, we shall assume now that the magnetic field **H** is constant in space. This is correct for essentially all experiments carried out up to now. Then the only external force acting on the molecules is gravitation: this adds to the integrand in eqn (3.98) a term $+u_\alpha\partial_\alpha F_g$.

(3) We must allow now for small changes in **n** at each point, $\mathbf{n} \rightarrow \mathbf{n} + \delta\mathbf{n}$, as in the equations defining the molecular field. This adds to the integrand the terms listed in eqn (3.20). We transform them by a partial integration, as was done after eqn (3.20), but now we are careful to keep the surface terms; they are explicitly

$$\int \pi_{\beta\gamma}\delta n_\gamma \text{ d}S_\beta$$

(where $\text{d}S_\beta$ is the vectorial surface element on the boundary).

Grouping all the contributions, we arrive at the total variation of f:

$$\delta f = \int \{\sigma_{\beta\alpha}^e \, \partial_\beta u_\alpha + u_\alpha \, \partial_\alpha F_g - h_\gamma \delta n_\gamma\} \, \mathrm{d}^3\mathbf{r} + \int \pi_{\beta\gamma} \, \delta n_\gamma \, \mathrm{d}S_\beta \quad (3.101)$$

We now integrate by parts the first term, and obtain

$$\delta f = \int \{-u_\alpha \phi_\alpha - h_\gamma \, \delta n_\gamma\} \, \mathrm{d}^3\mathbf{r} + \int \{\sigma_{\beta\alpha}^e u_\alpha + \pi_{\beta\gamma} \, \delta n_\gamma\} \, \mathrm{d}S_\beta \quad (3.102)$$

From eqn (3.102), the vector:

$$\phi_\alpha = \partial_\beta \sigma_{\beta\alpha}^e - \partial_\alpha F_g \quad (3.103)$$

appears as the *force per unit volume* in the bulk of the sample.

The condition for hydrostatic equilibrium is $\phi = 0$. We shall now show that this condition is automatically satisfied when eqn (3.91) holds, provided that the pressure distribution is chosen properly. Inserting the definitions (3.100) and (3.99) for the Ericksen stress in (3.103), we arrive at:

$$\phi_\alpha = -\pi_{\beta\gamma} \, \partial_\alpha \, \partial_\beta n_\gamma - \partial_\alpha n_\gamma (\partial_\beta \pi_{\beta\gamma}) - \partial_\alpha (p + F_g). \quad (3.104)$$

The quantity $\partial_\beta \pi_{\beta\gamma}$ in the second term may be related to the molecular field h_γ by eqn (3.89). This gives

$$\phi_\alpha = -\pi_{\beta\gamma} \, \partial_\alpha \, \partial_\beta n_\gamma - \partial_\alpha n_\gamma \, \frac{\partial F_d}{\partial n_\gamma} - \partial_\alpha (p + F_g)$$
$$+ \partial_\alpha n_\gamma \chi_\alpha H_\gamma n_\delta H_\delta - h_\gamma \, \partial_\alpha n_\gamma. \quad (3.105)$$

Recalling that $\pi_{\beta\gamma}$ is a partial derivative of F_d (eqn 3.90) we recognize that the first two terms of (3.105) add up to $-\partial_\alpha F_d$. The fourth term is $-\partial_\alpha F_m$. Finally, when eqn (3.91) holds, the last term reduces to

$$-\lambda n_\gamma \, \partial_\alpha n_\gamma = -\tfrac{1}{2}\lambda \, \partial_\alpha (\mathbf{n}^2) = 0.$$

Then ϕ_α reduces to

$$\phi_\alpha = -\partial_\alpha (F_d + F_m + F_g + p) \quad (3.106)$$

and the condition of hydrostatic equilibrium ($\phi_\alpha = 0$) imposes the following choice for p:

$$p(\mathbf{r}) = -(F_d + F_m + F_g) + \text{const.} \quad (3.107)$$

Returning to eqn (3.102), let us look now at the two surface terms. The first term represents the work done on the sample by the limiting walls, if the walls are displaced by u_α, but with no change in the director at the walls ($\delta \mathbf{n} = 0$). The second term represents another type of work, done by the walls, when their motion implies a change of $\mathbf{n}$ at the surface: we discussed an example of this effect in §3.1.5, where we had a nematic slab twisted between two polished glasses.

Other definitions of the stress. The Ericksen definition of a stress tensor is based on the work done when the molecules are displaced, but each molecule retains its initial orientation. However, this set of defining operations for the stress is not unique; a different set will lead to a different stress tensor. To illustrate this point, we shall now describe another stress tensor $\sigma_{\alpha\beta}^{M}$ (which is in fact related to the tensor used by the Harvard group [30]. The defining operations for $\boldsymbol{\sigma}^{M}$ are the following:

(1) The molecular centres of gravity are displaced by $\mathbf{u}(\mathbf{r})$. At the same time, the director is rotated by an amount equal to the local rotation of the centres of gravity; the corresponding rotation vector is $\frac{1}{2}$curl $\mathbf{u}$, and the change in $\mathbf{n}$ involved, which we shall call $\delta^{(1)}\mathbf{n}$, is thus given by

$$\delta^{(1)}\mathbf{n} = \tfrac{1}{2}(\text{curl } \mathbf{u}) \times \mathbf{n}$$

or in terms of the components

$$\delta^{(1)}n_{\alpha} = \tfrac{1}{2}(n_{\beta}\,\partial_{\beta}u_{\alpha} - n_{\beta}\,\partial_{\alpha}u_{\beta})$$

The change in the total free energy f resulting from the first operation is defined to be:

$$\delta f^{(1)} = \int \sigma_{\beta\alpha}^{M}\,\partial_{\beta}u_{\alpha}\,\mathbf{dr} + \int \pi_{\beta\gamma}\,\delta^{(1)}n_{\gamma}\,dS_{\beta}$$

Comparing with our earlier equations, the reader may check that

$$\sigma_{\alpha\beta}^{M} = \sigma_{\alpha\beta}^{e} + \tfrac{1}{2}(h_{\beta}n_{\alpha} - h_{\alpha}n_{\beta})$$

(2) The second operation amounts to an arbitrary rotation of the director at each point, giving rise to a change $\delta\mathbf{n}^{(2)}(\mathbf{r})$. The overall shift of $\mathbf{n}$ is thus split in two parts

$$\delta\mathbf{n} = \delta^{(1)}\mathbf{n} + \delta^{(2)}\mathbf{n}$$

The free-energy change associated with the second operation is

$$\delta f^{(2)} = \int (-\mathbf{h}.\delta^{(2)})\mathbf{n}\,\mathbf{dr} + \int \pi_{\beta\gamma}\,\delta^{(2)}n_{\gamma}\,dS^{\beta}$$

This new set of definitions may appear more complicated at first sight. In fact the separation of $\delta\mathbf{n}$ into $\delta\mathbf{n}^{(1)} + \delta\mathbf{n}^{(2)}$ has a very physical meaning for dynamical studies; $\delta\mathbf{n}^{(1)}$ corresponds to a non-dissipative motion, while $\delta\mathbf{n}^{(2)}$ is always associated with some friction.

9

3.5.3. *The balance of torques*

The external agents acting on our sample are the magnetic field $\mathbf{H}$ and the gravitational force $\rho \mathbf{g} = -\nabla F_g$. The resulting torques are

$$\int \{(\mathbf{M} \times \mathbf{H}) + \mathbf{r} \times \rho \mathbf{g}\}\, d_3 r.$$

In equilibrium, these torques must be balanced by the action of the walls: the above volume integral must be convertible into a surface integral, which we shall now derive.

Our starting point is a *rotational identity* satisfied by the deformation energy F_d. As already pointed out in section I, F_d is invariant if (and only if) we rotate both the centres of gravity and the director by the same angle ω. This means that if simultaneously

$$\left.\begin{aligned}\mathbf{u}(\mathbf{r}) &= \boldsymbol{\omega} \times \mathbf{r} \\ \delta\mathbf{n}(\mathbf{r}) &= \boldsymbol{\omega} \times \mathbf{n}\end{aligned}\right\} \tag{3.108}$$

the deformation energy must be unaltered.

Let us write (3.108) in tensor notation

$$\left.\begin{aligned}\partial_\beta u_\alpha &= \epsilon_{\alpha\mu\beta}\omega_\mu \\ \delta n_\gamma &= \epsilon_{\gamma\mu\rho}n_\rho\omega_\mu\end{aligned}\right\} \tag{3.108a}$$

where $\epsilon_{\alpha\mu\beta}$ is the alternant symbol ($\epsilon_{xyz} = -\epsilon_{yxz} = 1$, $\epsilon_{xx\beta} = 0$, etc). We now insert (3.108a) into δf_d (see eqn 3.101) and obtain

$$\delta f_d = \omega_\mu \int \left\{\sigma_{\beta\alpha}^d \epsilon_{\alpha\mu\beta} + \frac{\partial F_d}{\partial n_\gamma}\, \epsilon_{\gamma\mu\rho}n_\rho + \pi_{\beta\gamma}\epsilon_{\gamma\mu\rho}\, \partial_\beta n_\rho\right\} d^3 r = 0 \tag{3.109}$$

For each μ value (x, y, or z) the coefficient of ω_μ must vanish. This gives a rotational identity, which is satisfied by all scalar functions F_d (all functions of div $\mathbf{n}$, $\mathbf{n}$ curl $\mathbf{n}$ and ($\mathbf{n} \times$ curl $\mathbf{n}$)^2). Transforming the second term in eqn (3.109) by eqn (3.89), we obtain

$$\int \epsilon_{\alpha\mu\beta}\sigma_{\beta\alpha}^d\, d^3r + \int \epsilon_{\gamma\mu\rho}\{-h_\gamma n_\rho + n_\rho\, \partial_\beta \pi_{\beta\gamma} + \chi_a(\mathbf{n}.\mathbf{H})H_\gamma n_\rho + \pi_{\beta\gamma}\, \partial_\beta n_\rho\}\, d^3r = 0 \tag{3.110}$$

The term $\epsilon_{\gamma\mu\rho}n_\rho h_\gamma = (\mathbf{n} \times \mathbf{h})_\mu$ vanishes at equilibrium. The term involving χ_a is simply $(\mathbf{M} \times \mathbf{H})_\mu$ since $\mathbf{M} = \chi_\perp\mathbf{H} + \chi_a(\mathbf{n}.\mathbf{H})\mathbf{n}$. The other terms in the bracket add up to a total differential, giving

$$\int \{\epsilon_{\alpha\mu\beta}\sigma_{\beta\alpha}^d + (\mathbf{M} \times \mathbf{H})_\mu\}\, d^3r + \int \epsilon_{\gamma\mu\rho}\pi_{\beta\gamma}n_\rho\, dS_\beta = 0. \tag{3.111}$$

The first term in eqn (3.111) would vanish if the tensor $\boldsymbol{\sigma}^d$ was symmetric, but, as can be seen from eqn (3.99) and (3.15), $\boldsymbol{\sigma}^d$ is not symmetric in general.† We can transform the first term by writing that the bulk force ϕ_α vanishes at equilibrium:

$$0 = \int \epsilon_{\alpha\mu\rho} \phi_\alpha r_\rho \, d^3\mathbf{r}$$

$$= \int \epsilon_{\alpha\mu\rho} (\partial_\beta \sigma^e_{\beta\alpha}) r_\rho + G_\mu \, d^3\mathbf{r} \qquad (3.112)$$

where

$$G_\mu = \int \epsilon_{\alpha\mu\rho} (-\partial_\alpha F_g) r_\rho \, d^3\mathbf{r} \qquad (3.113)$$

is the torque of the gravitational forces.‡ Integrating by parts the first term in (3.112) we have in

$$\int \epsilon_{\alpha\mu\beta} \sigma^e_{\beta\alpha} \, d^3\mathbf{r} = \int \epsilon_{\alpha\mu\rho} \sigma^e_{\beta\alpha} r_\rho \, dS_\beta + G_\mu \qquad (3.114)$$

In the left hand side of (3.114) we can replace $\boldsymbol{\sigma}^e$ by $\boldsymbol{\sigma}^d$, since both tensors differ only by a symmetric tensor ($p\delta_{\alpha\beta}$). Inserting (3.114) into (3.111) we finally obtain

$$\int \epsilon_{\alpha\mu\rho} \{r_\rho \sigma^e_{\beta\alpha} + n_\rho \pi^e_{\beta\alpha}\} \, dS_\beta + \int \{(\mathbf{M} \times \mathbf{H})_\mu + G_\mu\} \, d^3\mathbf{r} = 0 \qquad (3.115)$$

Equation (3.115) shows that in a nematic sample, at equilibrium, the body torques ($\mathbf{M} \times \mathbf{H}$) and $\mathbf{G}$ are balanced by surface torques. There are two contributions to these surface torques, one deriving from the Ericksen stress $\boldsymbol{\sigma}^e$, and one deriving from the tensor $\boldsymbol{\pi}$. We shall now try to make the discussion of these torques less abstract, by a consideration of specific examples.

Problem (1): a floating object. We consider a single small object, of arbitrary shape, suspended in a nematic matrix. In practice the object may be a large solute molecule, or a colloidal grain—the case of a magnetic grain being of particular interest [16]. We want to know what are the long range distortions induced in the nematic by such an object.

Solution: Let us take a frame of reference with the origin at the center of gravity of the grain, and the z axis parallel to the unperturbed director $\mathbf{n}_0$. At large distances from the grain ($r \to \infty$) the director will deviate only slightly from $\mathbf{n}_0$:

$$\mathbf{n}(r) = \mathbf{n}_0 + \delta\mathbf{n}$$

$$\delta\mathbf{n} = (n_x, n_y, 0)$$

---

† In the one-constant approximation $\boldsymbol{\sigma}^d$ becomes symmetric, because F_d is then invariant separately for rotations of $\mathbf{n}$ and for rotations of $\mathbf{r}$.

‡ We could dispose of G_μ by choosing the origin of coordinates at the centre of gravity of the sample.

In the one constant approximation the free energy density (3.17) becomes

$$F_d = \tfrac{1}{2}K\{(\nabla n_x)^2 + (\nabla n_y)^2\}$$

and the local equilibrium equations are

$$\nabla^2 n_x = \nabla^2 n_y = 0$$

It is convenient to introduce a rotation vector $\boldsymbol{\omega}(\mathbf{r})$ such that $\delta\mathbf{n} = \boldsymbol{\omega} \times \mathbf{n}_0$. Then $\nabla^2\omega_x = \nabla^2\omega_y = 0$. Since ω_z is arbitrary we may also impose $\nabla^2\omega_z = 0$. The most general form for $\boldsymbol{\omega}$, vanishing at large $\mathbf{r}$, is

$$\boldsymbol{\omega} = \boldsymbol{\alpha}r^{-1} + \boldsymbol{\beta}.\nabla(r^{-1}) + \cdots$$

where $\boldsymbol{\alpha}$ is a vector and $\boldsymbol{\beta}$ a dyadic; $\boldsymbol{\alpha}$ and $\boldsymbol{\beta}$ are independent of $\mathbf{r}$; they depend, however, on the orientation of the grain. We shall now show that $\boldsymbol{\alpha}$ is directly related to the torque $\boldsymbol{\Gamma}_t$ which the grain applies to the nematic. Since nematics transmit torques, we may compute $\boldsymbol{\Gamma}_t$ by integration of eqn (3.115) over a large surface Σ surrounding the grain, and far enough from it to use the asymptotic for of $\delta\mathbf{n}$, or $\boldsymbol{\omega}$. To first order in $\delta\mathbf{n}$, the contributions come only from the $\boldsymbol{\alpha}$ term; only the π contribution remains, and gives:

$$\boldsymbol{\Gamma}_t = K\boldsymbol{\alpha}\int_{(\Sigma)} d\boldsymbol{\Sigma}.\nabla(r^{-1}) = 4\pi K\boldsymbol{\alpha}.$$

If the grain is subjected to other external forces (gravitational, magnetic, etc.†) it will in general feel, from these forces, a torque $\boldsymbol{\Gamma}_{ext}$. It feels from the nematic, a torque $-\boldsymbol{\Gamma}_t$. In equilibrium, the sum of these two torques must vanish:

$$\boldsymbol{\alpha} = \boldsymbol{\Gamma}_{ext}/4\pi K$$

Thus, when $\boldsymbol{\Gamma}_{ext} \neq 0$, there is a long range distortion in the nematic, decreasing only like $1/r$.

Problem (2): the 'magic spiral.' A nematic liquid is enclosed between two concentric cylinders, with the following boundary conditions: on the inner cylinder the molecules are normal to the wall; on the outer cylinder they are tangential. One possible equilibrium conformation is then the spiral shown on Fig. 3.23. Clearly each of the cylinders must feel no overall torque, when the nematic has settled to equilibrium (each cylinder can rotate without changing the energy stored in the nematic). On the other hand, the molecular conformation is bent and the molecules do experience some torques; for instance, if we relaxed slightly the boundary condition at the inner surface, the molecules would immediately rotate to decrease the bending. The question (raised by R. Meyer, and solved by O. Parodi) is to reconcile these two aspects, through the Ericksen equations.

---

† If a magnetic field H is applied to the grain, we assume H to be small, or $\xi(H)$ to be large in comparison of the distances r of interest: in this limit the distortions inside the nematic will not be affected by the direct action of H.

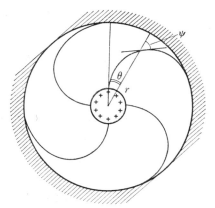

FIG. 3.23. The 'magic spiral': nematic contained between two cylinders with different boundary conditions.

Solution: Let us introduce cylindrical coordinates, r and θ, and call ψ the angle between the local optical axis and the radial direction, For the spiral, ψ will be a function of r only. In the one constant approximation, the free energy (per unit length along z) is found to be

$$\int_{r_0}^{r_1} 2\pi r \, dr \tfrac{1}{2} K \left(\frac{\partial \psi}{\partial_t}\right)^2.$$

Writing that this is a minimum with respect to all variations of $\psi(r)$ leads to the equation

$$\frac{\partial}{\partial r}\left(r\frac{\partial \psi}{\partial r}\right) = 0$$

and the solution (correctly fitted to the boundary conditions) is

$$\psi = \frac{\pi}{2}\cdot\frac{\ln(r/r_0)}{\ln(r_1/r_0)}.$$

Let us consider first the π contributions to the torque on a surface element dS of the inner cylinder. From the definition (3.90) of $\pi_{\beta\gamma}$ we find

$$\pi_{\beta\gamma}\,dS_\beta = K\,dS\,\frac{\partial n_\gamma}{\partial r}$$

The corresponding torque is thus, according to eqn (3.115)

$$dc_z(\pi) = K\,dS\left(n_x\frac{\partial n_y}{\partial r} - n_y\frac{\partial n_x}{\partial r}\right)$$

$$= K\,dS\,\frac{\partial \phi}{\partial r}$$

where $\phi = \theta + \psi$ is the angle between the director and the x-axis. The torque $C(\boldsymbol{\pi})$ is perfectly finite; per unit length along z it is

$$C(\boldsymbol{\pi}) = K2\pi r_0 \left(\frac{\partial \phi}{\partial r}\right)_{r_0} = \pi^2 K/\ln(r_1/r_0).$$

However, we must also consider the contribution from $\boldsymbol{\sigma}^e$ in eqn (3.115). We note first that the scalar pressure p (eqn 3.107) has cylindrical symmetry and thus does not contribute to the torque: the only term of interest in the stress is $\boldsymbol{\sigma}^d$. Using eqn (3.99) for $\boldsymbol{\sigma}^d$, we find:

$$\sigma_{ij}^d = -K \frac{\partial \phi}{\partial x_i} \frac{\partial \phi}{\partial x_j}$$

$(i, j) = (x, y)$

The corresponding force $\mathrm{d}\mathbf{f}$ on the element $\mathrm{d}S$ is given by

$$\mathrm{d}f_i = \sigma_{ij}^d \, \mathrm{d}S_j$$

or explicitly

$$\mathrm{d}\mathbf{f} = -K \, \mathrm{d}S \frac{\partial \phi}{\partial r} \nabla \phi$$

The vector $\nabla \phi = \nabla \psi + \nabla \theta$ has one radial component $(\nabla \psi)$ which does not contribute to the torque; the tangential component $(\nabla \theta)$ does contribute however, and gives

$$\mathrm{d}C_z(\boldsymbol{\sigma}) = -K \, \mathrm{d}S \frac{\partial \phi}{\partial r} . \frac{1}{r} . r = -K \, \mathrm{d}S \frac{\mathrm{d}\phi}{\mathrm{d}r}$$

Thus the sum of the two torques cancels exactly

$$\mathrm{d}C_z(\boldsymbol{\sigma}) + \mathrm{d}C_z(\boldsymbol{\pi}) = 0$$

This is gratifying, because, if the sum did not cancel, we would have discovered perpetual motion!

Problem (3): What are the conditions to be imposed on the Frank equations for the director, at the interface between a nematic and an isotropic phase, when the manifold of easy directions at the interface is continuously degenerate (i.e. = 'conical,' or 'tangential' situations)?

Answer: The bulk free-energy f must be stationary for any small rotation of the director $\mathbf{n}$ (at the surface) around the normal (z) to the interface, because the surface energy does not change in such a rotation. This leads to the condition, derived from eqn (3.115)

$$\epsilon_{\alpha z \rho} n_\rho \pi_{z\alpha} = 0,$$

or more explicitly

$$n_x \pi_{zy} - n_y \pi_{zx} = 0.$$

REFERENCES

CHAPTER 3

[1] OSEEN, C. W. *Trans. Faraday Soc.* **29**, 883 (1933).

[2] ZOCHER, H. *Trans. Faraday Soc.* **29**, 945 (1933).

[3] FRANK, F. C. *Discuss. Faraday Soc.* **25**, 19 (1958).

[4] ERICKSEN, J. L. *Archs. ration. Mech. Analysis* **23**, 266 (1966).

[5] See, for instance, LANDAU, L. D. and LIFSHITZ, E. M. *Theory of elasticity*, § 10, section on hexagonal crystals (for the elastic coefficients, the symmetries C_6 and C_∞ are equivalent). Pergamon, London (1959).

[6] ERICKSEN, J. L. *Archs. ration. Mech. Analysis* **10**, 189 (1962); see also OSEEN, C. W. *Ark. Math. Asho. Phys. K. Svenska, Vetenskapsakademien*, **A19**, 1–19 (1925).

[7] See LANDAU, L. D., and LIFSHITZ, E. M. *Electrodynamics of continous media*, § 39. Pergamon, London (1960).

[8] ERICKSEN, J. L. *Trans. Soc. Rheol.* **11**, 5 (1967).

[9] MAUGUIN, C. *Phys. Z.* **12**, 1011 (1911); *C.r. hebd. Séanc. Acad. Sci., Paris* **156**, 1246 (1913).

[10] GRANDJEAN, F. *Bull. Soc. fr. Minér.* **29**, 164 (1916).

[11] (a) This was noticed already by Lehmann: *Verh. d. Naturwiss. Vereins in Karlsruhe* **19Bd** Sonderdruck 275 (1906). But the precise recipe giving reproducible planar textures was established and utilized only much later by CHATELAIN, P. *Bull. Soc. fr. Minér.* **66**, 105 (1943); (b) LANNING, J. L. *Appl. Phys. Lett.* **21**, No. 4, 15 Aug. (1972).

[12] (a) HALLER, I. and HUGGINS, H. A. U.S. Pat. 3 656 834 (Cl. 350/150; G02f) (1972); HAAS, W., ADAMS, J., and FLANNERY, J. *Phys. Rev. Lett.* **25**, 1326 (1970). (b) CREAGH, L. T. and KMETZ, A. R. *Mol. Cryst. liquid Cryst.* (to be published); (c) PROUST, J. E., TER-MINASSIAN, L., and GUYON, E. *Solid State Commun.* **11**, 1227 (1972).

[13] BOUCHIAT, M. and LANGEVIN, D. *Phys. Lett.* A **34**, 331 (1971).

[14] For a discussion of weak anchoring and magnetic field effects, see RAPINI, A. and PAPOULAR, M. *J. Phys. (Fr.)* **30**, (Suppl. C4), 54 (1969).

[15] MAIER, W. and MEIER, G. *Z. Naturf.* A **16**, 470, 1200 (1961); GASPAROUX, H., REGAYA, B. and PROST, J. *C.r. hebd. Séanc. Acad. Sci., Paris, Ser. B* **272**, 1168 (1971); *J. Phys. (Fr.)* **32**, 953 (1971); ROSE, P. I. *Mol. Cryst. liquid Cryst.* (to be published).

[16] BROCHARD, F. and DE GENNES, P. G. *J. Phys. (Fr.)* **31**, 691, (1970).

[17] (a) FREDERIKS, V. and ZOLINA, V. *Trans. Faraday Soc.* **29**, 919 (1933); (b) FREDERIKS, V. and ZWETKOFF, V. *Sov. Phys.* **6**, 490 (1934); (c) ZWETKOFF, V. *Acta Phys.-chim. URSS* **6**, 865 (1937); (d) SAUPE, A. *Z. Naturf.* A **15**, 815 (1960); (e) GRULER, H., SCHEFFER, T., and MEIER, G. *Z. Naturf.* A **27**, 966 (1972).

[18] (a) GRULER, H. and MEIER, G. *Mol. Cryst. liquid Cryst.* **16**, 299, (1972); (b) CARR, E. F. *Mol. Cryst. liquid Cryst.* **7**, 269 (1969); (c) CARR, E. F. in *Ordered fluids and liquid crystals*, p. 76. Am. chem. Soc. (1967)

[19] GUYON, E., PIERANSKI, P. and BROCHARD, F. *C.r. hebd. Séanc. Acad. Sci.*, *Paris, Ser. B* **273**, 486 (1971). *J. Phys. (Fr.)* **33**, 681, (1972).

[20] MEYER, R. B. Private communication (1971).

[21] CLADYS, P. *Phys. Rev. Lett.* **28**, 1629 (1972).

[22] MEYER, R. B. *Phys. Rev. Lett.* **22**, 918 (1969).

[23] SCHMIDT, D., SCHADT, M. and HELFRICH, W. *Z. Naturf.* A **27**, 277 (1972).

[24] ORSAY LIQUID CRYSTAL GROUP, in *Liquid crystals and ordered fluids*, p. 195. Plenum Press, New York (1970).

[25] CHATELAIN, P. *Acta Cristallogr.* **1**, 315 (1948).

[26] DE GENNES, P. G. *C.r. hebd. Séanc Acad. Sci.*, *Paris*, **266**, 15 (1968).

[27] See, for instance, MARTIN, P. in *The many-body problem* (C. de Witt, R. Balian, eds.), Gordon and Breach, New York (1968).

[28] MARTINAND, J. L. and DURAND, G. *Solid State Commun.* **10**, 815 (1972).

[29] ERICKSEN, J. L., *Arch. rat. mech. Anal.* **9**, 371 (1972).

[30] FORSTER, D., LUBENSKY, T., MARTIN, P., SWIFT, J. and PERSHAN, P. *Phys. Rev. Lett.* **26**, 1016 (1971).

4

DEFECTS AND TEXTURES IN NEMATICS

'Il y a de certains défauts qui, bien mis en œuvre, brillent plus que la vertu même'

LA ROCHEFOUCAULD

4.1. Observation

IN CHAPTER 3 we restricted our attention to distortions of the nematic arrangement which involved continuous variations of the director $\mathbf{n}(\mathbf{r})$. But there are other, important, physical situations, where $\mathbf{n}(\mathbf{r})$ is not a smooth function of $\mathbf{r}$ at all points. Two current examples are described below.

4.1.1. Black filaments ('structures à fils')

In rather thick nematic samples, both with and without crossed polarizers, it is common to observe a system of dark, flexible filaments: as explained in Chapter 1, this thread-like structure is responsible for the name 'nematic' (see Fig. 4.4). Some of the filaments appear to float freely in the fluid. Others are attached by both ends to the walls. Some are less mobile and seem to stick entirely to the walls.

A first analysis of the optical properties of these filaments was carried out very early by Grandjean and refined by G. Friedel [1]. Both had already recognized that the filaments were not due to impurities or the like, but corresponded to lines of singularity in the molecular alignment: the word 'disclinations' was introduced later for these lines by Frank [2].

4.1.2. 'Structures à noyaux' ('Schlieren structure')

When the boundary conditions imposed by the glass walls to the nematic are *continuously degenerate* (tangential or conical, without any preferred axis in the plane of the walls), one often sees a system of singular points on the surface (Fig. 4.1; see between pp. 132–3). Between crissed Nicol prisms, the singular points are connected by black stripes, showing the regions where the optical axis (projected on the plane of the preparation) is parallel to one of the Nicols. The points have been called 'noyaux' by G. Friedel. The general texture resulting from those points is often called 'Schlieren texture' in the German literature.

The arrangement of the molecules in the vicinity of the limiting surface may be derived by various means. One is the polarizing microscope: the principle of the experiment is shown in Fig. 4.2. Alternatively, physico-chemical changes, such as precipitation [3] or bubble formation [4] can be induced at the limiting surface: the microcrystallites (or the bubbles) which appear show an oriented texture, revealing the local optical axis (Fig. 4.3 and 4.4; see between pp. 132–3). With all these methods

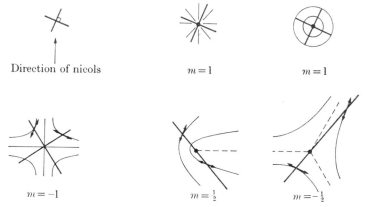

Direction of nicols $m = 1$ $m = 1$

$m = -1$ $m = \frac{1}{2}$ $m = -\frac{1}{2}$

FIG. 4.2. Geometrical arrangement of the molecules, at the sample surface, around a 'noyau'. The thick, black, lines give the regions of extinction between crossed nicols.

of observation, one detects four essential types of 'noyaux' with molecular arrangments as shown in Fig. 4.2. We shall classify these four types with an index m taking the values:

$$-1, \quad -\tfrac{1}{2}, \quad \tfrac{1}{2}, \quad 1$$

The two-dimensional structure of the various types (in the plane of the wall which is observed) may be summarized in the following way: assume that the director $\mathbf{n}$ is in the plane of the wall (tangential boundary condition) and define two orthogonal axes (x, y) in this plane. Call $\mathbf{r}$ the distance between the singularity and the observation point, and $\phi(\mathbf{r})$ the angle between $\mathbf{r}$ and x (tan $\phi = y/x$). The angle between $\mathbf{n}$ and x will be named $\theta(\mathbf{r})$. An approximate relation for θ, which is correct from the point of view of symmetries, is

$$\theta(\mathbf{r}) = m\phi(\mathbf{r}) + \text{const.} \tag{4.1}$$

Thus, if we follow a closed circuit around the singular point, and make one full turn ($\Delta\phi = 2\pi$), we find that the director has rotated by $\Delta\theta = 2\pi m$. This is acceptable only if m is an integer, or a half integer (since the states $(\mathbf{n})$ and $(-\mathbf{n})$ are indistinguishable).

4.1.3. Types of defects

4.1.3.1.   Disclination. The word disclination was invented by
F. C. Frank [2] and comes from the Greek *kline* = slope. A disclination
is a discontinuity in orientation, i.e. a discontinuity in the director-field
n(r). The discontinuity may be located at one point, on a line, or on a
surface, and is referred to as a point, line, or sheet disclination. However,
it is easy to see that sheet disclinations are completely unstable, and can
thus be omitted from the list. This instability can be understood from
Fig. 4.5: in an abrupt sheet singularity the energy stored per unit area

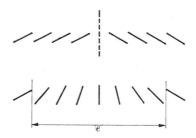

FIG. 4.5. Spontaneous smearing out of a sheet discontinuity.

on the surface $\sum$ of the sheet is of the order of U/a^2 (where U is a
molecular binding energy and a an average molecular dimension). If we
substitute for the discontinuous change of **n** at $\sum$, a gradual variation
taking place in a finite thickness $e \gg a$, the distortion energy (per unit
area) is of the order of $(K/e^2)e = K/e \simeq U/ea \ll U/a^2$. Thus smearing
out is favourable. The smearing operation is allowed, except possibly at
some points or lines on $\sum$† and the sheet discontinuity will fade out,
leaving at most a few disclination lines.

Thus we have only two types of disclinations in nematics: lines and
points. It is well established that the black filaments, described at the
beginning of this section, correspond to the lines. They will be discussed
in Section 4.2. The meaning of the 'noyaux' is less obvious. They can be
either point disclinations located at the boundary surface, or lines
normal to the plane of the surface: both cases occur in practice. A simple
experiment, quoted by G. Friedel, sometimes allows one to decide: if the
'noyaux' are observed at the boundary between the nematic and a
cover glass, the cover glass is displaced in its own plane. If the 'noyau'

---

† E.g. at the intersection of Σ and of the wall of the container, if the latter imposes
a well defined easy direction. After smearing out, this intersection may remain as a
disclination line.

corresponded to a line which was initially vertical, the line becomes tilted and is then observable as a black thread in the preparation. It is found that the 'noyaux' of half integral index ($m = \pm\frac{1}{2}$) are always associated with a line, while the noyaux of integral index may be either lines, or point disclinations, and are nearly always of the latter type in usual materials. We shall discuss this in Section 4.3.

4.1.3.2. Walls.   We have seen that a sheet singularity always tends to smear out into a continuously distorted region of finite thicknesses. With our simple argument (which ignored all effects due to boundaries or magnetic fields) the energy was minimized when the thickness e became very large. In practice, e is limited by size effects, or by the magnetic coherence length $\xi(\mathbf{H})$. It is thus possible to find, in nematics under fields, a *wall* separating two regions which are optically aligned in the field $\mathbf{H}$. These walls will be analysed in Section 4.4.

4.2. Disclination lines

4.2.1. Definition of the 'strength'

A simple geometrical process, generating a closed disclination line L (or 'loop'), in a nematic single crystal, is the following (Fig. 4.6): choose a surface $(\sum)$ limited by the loop L. Call the two sides of $(\sum)$ $\sum^+$ and $\sum^-$. Take the molecules which are in contact with $\sum^+$; by some 'external force,' rotate their director around an axis $(\boldsymbol{\Omega})$ normal to the optical axis $\mathbf{n_0}$ of the unperturbed crystal. $\boldsymbol{\Omega}$ defines both the axis and the magnitude of the rotation. On the other side $(\sum^-)$, maintain the unperturbed director n_0.

At finite distances from $(\sum)$ the director $\mathbf{n(r)}$ will then adjust itself and vary continuously with $\mathbf{r}$. The resulting configuration is continuous except on $(\sum)$; in general we have on $(\sum)$ a sheet of discontinuity, which can be maintained only in the presence of the external forces. However, if the rotation angle Ω is such that the optical axes above and below the cut coincide, we may switch off the external forces and retain a non-trivial conformation. The condition for this is that Ω be a multiple of π

$$\Omega = 2\pi m \qquad \text{(m integer or half integer).} \qquad (4.2)$$

The resulting conformation is everywhere continuous, except on the line L. We say that L is a disclination of 'strength' m (following a notation proposed by Friedel and Kleman [3]).

In Fig. 4.7 and 4.2 we show the molecular arrangement around a straight portion of a disclination line, for two particular cases of

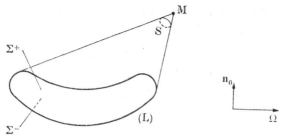

FIG. 4.6. Generation of a disclination loop (L) by the 'Volterra process'.

importance: Ω parallel to the line ('wedge-disclination': Fig. 4.7); and Ω normal to the line ('twist-disclination': Fig. 4.2.).

Note that different parts of one, curved, disclination line L, may be successively wedge-like and twist-like, as in Fig. 4.6; the quantity which is really characteristic of each line, and conserved all along its length, is the strength m.

4.2.2. Distortion field around one line

4.2.2.1. Straight lines. Just as for dislocations in solids, the calculation of the distortions around one line is often difficult. Here, as a first step towards simplification, we shall always use the one-constant approximation for the distortion energy. We start with simple wedge disclinations (Fig. 4.7). The z axis is set along the line. The director $\mathbf{n}$ is

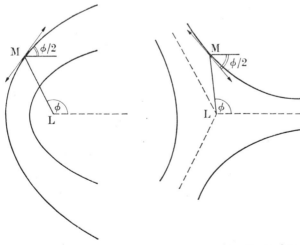

FIG. 4.7. Shape of a wedge disclination of strength $m = \pm\frac{1}{2}$. (The line L is normal to the sheet.)

in the (x, y) plane and makes an angle $\theta(x, y)$ with the x-axis. The distortion energy (eqn 3.17) reduces to

$$F_{\mathrm{d}} = \tfrac{1}{2}K(\nabla\theta)^2. \tag{4.3}$$

The minimization of $\mathscr{F}_{\mathrm{d}} = \int F_{\mathrm{d}}\, \mathbf{dr}$ leads to the equilibrium condition (first formulated by F. C. Frank [2]):

$$\nabla^2\theta = 0. \tag{4.4}$$

The solutions of eqn (4.4) which are discontinuous on the z-axis are precisely of the form found empirically by G. Friedel (eqn 4.1). At the

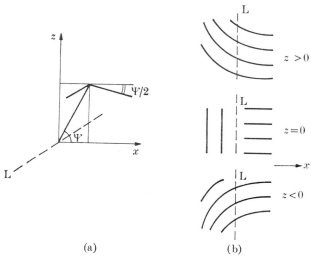

(a)   (b)

FIG. 4.8. Shape of a twist disclination of strength $\tfrac{1}{2}$. (a) geometrical construction: (b) molecular pattern: the molecules are always horizontal. The arrangement in three successive horizontal planes is shown. The pattern shown is only one particular example; to obtain other allowed patterns one may rotate the molecules, inside each horizontal plane, by a constant angle.

'noyaux,' θ is a linear function of the angle $\phi = \tan^{-1} y/x$. Rotation by 2π around the line restores the initial direction of the optical axis: θ can change at most by π, or $2\pi,\ldots$ Thus the coefficient $m = \mathrm{d}\theta/\mathrm{d}\phi$ must be an integer or a half integer.

To check that eqn (4.1) is indeed compatible with eqn (4.4) it may be convenient to investigate the vector $\nabla\theta$. According to eqn (4.1), this vector is everywhere tangential and of magnitude

$$|\nabla\theta| = m/\rho, \tag{4.5}$$

where $\rho = \sqrt{x^2+y^2}$ is the distance to the line. It is easy to verify that such a vector field has no divergence $\operatorname{div}(\nabla\theta) \equiv \nabla^2\theta = 0$.

Let us now discuss the energy $\mathcal{T}$ (per unit length of line) associated with these distortions. According to eqn (4.3) and (4.5) this is given by

$$\mathcal{T} = \int_a^{\rho\max} 2\pi\rho \, \mathrm{d}\rho \, \tfrac{1}{2}Km^2/\rho^2 \qquad (4.6)$$

We have set a lower limit a (of the order of molecular dimensions) to the integral, and also an upper limit $\rho_{\max}$: the cut off $\rho_{\max}$ may be given either by the distance between the line and the walls of the container, or by the distance to other disclinations screening out the distortion (eqn 4.5)—whichever is the smallest [6]. The precise value of $\rho_{\max}$ is not very important, because it enters only in a logarithm

$$\mathcal{T} = \pi K m^2 \ln(\rho_{\max}/a) \qquad (4.7)$$

Thus $\mathcal{T}$ is proportional to the average elastic constant K. The (log) factor in eqn (4.7) is typically of order 10, and $\mathcal{T} \sim 30K$.

Equation (4.7) does not include the contributions of the inner region or 'core' ($\rho < a$). These contributions are hard to calculate with any accuracy (except near the nematic–isotropic transition point) but we may guess that they will be of the order of $U/a \sim K$: thus they may be lumped together in eqn (4.7) by a slight change in the argument of the logarithm.

Problem: compute the energy of two parallel wedge disclinations (L_1 and L_2) of opposite strengths, separated by a distance d.

Solution: if the strengths are m and $-m$ respectively, the angle θ defining the director still obeys eqn (4.4) and may be taken to be of the form:

$$\theta = m(\phi_1 - \phi_2) + \text{const.},$$

ϕ_1 and ϕ_2 being the azimuthal angles relative to both lines (see Fig. 4.9). To integrate the distortion energy (eqn 4.3) it will be convenient to arrange that θ, or $\phi_1 - \phi_2$, be a single-valued function. To ensure this, we must introduce a cut from L_1 to L_2; crossing the cut from below decreases $\phi_1 - \phi_2$ by 2π. (on the other hand, if we follow a large circle surrounding both lines, which does not cross the cut, we recover the same value of $\phi_1 - \phi_2$ after one turn). The energy (4.3) may then be integrated by parts:

$$\tfrac{1}{2}K\int (\nabla\theta)^2 \, \mathrm{d}\mathbf{r} = -\tfrac{1}{2}K\int \theta\nabla^2\theta \, \mathrm{d}\mathbf{r} + \tfrac{1}{2}K\int \theta\nabla\theta \, \mathrm{d}\boldsymbol{\sigma}$$

The first integral vanishes since $\nabla^2\theta = 0$. The second integral is taken (a) on a very remote surface (distance R from the lines); this part is easily seen to be of order $R(1/R^2) = (1/R) \to 0$; (b) on the surface of the cut, where θ has a discontinuity

$$\{\theta\} = 2\pi m.$$

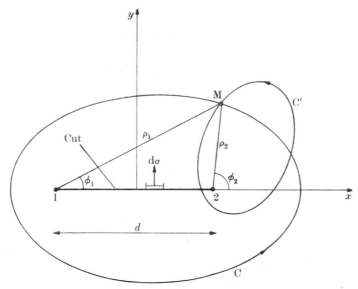

FIG. 4.9. Configurations with two parallel wedge disclinations of opposite strengths $\pm\frac{1}{2}$. If the observation point M makes a closed circuit C without crossing the cut, the angle θ defining the director is unchanged. If the circuit (C′) crosses the cut, θ changes by π on every turn. In the one constant approximation $\theta = (\phi_1 - \phi_2)/2$.

On the cut both the vector $d\boldsymbol{\sigma}$ and $\nabla\theta$ are parallel to the y-axis, and

$$|\nabla\theta| = \frac{m}{\rho_1} + \frac{m}{\rho_2} = m\left\{\frac{1}{\rho_1} + \frac{1}{d - \rho_1}\right\}.$$

Thus the distortion energy (per unit length of the lines) reduces to

$$\mathscr{F}_{12} = \frac{K}{2}\int_a^{d-a} \{\theta\} m\left\{\frac{1}{\rho_1} + \frac{1}{d - \rho_1}\right\} d\rho_1$$

(where we have introduced suitable cut-offs corresponding to the core radius). This leads to

$$\mathscr{F}_{12} = 2\pi K m^2 \ln\left(\frac{d}{a}\right).$$

$\mathscr{F}_{12}$ increases with d; the interaction between two lines of opposite strength is *attractive*, and the attraction force per unit length is $2\pi K m^2/d$.

Problem: a wedge disclination of strength m is parallel to a wall, at a distance h from it. The wall imposes tangential (or normal) boundary conditions. What is the force acting on the line?

Solution: Introduce an image (L_2) of the line, at a distance $d = 2h$ from it (Fig. 4.10). The form of θ, satisfying to eqn (4.4) and to the boundary conditions,

F I G. 1.8. The three-dimensional structure of cholesterol nonanoate. In a zero-order approximation, the molecule is equivalent to one simple rod, and the differences between the molecule and its mirror image drop out. Looking closer, one notices that the steroid nucleus (bottom part of the figure) is really a twisted rod. Also the aliphatic part (upper part of the figure) is not collinear with the steroid rod; both effects contribute to the chirality.

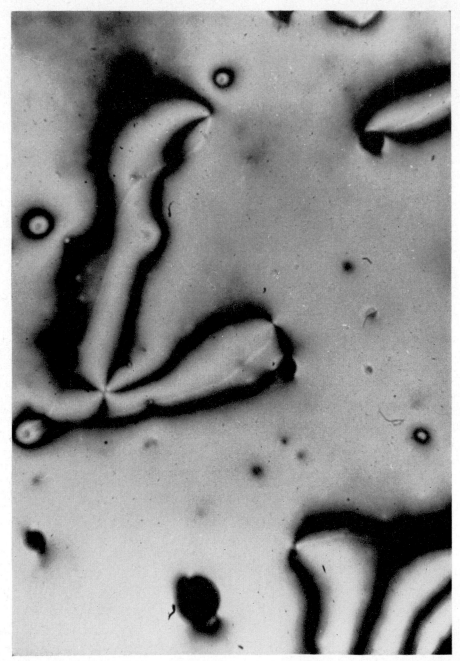

F𝗂ɢ. 4.1. Schlieren texture, observed between cross nicols, with a nematic giving tangential boundary conditions on the glass slide (Courtesy J. Dreyer). This particular photograph is remarkable, because it also shows the disclination lines in the bulk (white lines), which connect to the 'noyaux': note that all 'noyaux' with two black lines have one disclination line, while those with four black lines have none.

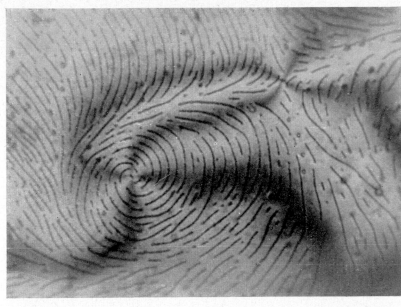

Fig. 4.4. Bubble decoration technique, displaying the arrangement of the molecules at a free surface (courtesy P. Pieranski). In the present example we see two singular points of strengths $(+1)$ and (-1) between crossed polarizers.

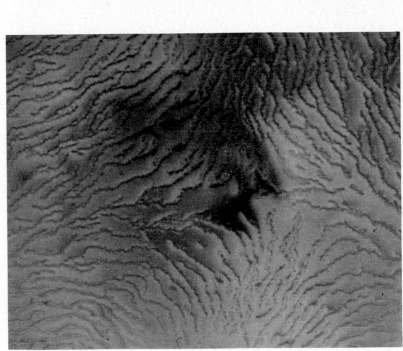

Fig. 4.3. Microprecipitate decoration technique displaying the arrangement of the molecules at a surface (Courtesy J. Rault). This particular example is a (-1) defect.

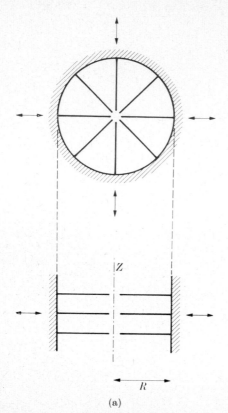

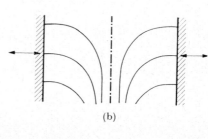

FIG. 4.13. (a) Wedge disclination of strength $+1$ in a cylinder: the arrangement near the centre has a discontinuity and must involve a core. (b) 'Escape' of the disclination in the third dimension: the arrangement is now continuous (no core). (c) Visualisation of the 'escaped' arrangement in a capillary (courtesy C. Williams).

(a)

(b)

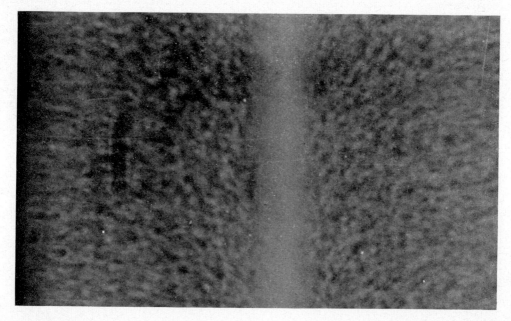

(c)

is

$$\theta = m(\phi_1 + \phi_2) + \theta_0.$$

On the wall $\phi_1 + \phi_2 = \pi$ and θ has a prescribed value $\theta_{wall} = m\pi + \theta_0$. This defines θ_0. To compute the energy we again use the equation

$$\mathscr{F} = \tfrac{1}{2} K \int \theta \nabla \theta \cdot d\boldsymbol{\sigma}.$$

The contour is represented in Fig. 4.10. The part of the integral taken along the wall vanishes, because $\theta = \theta_{wall}$ is constant while $\nabla \theta$ is an odd function of y. The

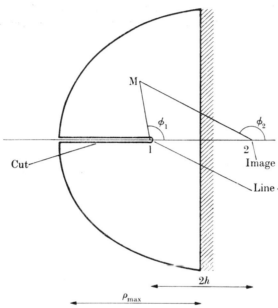

FIG. 4.10. A wedge disclination (1) parallel to a wall in a nematic. The line is repelled by its image (2). Also shown is the contour used to calculate the elastic energy of the system.

integration along the large semi-circle does not contribute, because $\nabla \theta$ is tangential in this region. Finally we are left with the contribution from the cut

$$\mathscr{F} = \tfrac{1}{2} K \{\theta\} \int_a^{\rho_{max}} m \left[\frac{1}{\rho} + \frac{1}{\rho + 2h} \right] = \frac{\pi m^2 K}{2} \left[\ln \frac{\rho_{max}}{a} + \ln \frac{\rho_{max}}{2h} \right]$$

$$\mathscr{F} = \mathscr{T} - \frac{\pi m^2 K}{2} \ln \frac{2h}{a}.$$

Thus the line is *repelled* from the wall, with a force per unit length

$$-\frac{\delta \mathscr{F}}{\delta h} = \frac{\pi K m^2}{2h} = \frac{2\pi K m^2}{d}.$$

This is called the image force.

10

Problem: a field H is applied horizontally to the free surface of a nematic: discuss the shape of this free surface, including possible disclination lines at the boundary [7].

Solution: As shown in the problem, page 93–95, the free-energy per unit area of surface is

$$F_{\text{surf}} = \tfrac{1}{2}A \left\{ \left(\frac{\mathrm{d}\zeta}{\mathrm{d}x}\right)^2 + \left(\frac{\zeta}{\lambda}\right)^2 \right\} \pm \frac{K}{\xi} \sin \psi \, \frac{\mathrm{d}\zeta}{\mathrm{d}x}$$

where $\zeta(x)$ gives the vertical displacement of the surface (and is assumed to vary in one direction x only), A is the surface tension, $\lambda = \sqrt{(A/\rho g)}$, ρ is the density-difference between the two phases in contact, and ψ is the angle of the director

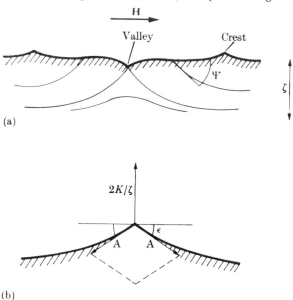

FIG. 4.11. (a) Valleys and crests induces by a horizontal magnetic field on the free surface of a nematic with conical or normal boundary conditions. (b) Determination of the slope ϵ near a crest (typically $\epsilon \sim 10^{-3}$).

with the plane of the surface. The ($\pm$) sign refers to two possible types of distortion below the surface.

One lowers the energy F by choosing a suitable, non-zero value for $\mathrm{d}\zeta/\mathrm{d}x$, provided that this does not imply too large values for the gravitational term (ζ^2/λ^2). As seen in the former problem, this term forces the surface to remain flat if the type of distortion is the same everywhere and the dimensions of the surface are $\gg \lambda$. On the other hand, if we use alternatively the two types of solutions, as shown on Fig. (4.11a), we may benefit from the last term in P_{surf}, while retaining small ζ values; the regions with different types of distortions will be separated by disclination lines, lying in the surface, parallel to the y-axis.

Locally, at all regular points, the minimum of F_{surf} still gives the usual equation $\lambda^2(\mathrm{d}^2\zeta/\mathrm{d}x^2) = \zeta$. The solution for a regular array of lines, separated by the distance L, is

$$\zeta(x) = \epsilon \frac{\sin h(x/\lambda)}{\cos h(L/2\lambda)} \qquad -\frac{L}{2} < x < \frac{L}{2}$$

$$\zeta(x+L) = -\zeta(x)$$

where 2ϵ is the angle of the cusp on a disclination line.

One may derive ϵ either by minimizing the free energy or, more physically, by an argument based on capillary forces (Fig. 4.11(b)); integrating the last term in F_{surf} on one side of a cusp gives an energy contribution $-(K/\xi) \sin \psi \zeta$ (where ζ is the altitude of the cusp) and a vertical force $(K/\xi) \sin \psi$. The other side doubles this force. Balancing these forces against the surface tensions gives

$$\epsilon = \frac{K}{\xi A} \sin \psi.$$

The free energy obtained for this solution is, per cm^2

$$\bar{F} = -\frac{\lambda A \epsilon^2}{L} \tanh u + \frac{\mathscr{T}}{L}$$

$$u = \frac{L}{2\lambda}.$$

where $\mathscr{T}$ is the line energy of the disclination at the cusp. We may estimate roughly $\mathscr{T}$ for this problem as for a wedge disclination of strength 1, with distortions limited to a half space and a cut-off at thickness ξ

$$\mathscr{T} \simeq \tfrac{1}{2} K \int_a^\xi \pi \rho \, d\rho \, \frac{1}{\rho^2} = \frac{\pi}{2} \ln \frac{\xi}{a} K = c_1 K,$$

where $c_1 \sim 10$ and is nearly independent of the magnitude of H. One must finally write that $\bar{F}$ is a minimum with respect to variations of L. This gives the condition

$$\left(\frac{H_{\frac{1}{2}}}{H}\right)^2 = \tanh u - \frac{u}{\cosh^2 u} \qquad H_1 = \left(\frac{c_1}{\chi_a}\right)^{\frac{1}{2}} (\rho g A)^{\frac{1}{4}},$$

which defines implicitly u and L as a function of H. The right hand side is always smaller than unity. Thus the field H_1 is the threshold field, above which the solution with crests and vallays is thermodynamically more stable than a flat free surface. Typically $H_1 \sim 10^4$ gauss. When H increases above H_1, L decreases from ∞ to 0 but is typically of order λ. Domain structures under horizontal magnetic fields have indeed been observed by Williams on PAA [8a], with inter-line distances of a few millimetres. They might correspond to the situation described in the problem, if the contact angle ψ was different from zero. (It is commonly accepted that $\psi = 0$ for the PAA–air interface, but it is also known in similar cases that ψ may be quite sensitive to chemical contaminants.)

More recent work on MBBA by the same group [8b] tends to confirm the effect and the present interpretation.

It might also be possible to detect the slight deformation of the surface (vertical displacements of order 1 μm) which is predicted by the model, using an inter-ferometric technique (Newton's rings). However, to reach a significant result, it is important to use light polarized perpendicular to the field direction (i.e. normal to the plane of Fig. 4.11); for this particular polarization, the refracting index inside the liquid is constantly equal to the 'ordinary' index n_0, and the optical paths for vertical beams are not complicated by the deformations of the director field which take place under the interface.

4.2.2.2. *'Planar' distortions for arbitrary line shapes* [9].

We shall now discuss the shape of the distortion for somewhat more general line structures (always keeping, however, the one-constant approximation). Let us *assume* that at all points the director σ is parallel to a constant plane (which we take as the xy plane). We retain the same notation and call $\theta(\mathbf{r})$ the angle between $\mathbf{n}(\mathbf{r})$ and the x-axis. Eqns (4.3) and (4.4) remain valid. It is possible to check first that the solutions of eqn (4.4) do give admissible configurations; the molecular field:

$$\mathbf{h} = K\nabla^2\mathbf{n}$$

is given explicitly by

$$\left.\begin{aligned}
h_x &= -K\nabla^2\theta \sin\theta - K(\nabla\theta)^2 \cos\theta = -K(\nabla\theta)^2 \cos\theta \\
h_y &= +K\nabla^2\theta \cos\theta - K(\nabla\theta)^2 \sin\theta = -K(\nabla\theta)^2 \sin\theta \\
h_z &= 0.
\end{aligned}\right\} \quad (4.9)$$

Thus $\mathbf{h} = -K(\nabla\theta)^2\,\mathbf{n}$ is correctly collinear with $\mathbf{n}$. We call these configurations, with $n_z \equiv 0$, 'planar' configurations.

Let us now restrict our attention to planar configurations, and construct a loop (L) of arbitrary shape, and strength m (see Fig. 4.6). We wish to describe a planar distortion around such a loop; this means that we must find a function $\theta(\mathbf{r})$ with the following properties:

(1) $\theta(\mathbf{r})$ satisfies the eqn $\nabla^2\theta = 0$ and is regular except on the contour L;

(2) $\theta(\mathbf{r})$ does not diverge far from the loop;

(3) $\theta(\mathbf{r})$ increases by $2\pi m$ if, starting from point $\mathbf{r}$, we make one turn around the line L and come back to $\mathbf{r}$.

This problem has a direct magnetic analogue: we may interpret $\nabla\theta$ as the magnetic field $\mathbf{b}$ due to a current loop L with current $I = m/2$.† The solution for θ (i.e. for the magnetic potential) is

$$\theta(\mathbf{r}) = \tfrac{1}{2}mS(\mathbf{r}) \quad (4.10)$$

where $S(\mathbf{r})$ is the solid angle subtended by the loop at the observation point $\mathbf{r}$ (see Fig. 4.6). If we make one turn around the line S increases by 4π and condition (3) is correctly obeyed. Eqn (4.10) is often helpful to visualize the molecular arrangment around a loop.

† We use electromagnetic c.g.s. units.

The main interest of this magnetic analogy is to relate the distortion energy to a self-induction coefficient L_S for the magnetic loop; formulae and tables of L_S values are available for all sorts of loop shapes. The connection is given by the equation

$$\tfrac{1}{2}L_S I^2 = \int \frac{b^2}{8\pi} \, d\mathbf{r} \qquad (4.11)$$

Thus, putting $\mathbf{b} = \nabla\theta$ and $I = m/2$, we get

$$\tfrac{1}{2}K \int (\nabla\theta)^2 \, d\mathbf{r} = \frac{\pi}{2} K L_S m^2 \qquad (4.12)$$

For instance, a circular loop of radius R (with a core of radius a) has a self-induction coefficient

$$L_S = 4\pi R \log(R/a) \qquad (4.13)$$

It is worthwhile to note that (apart from weak differences due to the logarithmic factors) the energy derived from eqns (4.12) and (4.13) is the product of the line tension $\mathscr{T}$ (eqn 4.7) by the length of the loop $2\pi R$.

Problem: discuss the energy of a loop in a uniformly twisted nematic [9].

Solution: Let us take the helical axis of the twist along z, and consider planar solutions with $n_z = 0$. Keeping the same notations, a suitable solution $\theta(\mathbf{r})$, satisfying the equation $\nabla^2\theta = 0$, is

$$\theta(\mathbf{r}) = qz + \tfrac{1}{2}mS(\mathbf{r}) = \theta_0 + \theta_1,$$

where q is the rate of twist. The energy is

$$\mathscr{F} = \mathscr{F}_0 + K \int \nabla\theta_0 \cdot \nabla\theta_1 \, d\mathbf{r} + \tfrac{1}{2}K \int (\nabla\theta_1)^2 \, d\mathbf{r}.$$

The first term is the energy in the absence of disclinations. The last term is given by eqn (4.12). The cross term may be integrated by parts:

$$\int \nabla\theta_0 . \nabla\theta_1 \, d\mathbf{r} = \int \{\theta_1\} \nabla\theta_0 \, d\boldsymbol{\sigma} - \int \theta_1 \nabla^2\theta_0 \, dr$$

The volume integral on the right-hand side vanishes; the surface integral is taken on a cut surface (Σ) limited by the loop L. The discontinuity $\{\theta_1\}$ across Σ is equal to $m\pi$. This gives

$$\mathscr{F} - \mathscr{F}_0 = \pi K m \{ \tfrac{1}{2} D_S m - q\Sigma \}$$

where Σ is the projected area of the loop on the xy plane.† For instance, with a circular loop parallel to the xy plane, of radius R, of strength $m = \tfrac{1}{2}$:

$$\mathscr{F} - \mathscr{F}_0 = \tfrac{1}{2}\pi^2 K \{ R \log (R/a) - qR^2 \}$$

† This equation is the analogue of a classic theorem for dislocations under stress in solids.

$\mathscr{F}$ first increases with R, reaches a maximum for $R = R^* \sim (1/4q) \log (1/qa)$, and then decreases; large loops are favourable, but there is a very high energy-barrier (of order K/q) for loop nucleation; in practice nucleation never takes place in the bulk, but on defects at the wall surfaces etc.

4.2.3. The concept of line tension

For a line of arbitrary shape, creating distortions which may be planar or non-planar, the calculation of the energy becomes more delicate. However, in the one constant approximation, the result is a simple generation of eqn (4.7). The energy E of the line is related to

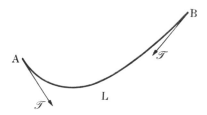

Fɪɢ. 4.12. Typical application of the line tension concept. A line L is anchored at points A and B on the sample boundary, and urged towards the bottom part of the figure by some external agent (for instance a flow). The forces on points A and B, due to the deformed line, are shown on the figure. They are equal in magnitude to the line tension $\mathscr{F}$.

the length L by:

$$E/L = \mathscr{F} = \pi K m^2 l, \tag{4.14}$$

where l is a logarithmic factor, which is nearly independent of line size or line shape. Thus we may also write, with logarithmic accuracy that

$$\delta E/\delta L = \mathscr{F}.$$

This later definition of $\mathscr{F}$ coincides with the usual definition of a *line tension*, as it is used for instance in connection with vibrating strings. Many concepts which we have been taught concerning vibrating strings in high school may thus be transposed to disclination lines; for instance the forces at both ends of a bent disclination are obtained by the construction shown on Fig. 4.12.

Line tensions can be measured; we shall describe here one typical set up used by R. B. Meyer in this connection [10]. A nematic slab, of thickness D is contained between two polished glasses. The easy axes of the two plates are at right angles. (We shall call x the easy direction of the lower plate, and y the easy direction of the upper plate.) This leads to the '*plages tordues*,' first observed long ago by Mauguin, and mentioned in Chapter 3. In one '*plage*' we have a uniform twist, described by an angle

$$\theta(z) = \pm \pi z/2D, \tag{4.15}$$

whose z measures the level counted from the lower plate. The two possible signs for the twist give equal energies; some 'plages' are $(+)$ and some are $(-)$. Two $(+)$ and $(-)$ regions are separated by a line (lying more or less at $z = D/2$), of strength $\frac{1}{2}$. Most of these lines are closed into loops.

When no special precaution is taken, a loop tends to shrink (to decrease its line-like energy $2\pi R \mathscr{T}$) and disappears rapidly. However, when a given loop is observed under the microscope, it can be stabilized by the following trick: let us assume for instance that the inside of the loop is $(+)$. To avoid shrinkage, we apply a weak magnetic field $\mathbf{H}$ at an angle θ_H from the x axis such that $0 < \theta_H < \pi/2$. Then the $(+)$ region is slightly favoured; by a suitable choice of θ_H (in practice by rotating the sample in a fixed field) one can achieve an exact balance between the magnetic effect and the line effect; the loop is then immobile, and its radius R may be measured accurately.

Let us write down the corresponding equations, assuming that. The field is weak ($\xi(H) \gg D$) so that the local twist in each $(+)$ or $(-)$ is not altered; and that the radius R of the loop is large compared with D. Then, each $(+)$ or $(-)$ region is macroscopic, and has a magnetic energy (per unit area in the xy-plane)

$$F_{\text{mag}} = -\tfrac{1}{2}\chi_a H^2 \int_0^D dz \, \cos^2(\theta(z) - \theta_H)$$
$$= -\tfrac{1}{2}\chi_a H^2 D \{ \overline{\cos^2\theta(z)}\cos^2\theta_H + \overline{\sin^2\theta(z)}\sin^2\theta_H +$$
$$+ \tfrac{1}{2}\sin 2\theta_H \, \overline{\sin 2\theta(z)} \}. \tag{4.16}$$

The bars represent averages over the slab thickness. The averages of $\cos^2\theta$ (or $\sin^2\theta$) are the same for $a(+)$ or $a(-)$. The average

$$\overline{\sin 2\theta(z)} = 1/D \int_0^D dz(\pm\sin \pi z/D) = \pm 2/\pi \tag{4.17}$$

differs for the two signs of the twist. The difference in magnetic energies between the two regions is thus:

$$\Delta F_{\text{mag}} = F_{\text{mag}}^{(-)} - F_{\text{mag}}^{(+)} = -(\chi_a H^2 D/\pi)\sin 2\theta_H. \tag{4.18}$$

Consider now the loop of radius R, surrounding an area πR^2 of the $(+)$ region. The overall energy is of the form:

$$\mathscr{F} = -\pi R^2 \, \Delta F + 2\pi R \mathscr{T} + \text{constant} \tag{4.19}$$

and the loop will be in equilibrium when $\partial \mathscr{F}/\partial R = 0$, or

$$\mathscr{T}/R = \Delta F_{\text{mag}}$$

Eqn (4.20) could also have derived in terms of known rules for vibrating strings; $\mathscr{T}/R$ is the inward restoring force (per unit length) for a bent line: this is balanced by the magnetic force ΔF_{mag}. Thus, if R is measured and if the magnetic parameters (H, θ_H, χ_a) are known, one can derive $\mathscr{T}$ from eqn (4.20). For a more detailed discussion of these measurements, see ref. [10].

Remark; anisotropy of the line tension $\mathscr{T}$. When the differences between K_1, K_2, and K_3 are taken into account, a twist and a wedge disclinations will have different line tensions. If this anisotropy of $\mathscr{T}$ is very strong, it can lead to angular cusps in the shape of the line.† In practice, however, these cusps do not seem to have been observed in nematics; the angular dependence of $\mathscr{T}$ is probably too weak.

4.3. Point disclinations

4.3.1. Instability theorem for lines of integral strength

The distortions around a line of integral strength ($m = \pm 1$, ± 2, ...) may always be continuously transformed into a smooth structure with no singular line. A typical example of this smoothing process is shown on Fig. 4.13(a, b) (see facing p. 133); the sample is cylindrical, with a large radius R, and normal boundary conditions. Fig. 4.13a shows the simplest arrangement, where the director is everywhere radial; the deformation is pure splay; on the axis of the cylinder we have a disclination line of strength $m = \pm 1$. The energy per unit length of line easily calculated to be

$$\mathscr{T} = \pi K_1 \ln(R/a). \tag{4.21}$$

Fig. 4.13(b) shows another possible conformation, involving both splay and bend. With cylindrical coordinates (ρ, ϕ, z) this is described by a director field of the form

$$\begin{aligned} n_z &= \cos u(\rho) \\ n_\rho &= \sin u(\rho) \\ n_\phi &= 0, \end{aligned} \tag{4.22}$$

where $u(R) = \pi/2$ and $u(0) = 0$. In the one-constant approximation the form of $u(\rho)$ turns out to be very simple [11, 12]:

$$\tan \tfrac{1}{2}u = \rho/R \tag{4.23}$$

We can see on eqn (4.23) that u vanishes linearly when $\rho \to 0$ and that the gradients of **n** are not singular on the axis; thus, in this splay–bend solution, the disclination line has vanished.

† This was pointed out to the author by J. Friedel.

The calculation can also be performed for the more realistic case where $K_1 \neq K_3$. The functional form of $u(\rho)$ is then modified, but the qualitative features shown by eqn (4.23) remain unchanged. The energy can be derived exactly, for instance, when $K_3 > K_1$, we have an energy per unit length [13]

$$\mathscr{T}' = \pi(2K_1 + K_3(k/\tan k)) \qquad (4.24)$$

where

$$\tan^2 k = K_3/K_1 - 1 \qquad (4.25)$$

In most practical cases K_3/K_1 is close to unity and

$$\mathscr{T}' \cong 3\pi K$$

Then the splay–bend conformation is more favourable than the line, provided that $\ln R/a > 3$ or $R > 20a \sim 40$ nm. Thus for all physical sizes R, the line is not stable. This has been noticed independently by R. B. Meyer [11] and by Cladys and Kleman [12]. Meyer describes the effect as an 'escape in the third dimension.' On the experimental side, the 'rule of escape' had also been recognized independently by Rault (doctoral dissertation, Orsay 1971).

The line could become stable only if K_1 was much smaller than K_3 (and K_2). Consider for instance the limit $K_3 \gg K_1$ in eqn (4.24). Then $k \rightarrow \pi/2$ and the energy of the 'escaped' configuration becomes

$$\mathscr{T}' \rightarrow \tfrac{1}{2}\pi^2 (K_3 K_1)^{\frac{1}{2}} \qquad (K_3 \gg K_1) \qquad (4.27)$$

Comparing this with eqn (4.21), we see that the disclination line may become stable if

$$\ln(R/a) < \tfrac{1}{2}\pi(K_3/K_1)^{\frac{1}{2}} \qquad (4.28)$$

As we shall see later, in Chapter 7, there is one case where the ratio K_3/K_1 may be very large—namely in the vicinity of a transition from nematic to smectic A. For instance if $K_3/K_1 = 16$ the condition (4.28) corresponds to $R \lesssim 400a \simeq 1$ μm. Thus, in such a case, and using fine capillaries, one might possibly stabilize a line with integral m. But, excluding such extreme cases, we see that the line will escape in practice.

The argument may be extended to cover other types of 'escape' (involving twist) and other types of lines; in all cases, when the elastic constants are comparable, the lines of integral m are found to be unstable. This theorem will be very important for our discussion of the 'noyaux.'

On the other hand, the lines of half-integral strength ($m = \pm\tfrac{1}{2}$) are stable, simply because there is no way for them to 'escape' into a smooth structure, while retaining the same m; if we follow a closed contour C surrounding the distorted region, starting from one point A with director $\mathbf{n}$(A), we reach point A again after one turn, but with a director $-\mathbf{n}$ (A). Now if we decrease the contour C we must at some moment hit a point I where $\mathbf{n}$ is discontinuous. The locus of these points I defines a disclination line, which is thus present whatever efforts we make.

4.3.2. Interpretation of the 'noyaux'

The 'structure à noyaux' or 'schlieren texture' has been presented in Section 4.1. We have seen that each 'noyau' (node) has a characteristic strength m which can be integral ($m = \pm 1$) or half integral ($m = \pm \frac{1}{2}$).

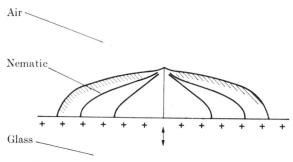

FIG. 4.14. One singular point at the free surface of a flat nematic droplet. Boundary conditions: homeotropic at the glass nematic interface; tangential or conical at the nematic–air interface.

We also noted that the nodes could be interpreted in terms of two different models; either as point disclinations, or as line disclinations normal to the plane of the slab. We are now in a position to discuss this more fully.

Since all lines with integral m are unstable, the nodes with $m = \pm 1$ *must* be disclination points; a typical geometry for such points is shown in Fig. 4.14. Any node with $m = \pm \frac{1}{2}$ must be the end point of a disclination line of the same strength, since (as explained at the end of the last paragraph) it is not possible to eliminate such a line by continuous deformations.

This very fundamental distinction between two types of nodes was not appreciated by G. Friedel. It has been established mainly through the work of R. B. Meyer [11]. All the existing observations appear to agree with it. A typical photograph showing both types of node, and the lines emerging only from the $\pm \frac{1}{2}$ species, is shown in Fig. 4.1.

4.3.3. Other observations on point defects

4.3.3.1. Points on an interface. A number of point defects, geometrically similar to 'noyaux' of integral strength ($m = \pm 1$) can be observed at the *free surface* of a nematic, or at the *interface* between a

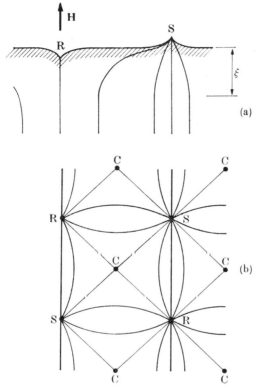

FIG. 4.15. Ideal array of point disclinations at the free surface of a nematic liquid: (a) view in a vertical plane; (b) horizontal pattern—sketch of one possible lattice (square lattice) S = peak, R = hole, C = saddle point. In practice, only disordered points have been observed up to now.

nematic and an isotropic phase, provided that the boundary conditions at the interface are tangential or conical. Typical geometries are shown in Figs. 4.14 and 4.15. A detailed study of these points has been carried out by R. B. Meyer [7a].

In most cases, the distortions taking place below the surface will react on the shape of the surface itself [7b]. The singular points are associated with cusps (or dips) on the surface. These cusps are expected to be very weak (typical height of order 1 μm), but the energy of the singular point does contain significant contributions from surface tensions and gravitational potentials, associated with the cusp.

4.3.3.2.   Points in the bulk.   When a line, such as the wedge disclination of Fig. 4.13(a), with $m = 1$, 'escapes' in the third dimension, it may do it in two ways ('upwards' or 'downwards'). An upwards portion and a downwards portion will be linked by a singular point, as shown in Fig. 4.16 (see facing p. 148). A systematic classification of the singular points has been constructed by Nabarro [13].

At first sight, the simplest geometry allowing for one singular point in the bulk corresponds to spherical nematic droplet, floating in an isotropic liquid, with normal boundary conditions at the interface: In the most naive solution for this problem, the director is everywhere radial, the singular point being at the centre of the droplet, and the deformation being pure splay [14a]. However, this conformation is not usually observed. A much more complex arrangment, involving a strong twist in the central region, is preferred [14b]. This is probably due to the fact that the twist constant K_2 is significantly smaller than K_1 in typical nematics.†

4.4. Walls under magnetic fields

In Section 4.1 we saw that sheet singularities are not stable in nematics, but that diffuse walls are allowed under a magnetic field. We now discuss these walls in more detail, starting from a simple example—which unfortunately is hard to realize experimentally—and then progressing towards more realistic situations.

4.4.1. 180° walls

4.4.1.1.   Structure of the wall.   Under a field **H**, a bulk nematic will align with the director **n** parallel or antiparallel to **H**; both situations are physically equivalent, and lead to equal energies. It may happen that in one domain **n** and **H** are parallel, while in another domain they are antiparallel; the border region between the two domains is then called a 180° wall. It is similar to a Bloch wall in a ferromagnet with uniaxial anisotropy; the analogue of the easy axis in the ferromagnet is the field axis in the nematic.

For the nematic problem, these walls were first considered by Helfrich [15]. The wall surface may be normal or parallel to **H**. In the

† Near a transition nematic–smectic A the ratio K_2/K_1 is expected to increase, and the 'naive' solution might prevail.

latter case, the only deformation involved is twist, and, since in usual materials K_2 is the smallest elastic constants, the twist wall is the least expensive from an energy point of view. It is also the simplest one to compute, and we shall restrict our attention to this type.

Let us put the field $\mathbf{H}$ along the X axis, and assume that the director $\mathbf{n}$ depends only on y, with the form:

$$\left.\begin{array}{l} n_x = \cos u(y) \\ n_z = \sin u(y) \\ n_y = 0. \end{array}\right\} \qquad (4.29)$$

At $y = +\infty$ we impose $u = 0$, and at $y = -\infty$, $u = \pi$. The calculation of $u(y)$ coincides with an earlier one, described in Chapter 3 (eqns 3.50–3.56) to introduce the concept of a coherence length; the only change is in the domain of variation of y. The implicit equation for $u(y)$ is:

$$\tan \tfrac{1}{2}u = \exp(-y/\xi_2(H)) \qquad (4.30)$$

where ξ_2 is defined as usual as $(K_2/\chi_a)^{\frac{1}{2}}H^{-1}$. We can check that eqn 4.30 is compatible with the boundary conditions. Most of the variation of u takes place in a thickness $2\xi_2$, centred at $y = 0$. Typically for $H = 5\,000$ G, $2\xi_2 \sim 10$ μm. The energy per unit area, or 'surface tension' σ, of the wall may be derived from eqns (3.47) and (3.54). The result is

$$\sigma = 2K_2/\xi_2; \qquad (4.31)$$

with the above values for H, $\sigma \sim 4 \times 10^{-3}$ erg cm^{-2}. The wall is a very smooth structure, and requires only low energies.

4.4.1.2. Boundary effects. In practice, it is often important to understand the interactions of the wall with the boundaries of the sample. If the wall intersects the boundary surface the situation is usually rather complex (in particular disclination lines may nucleate at the intersection). If the wall is parallel to the surface, the situation is somewhat simpler. Let us further assume that the boundary conditions at the surface are tangential, and that the distance d between the wall and the surface is much larger than ξ_2. Then a generalization of the above calculations leads to the following predictions [16].

(1) If the surface is rubbed, with an easy axis parallel to $\mathbf{H}$, the wall is *repelled* from it. The repulsive force per unit area, or pressure, is given by:

$$p = 8\chi_a H^2 \exp(-2d/\xi_2). \qquad (4.32)$$

(2) If the boundary condition is degenerate (e.g. at a free surface, the molecules being tangential) the wall is *attracted* towards the surface; the pressure is equal in magnitude, but opposite in sign, to eqn (4.32). In both cases these pressures may be understood as resulting from the interaction between the wall and its image, separated from it by a distance $2d$.

In principle, relying on statement (1), it appears possible to stabilize one wall between two anchored plates, which would both repel it. One might think of the following sequence to generate the wall: (*a*) in zero field, starting from a planar texture, rotate one of the glass plates by 180°, thus creating a 'plage tordue.' Apply **H** along the easy axis of the plates; the 'plage tordue' should shrink into a wall at the midplane of the slab. However, in practice, as soon as we twist the sample by more than 90°, disclination loops nucleate at the edges of the slab, and relax the torsion. Thus operation (*a*) can be performed only during a short transient period. (We shall come back to a discussion of these transient effects in Chapter 5.)

4.4.2. *Walls associated with a Frederiks transition*

The Frederiks transitions have been discussed in Section 3.2. In all these transitions we start with a nematic monodomain, apply to it a field

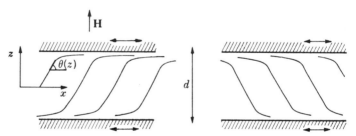

FIG. 4.18. Two types of domains above the Frederiks critical field.

normal to the optical axis, and observe a distortion above a certain threshold field H_c. For a given $H > H_c$, the system may choose between two different (but equivalent) distortion patterns: one example is shown in Fig. 4.18, corresponding to the case 1 in our classification of the various Frederiks transition —i.e, a transition from planar towards homeotropic texture.

For $H > H_c$ the nematic slab will thus break into domains, corresponding to either of the two distortion patterns. The border between

two domains of opposite distortions involves a wall.† The theoretical
structure of these walls has been analysed recently by F. Brochard [17].
It is significantly different from the structure of the Helfrich walls
described above.

Again the walls may be either parallel or normal to the easy axis; we
choose to discuss the parallel case (twist wall). The thickness of the wall
is *very large* when H is close to H_c. It is of order $2\kappa_2^{-1}$ where the inverse
length κ_2 is defined by

$$\xi_2^2 \kappa_2^2 = 1 - (H_c/H)^2 \qquad (H \to H_c)$$

From the point of view of the general theory of phase transitions, the
divergence of κ^{-1} when $H \to H_c$ is natural. The Frederiks transition is
of second order, and near a usual second order transition point
the correlation lengths diverge [18]. Fig. 4.19 (see between pp. 148–9)
shows the increase in size of a wall when $H \to H_c$.

When the field $H > H_c$ is suddenly applied, a number of domains
appear. After some time, many small domains shrink to zero, and we
are left with a few large loops. By slightly tilting the field, it is possible
to favour the inside of a loop, and to make it stationary (just as in the
Meyer technique for disclinations). The loop is not circular; the portions
of the wall which are parallel to the easy axis (twist wall) are less
expensive in energy than the normal positions (bend wall for case 1).
Thus, from a study of the ellipticity of the loops, it is possible to infer
the ratio of two elastic constants. Experiments of this type (but using
case 2 in the classification of section 3.2) have been performed by
L. Leger [19].

On the whole, the most interesting feature of the walls (as opposed to
the disclination lines) is that all their properties can be neatly analysed
in terms of the continuum theory—while for the lines, the core effect are
often non-negligible. For instance, the nucleation of a wall at a surface
involves large regions (or the order of the wall thickness) and is thus
rather unsensitive to atomic defects on the surface. On the other hand,
the nucleation of a line takes place on the molecular scale, and is not
easily controlled.

4.4.3. Transformation from walls to lines ('pincement')

Consider for instance the wall of Fig. 4.20 (see between pp. 148–9)
and increase the field H; the distortion in the wall region becomes
progressively larger (κ^{-1} decreases) and finally, at a certain field H',

† At least for H not too large. At high fields the situation is more complex, and will be
discussed later in this section.

a set of two disclination (of strengths $m = -\frac{1}{2}$ and $m = +\frac{1}{2}$) appears. This process has been called 'pincement' by Y. Bouligand. The field H depends on the choice of the system, but is typically of the order of $2H_c$. This transition from walls to lines has been observed first by R. B. Meyer [20]. A similar transition in cholesterics (with a more complex wall structure) has been found by Rault [21].

4.5. Umbilics

Another type of Frederiks transition is of interest in generating smooth defects. It is obtained with materials of negative dielectric anisotropy, prepared in a homeotropic texture, and subjected to a field E normal to the plates (direction z) (Fig. 4.21). Above a certain

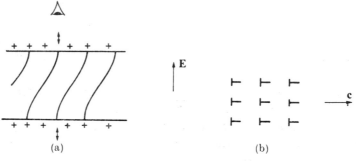

FIG. 4.21. 'Degenerate' Frederiks transition, under a field **E**, with homeotropic boundary conditions and negative dielectric anisotropy $\epsilon_a < 0$ (all electrolydrodynamic effects are assumed to be absent). Above the threshold field E_c the molecules tilt towards an arbitrary direction **c** of the slab plane (a) side view, (b) the alignment as seen by an observer looking through the plates.

Frederiks threshold E_c the molecules in the mid plane of the slab tend to tilt, but they can tilt towards any direction in the (x, y) plane. One can define the chosen tilt direction by a unit vector **c**. When observed with vertical light beams, a region tilted along **c** behaves as a birefringent slab with one neutral line parallel to **c**. It must be noticed however that **c** and $-$**c** are not equivalent.

In an ideal sample **c** would be independent of x and y. But in practice the lateral boundary conditions impose some distortions to the **c** field. One can in fact construct a two-dimensional Frank elasticity for this problem, involving two elastic constants ('c splay' and 'c bend'). Also one can find *singular points* in the **c** field (Fig. 4.22). These singular points have been called umbilics by the Orsay group [22]. They can have only integral strength (± 1). An interesting feature of the umbilics is

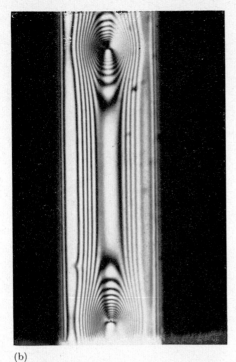

(b)

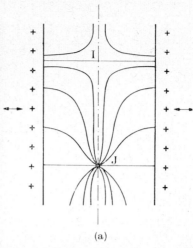

(a)

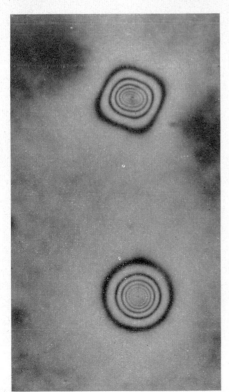

(c)

FIG. 4.16(a). Singular points in a capillary with normal boundary conditions. The diagram is schematic: in the vicinity of the points (I) or (J), a more complicated distortion may occur, as in Fig. 4.17. (b) Optical observation of a pair of defects in a capillary (crossed nicols at 45° from the capillary axis). The director field corresponds to (a). (courtesy H. Gruler and T. Scheffer.) (c) Singular points in a slab of MBBA with normal boundary conditions (courtesy R. B. Meyer). The drawing of Fig. 4.16a applies: point J gives circular rings. Point I gives elliptical rings. (Observation with crossed circular polarizers.)

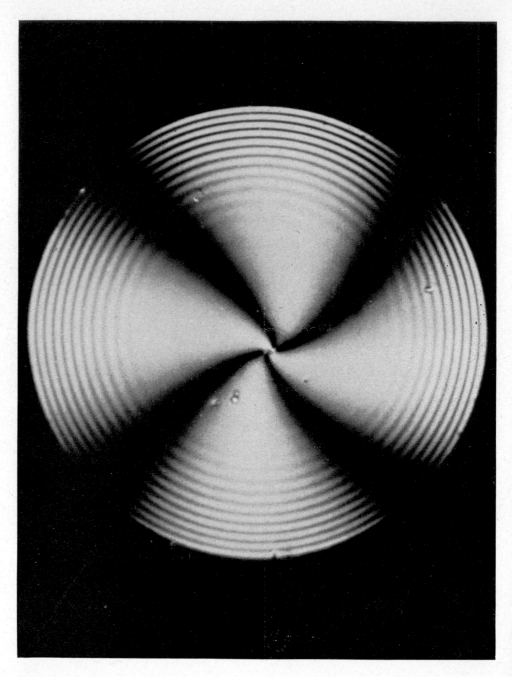

FIG. 4.17. Point defect at the centre of a nemalic droplet (courtesy S. Candau). The boundary conditions at the surface of the droplet are normal (or nearly so). Instead of a simple radial arrangement (of the type shown at point J in Fig. 4.16), one finds a twisted arrangement, clearly shown here from the black cross observed under crossed nicols.

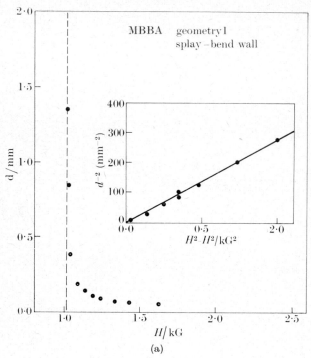

FIG. 4.19. (a) Thickness d of a wall as a function of the field H (d is derived from a study of interference fringes for normally incident monochromatic light). Note the divergence of d at the Frederiks threshold ($H/H_c \to 1$) (courtesy L. Leger). (b; see opposite) A closed wall in a homeotropic $\to$ planar transition. The long axis of the ellipse is along the field H (courtesy L. Leger).

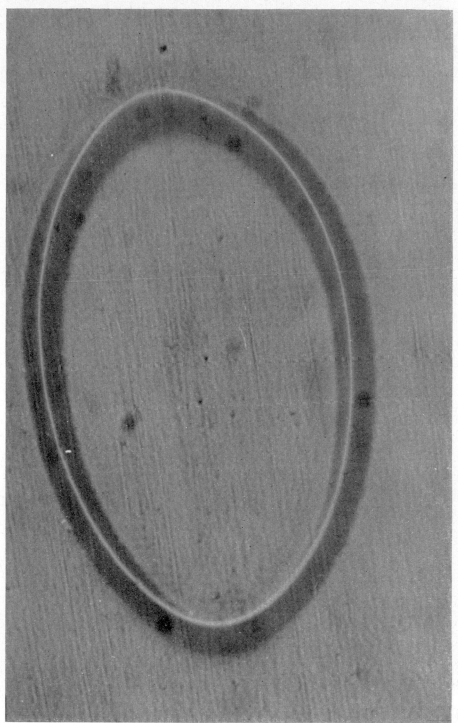

(b)

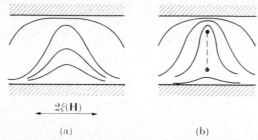

$2\xi(H)$

(a)   (b)

FIG. 4.20. 'Pincement': transformation of a wall (a) into a system of lines (b). The wall as seen under polychromatic light is shown in (c). In (d) the same region is observed at higher H. The image is focused on the upper line, and the lower line is blurred. In (e) the image is focused on the lower line. The use of a bent wall is convenient to avoid exact superposition of the two lines (courtesy L. Leger).

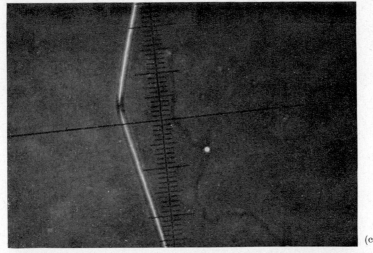

(c)

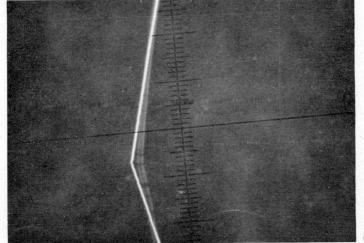

(d)

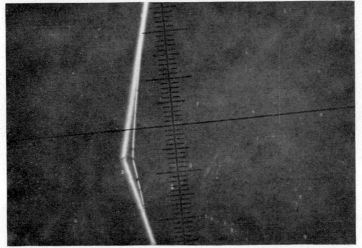

(e)

(a)

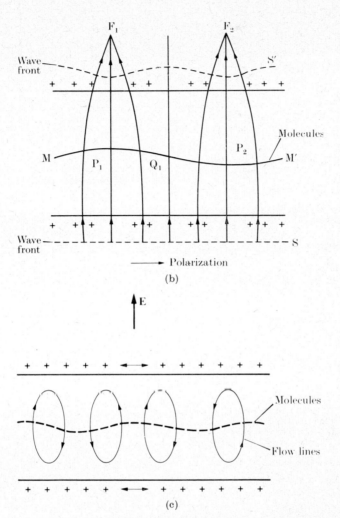

Wave front

F_1 F_2

S'

+   +   +   +   +   +   +   +   +   +

Molecules

M P_1 Q_1 P_2 M'

+   + +   + +   + +   + +   +

Wave front   S

→ Polarization

(b)

E

+   +   +   +   +   ← →   +   +   +   +   +   +

Molecules

Flow lines

+   +   +   +   +   ← →   +   +   +   +   +   +

(c)

FIG. 5.8. (a) The 'Williams domains' in MBBA. Glass plates impose tangential boundary conditions. The stripes are normal to the easy axis of the plates. (Courtesy G. Durand). (b) Light focusing by a distorted nematic arrangement. The molecular arrangement is represented by the line MM'. Plane parallel light is sent through the bottom plate: the initial wave front S is flat. Inside the sample the refractive index is larger at P_1 than at Q_1. Thus the light velocity is larger at Q_1 and the outgoing wave front S' is bent: this corresponds to a focusing of the rays (normal to S') at points F_1, F_2, etc. (c) Flow pattern associated with the distortion in the Williams domains.

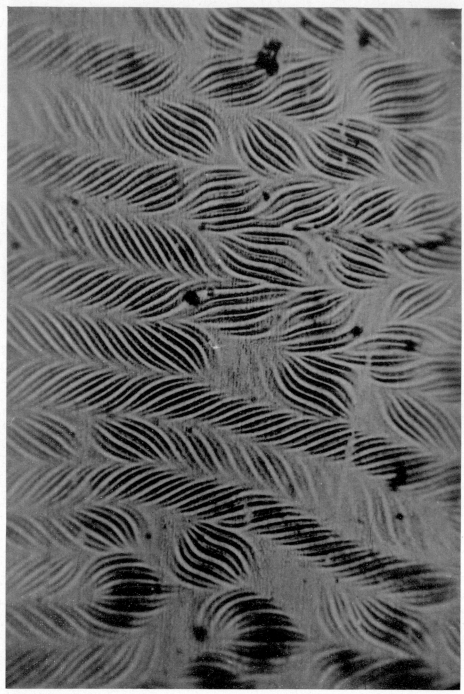

FIG. 5.11. The 'chevron' domains, characteristic of the unstabilities at high frequencies (courtesy G. Durand).

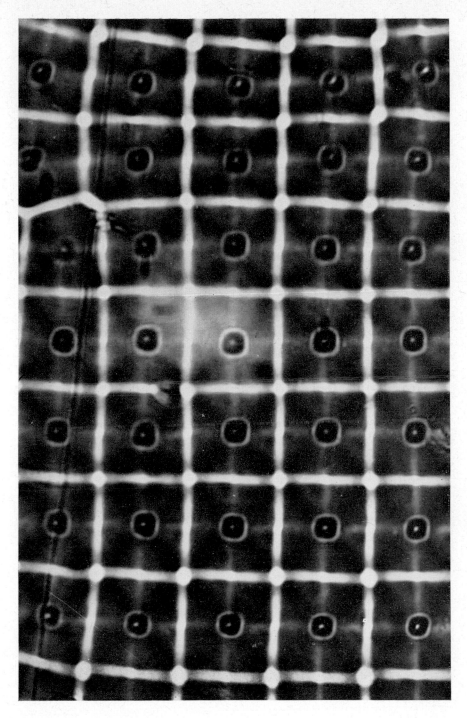

FIG. 6.17. Square lattice distortion observed in a planar cholesteric texture by application of an electric field E. Note that this distortion is visible optically only with materials of large pitch P. The field E is here roughly 10% higher than the threshold. At higher fields a very different pattern is observed. (Courtesy F. Rondelez).

FIG. 6.18. Focal conics in a cholesteric (courtesy Y. Bouligand). Compare with Fig. 7.2, which is the analogue for smectics A. The geometry is explained in Fig. 7.1. The ideal arrangement shown here is rather rare; usually one finds distorted focal conics—see Y. Bouligand, *J. Phys. (France)*, **33**, 715 (1972).

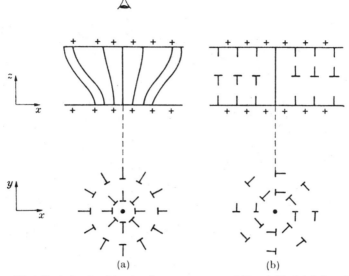

FIG. 4.22. 'Umbilics' obtained in the degenerate case of Fig. 4.21. (a) Splay dominant in the core. (b) Twist dominant in the core.

that (at least for fields E that are not too high) their core is continuous; the size of the core is essentially

$$\kappa^{-1} = \frac{DE_c}{\pi(E_c^2 - E^2)^{\frac{1}{2}}}$$

(D being always the sample thickness), and the director $\mathbf{n}$ varies smoothly inside the core [22]: thus the static and dynamic properties of the umbilics can be calculated accurately.

11

REFERENCES

CHAPTER 4

[1] FRIEDEL, G. *Annales. Phys.* **19**, 273 (1922).

[2] FRANK, F. C. *Discuss. Faraday Soc.* **25**, 1 (1958).

[3] RAULT, J. *C.r. Acad. Sci., Paris* **272**, 1275 (1971).

[4] PIERANSKY, P. Thèse 3e cycle, Orsay (1972).

[5] FRIEDEL, J. and KLEMAN, M. in *Fundamental aspects of the theory of dislocations*, p. 607. Natn. Bur. Stand. spec. pub. **317**, I (1970); KLEMAN, M. and FRIEDEL, J. *J. Phys. (Fr.)* **30**, (Suppl. C4) 43 (1969).

[6] Detailed studies on groups of disclinations have been carried out by ERICKSEN J.L. *Liquid crystals and ordered fluids* (Porter and Johnson, eds.), p. 181. Plenum Press, New York (1970); DAFERMOS, C. *Quart. J. Mech. app. Math.* **33**, 49 (1970).

[7] (a) MEYER, R. B. *Mol. Cryst. liquid Cryst.* **16**, 355 (1972); (b) DE GENNES, P. G. *Solid State Commun.* **8**, 213 (1970).

[8] (a) WILLIAMS, R. *J. chem. Phys.* **50**, 1324 (1969). The optical rotations observed by Williams are due to his use of a binocular microscope with beams tilted from the vertical: when such a beam propagates, for instance, in the (yz) plane, it enters into the thickness ξ below the surface, through regions where the optical axis is progressively rotating; (b) WILLIAMS, R., *et al. J. appl. Phys.* **43**, 3685 (1972).

[9] FRIEDEL, J. and DE GENNES P. G. *C.r. hebd. Séanc. Acad. Sci., Paris* **268**, 257 (1969).

[10] MEYER, R. B. *Phys. Rev. Lett.* **22**, 918 (1969); Thesis, Harvard University (1969); see also MEYER R. B. (to be published).

[11] MEYER, R. B. *Phil. Mag.* **27**, 405 (1973).

[12] CLADYS, P. KLEMAN, M. *J. Phys. (Fr.)* **33**, 591, (1972).

[13] NABARRO, F. R. N. *J. Phys. (Fr.)* **33**, 1089, (1972).

[14] (a) DUBOIS VIOLETTE, E. and PARODI, O. *J. Phys. (Fr.)* **30**, (Suppl. C4) 57 (1969); (b) CANDAU, S. *Mol. Cryst. liquid Cryst.* (to be published); BILLARD, J. Private communication. (c) SAUPE, A. *Mol. Cryst.* **21**, 211 (1973); (d) WILLIAMS, C., CLADYS, P., KLEMAN M. *Mol. Cryst.* **21**, 355 (1973).

[15] HELFRICH, W. *Phys. Rev. Lett.* **21**, 1518 (1968).

[16] DE GENNES, P. G. *J. Phys. (Fr.)* **32**, 789 (1971).

[17] BROCHARD, F. *J. Phys. (Fr.)* **33**, 607 (1972).

[18] See, for instance, STANLEY, H. E. *Introduction to phase transitions*, Oxford University Press (1971).

[19] LEGER, L. *Mol. Cryst. liquid Cryst.* (to be published).

[20] MEYER, R. B. Private communication.

[21] RAULT, J. *Mol. Cryst. liquid Cryst.* **16**, 143 (1972).

[22] RAPINI, A. *J. Phys.* **34**, 629, (1973).

DYNAMICAL PROPERTIES OF NEMATICS

'Before us the thick dark current runs. It talks to us in a murmur become ceaseless and myriad.'   W. FAULKNER

5.1. The equations of 'nematodynamics'

5.1.1. Coupling between orientation and flow

A NEMATIC flows very much like a conventional organic liquid with molecules of similar size. However, the flow regimes are more complex, and more difficult to study experimentally than in isotropic liquids, for the following reasons.

The translational motions are coupled to inner, orientational motions of the molecules; in most cases, the flow disturbs the alignment. Conversely, a change in the alignment (e.g. by application of an external field) will, in many instances, induce a flow in the nematic.

To measure these effects quantitatively, optical observations of the local state of affairs under flow would be extremely helpful. Unfortunately, nematics are turbid (for reasons which are not accidental; see §3.4). In practice, optical studies are restricted to thin samples (below 300 μm). Conventional viscometric equipment (based on capillaries, falling spheres, rotating cylinders. etc) is thus not adequate for optical observations. Also, the boundaries (e.g. at the inner surface of the capillary) are not controlled, and the state of nematic alignment is unknown. For these reasons, many of the existing data in the literature cannot be used for quantitative purposes.

It is possible however to improve on this situation, either by imposing an alignment by an external field, or by selecting some more refined probes: attenuation of acoustic shear waves, inelastic scattering of light, etc. (We shall discuss these experiments on "nematodynamics" in Section 5.2.)

From a theoretical point of view, the coupling between orientation and flow is a very delicate matter. It has been analysed essentially by two groups, which represent very different schools of thought:

(1) A macroscopic approach, based on classical mechanics, has been used by Ericksen [1], Leslie [2], and Parodi [3] (hereafter referred to as ELP). Most of the existing data have been analysed in this language.

(2) A microscopic approach, based on a study of correlation functions, has been set up more recently by the Harvard group [4a]. The results have been later rewritten directly in macroscopic terms, and extended to other mesomorphic phases [4b].

The two approaches are essentially identical in content. We shall present here a (slightly amended) version of the ELP theory, and then show how, by a slight change in the choice of variables, one can obtain the Harvard formulation.

5.1.2. Choice of dynamical variables

The first assumption of the ELP approach is that in a nematic liquid crystal a dynamical situation must be specified by:

(1) a velocity field $v(r)$ giving the flow of matter

(2) a unit vector $n(r)$ (the director) describing the local state of alignment.

This formulation was questioned in an early contribution by the Harvard group [5]. They proposed a completely different theory, where it was assumed that the velocity field $v(r)$ is enough to specify the state of affairs; in this theory the director is not an independent variable, but its orientation is deduced from the gradients of v. Thus, in this picture, a rotation of the optical axis can occur only if there is a macroscopic (non-uniform) flow.

Experimentally, it is in fact possible, in suitable geometries, to rotate the director n by an external field, without any macroscopic displacement of the molecules ($v \equiv 0$). A good example of this property is found with planar distortions of pure twist (Fig. 5.1); starting with a nematic single crystal between two polished plates, one applies a magnetic field H, in the plane of the plates, at a finite angle ψ ($< \pi/2$) to the unperturbed optical axis. If H is large enough ($\xi(H)$ much smaller than the sample thickness d) the molecules in the central part of the slab rotate by an angle ψ. To see if this is accompanied by backflow, one watches a dust particle floating in the nematic (Y. Galerne, unpublished). The result is negative; in this geometry a rotation of the optical axis is achieved without any macroscopic displacements. More refined versions of this experiment have also been carried out by inelastic scattering of light [6], and they lead to the same conclusion: the choice of state variables proposed in ref. [5] was not sufficient to describe a nematic, while the ELP choice is adequate.

We should also add one word of comment on the use of a director $n(r)$ of *constant length* ($n^2 \equiv 1$). Of course, as already mentioned in connection with the static continuum theory (Section 3.1), in a distorted state, the magnitude of the birefringence is slightly modified. However, this effect is very small at all temperatures below the nematic isotropic transition T_c. Also, in the ordered phase, the characteristic frequencies associated with a change of magnitude of the order parameter are high, and thus not amenable to a hydrodynamic description. Thus we must omit these changes.†

5.1.3. The entropy source for a flowing nematic

5.1.3.1. Nature of the losses.
We wish to write down equations of motion for the variables **v** and **n** in the limit of slow space-variations and low frequencies ('hydrodynamic limit'). The first step is to construct a formula giving the dissipation or, as it is called, the entropy source, due to all friction processes in the fluid. For simplicity we shall restrict our attention to *isothermal* processes (no thermal gradients); then we shall find two types of dissipative losses; conventional viscosity effects and losses associated with a rotation of the optical axis with respect to the background fluid.

Our derivation of the entropy source follows rather closely the approach of de Groot and Mazur [7] for isotropic fluids. The free energy stored in our nematic has the form

$$\int \{\tfrac{1}{2}\rho v^2 + F_0 + F_d + F_{mag}\} \, d^3\mathbf{r}$$

The first term represents kinetic energy (ρ is the density). F_0 is an internal free energy, depending on the density. F_d is the Frank free-energy for distortions. F_{mag} represents the coupling between the director and an external magnetic field **H**. Note that we did not include any surface terms in the free energy; this means that we restrict ourselves to situations with strong anchoring.

In all what follows we shall assume (as in Chapter 3) that H is uniform in space. Also, for simplicity, we have not included in our discussion the coupling with an electric field, or the effect of bulk forces such as gravitation.

For an isothermal process, the dissipation $T\dot{S}$ is equal to the decrease in stored free energy

$$T\dot{S} = -\frac{d}{dt}\int \{\tfrac{1}{2}\rho v^2 + F_0 + F_d + F_M\} \, d^3\mathbf{r} \qquad (5.1)$$

† Above T_c the situation is quite different, and it will be discussed separately in Section 5.4.

We shall now write down explicitly the various contributions to eqn (4.1); the kinetic term will be derived from an equation for the local acceleration; the internal free-energy terms will be taken directly from our discussion of hydrostatics in Chapter 3.

5.1.3.2.   Definition of the total stress and of the viscous stress.   We shall write the acceleration equation for the fluid in the form

$$\rho \frac{\mathrm{d}}{\mathrm{d}t} v_\beta = \partial_\alpha \sigma_{\alpha\beta} \tag{5.2}$$

where $\sigma_{\alpha\beta}$ is called the *stress tensor*. It must be emphasized that $\sigma_{\alpha\beta}$ is not uniquely defined by eqn (4.2). Changes of the form $\sigma_{\alpha\beta} \rightarrow \sigma_{\alpha\beta} + g_{\alpha\beta}$ leave the acceleration unchanged, provided that $\partial_\alpha g_{\alpha\beta} \equiv 0$.

Inserting eqn (5.2) into eqn (5.1) and integrating by parts we obtain

$$-\frac{\mathrm{d}}{\mathrm{d}t} \int (\tfrac{1}{2}\rho v^2) \, \mathrm{d}^3\mathbf{r} = \int \sigma_{\alpha\beta} \, \partial_\alpha v_\beta \, \mathrm{d}^3\mathbf{r} + \text{surface terms.} \tag{5.3}$$

Let us now turn to the contribution of $F_0 + F_d + F_m$ to the entropy source: it can be obtained from eqn (3.101) (putting $F_g = 0$) and is

$$-\frac{\mathrm{d}}{\mathrm{d}t} \int \{F_0 + F_d + F_m\} \, \mathrm{d}^3\mathbf{r} = \int (-\sigma^e_{\alpha\beta} \, \partial_\alpha v_\beta + \mathbf{h}.\,\dot{\mathbf{n}}) \, \mathrm{d}^3\mathbf{r} + \text{surface terms.} \tag{5.4}$$

where $\boldsymbol{\sigma}^e$ is the *Ericksen stress* (eqn 3.100), $\mathbf{h}$ is the molecular field (eqn 3.89) and

$$\dot{\mathbf{n}} = \frac{\mathrm{d}\mathbf{n}}{\mathrm{d}t} = \frac{\partial\mathbf{n}}{\partial t} + (\mathbf{v}.\nabla)\mathbf{n} \tag{5.5}$$

is the *material derivative* of $\mathbf{n}$ (change of $\mathbf{n}$ per unit time as experienced by a moving molecule).

Adding the contributions from eqn (5.3) and (5.4) we arrive at

$$T\dot{S} = \int \{(\sigma_{\alpha\beta} - \sigma^e_{\alpha\beta}) \, \partial_\alpha v_\beta + \mathbf{h}.\dot{\mathbf{n}}\} \, \mathrm{d}^3\mathbf{r} + \text{surface terms} \tag{5.6}$$

The difference $\boldsymbol{\sigma}'$ between the actual stress $\boldsymbol{\sigma}$ and the equilibrium stress $\boldsymbol{\sigma}^E$ will be called the *viscous stress*

$$\sigma'_{\alpha\beta} = \sigma_{\alpha\beta} - \sigma^e_{\alpha\beta}. \tag{5.7}$$

5.1.3.3.   The antisymmetric part of the viscous stress.   The tensor $\sigma'_{\alpha\beta}$ is not symmetric in general. To characterize the antisymmetric part, we shall introduce a vector $\mathbf{\Gamma}$:

$$\Gamma_z = -\sigma'_{xy} + \sigma'_{yx} \quad \text{etc.} \tag{5.8}$$

Our first task now is to derive an explicit expression for $\mathbf{\Gamma}$. Consider the integral

$$\dot{L} = \frac{d}{dt} \int d^3r \, \mathbf{r} \times \rho \mathbf{v} \tag{5.9}$$

This represents the rate of change of the angular momentum of the sample, i.e. the *total torque* applied to our liquid. In our case, $\dot{L}$ contains the following contributions: 1) magnetic torques

$$\int d^3r \mathbf{M} \times \mathbf{H} \tag{5.10}$$

where $\mathbf{M(r)}$ is the local magnetization in the nematic (see eqns 3.45 and 3.48); 2) torques due to external stresses ($\boldsymbol{\sigma}$) acting on the sample boundary

$$\int \mathbf{r} \times (d\mathbf{S}:\boldsymbol{\sigma}), \tag{5.11}$$

where $d\mathbf{S}$ is the surface element vector (pointing outwards from the fluid) on the boundary, and $(d\mathbf{S}:\boldsymbol{\sigma})_\alpha = dS_\beta \sigma_{\beta\alpha}$; 3) torques on the director at the boundary. For equilibrium situations, these have been discussed in Chapter 3 (eqn 3.115), and are given by

$$\int \mathbf{n} \times (d\mathbf{S}:\boldsymbol{\Pi}) \tag{5.12}$$

where $\Pi_{\alpha\beta}$ is defined in eqn 3.90. Here we have chosen to discuss 'strong anchoring' situations for which there are no specific losses at the surface. Then eqn (5.12) for the surface torques remains true even out of equilibrium.

Collecting the contributions from eqns 5.10–5.12 we arrive at

$$\frac{d}{dt} \int (\mathbf{r} \times \rho \mathbf{v}) \, d^3r = \int (\mathbf{M} \times \mathbf{H}) \, d^3r + \int \{\mathbf{r} \times (d\mathbf{S}:\boldsymbol{\sigma}) + \mathbf{n} \times (d\mathbf{S}:\boldsymbol{\Pi})\}. \tag{5.13}$$

We now make use of eqn (5.2) to transform the left-hand side of eqn (5.13), and integrate the result by parts, to obtain

$$\int (\sigma_{yx} - \sigma_{xy}) \, d^3r = \int (\mathbf{M} \times \mathbf{H})_z \, d^3r + \int \{\mathbf{n} \times (d\mathbf{S}:\boldsymbol{\Pi})\}_z. \tag{5.14}$$

In the left-hand side of (5.14) we write $\boldsymbol{\sigma} = \boldsymbol{\sigma}^e + \boldsymbol{\sigma}'$. The contribution from $\boldsymbol{\sigma}'$ is simply Γ_z. The contribution from $\boldsymbol{\sigma}^e$ has already been calculated in eqn (3.110)†

$$\int (\sigma^e_{yx} - \sigma^e_{xy})\, d^3r = \int \{\mathbf{M} \times \mathbf{H} - \mathbf{n} \times \mathbf{h}\}_z\, d^3r + \int \{\mathbf{n} \times (d\mathbf{S}:\mathbf{\Pi})\}_z. \quad (5.15)$$

Inserting these results into eqn (5.14) we see that the magnetic term and the $\mathbf{\Pi}$ term drop out, leaving us with

$$\int \{\mathbf{\Gamma} - \mathbf{n} \times \mathbf{h}\}\, d^3r = 0. \quad (5.16)$$

Thus we conclude that

$$\mathbf{\Gamma} = \mathbf{n} \times \mathbf{h}. \quad (5.17)$$

Physically we may say that $\mathbf{\Gamma}$ is the torque (per unit volume) exerted by the internal degree of freedom ($\mathbf{n}$) on the flow. This torque is non-zero only out of equilibrium, since we know from Chapter 3 that $\mathbf{n}$ is collinear to $\mathbf{h}$ at equilibrium. The lack of symmetry of the tensor $\boldsymbol{\sigma}'$ (eqn 5.8) simply reflects the presence of these bulk torques.

5.1.3.4. Final formula for the entropy source. Let us now return to eqn (5.6) giving the dissipation; it is convenient at this stage to separate the velocity gradient tensor $\partial_\alpha v_\beta$ into a symmetric part

$$A_{\alpha\beta} = \tfrac{1}{2}(\partial_\alpha v_\beta + \partial_\beta v_\alpha) \quad (5.18)$$

and an antisymmetric part, associated with the vector

$$\boldsymbol{\omega} = \tfrac{1}{2}\mathrm{curl}\, \mathbf{v}. \quad (5.19)$$

Similarly we shall call $\boldsymbol{\sigma}^s$ the symmetric part of $\boldsymbol{\sigma}'$. Making use of these definitions, and of eqn (5.8) for $\mathbf{F}$, we can transform eqn (5.6) and obtain

$$T\dot{S} = \int \{\mathbf{A}:\boldsymbol{\sigma}^s - \mathbf{\Gamma}.\boldsymbol{\omega} + \mathbf{h}.\dot{\mathbf{n}}\}\, d^3r \quad (5.20)$$

We now substitute $\mathbf{\Gamma} = \mathbf{n} \times \mathbf{h}$ (eqn (5.17)) into this result, and arrive at

$$T\dot{S} = \int \{\mathbf{A}:\boldsymbol{\sigma}^s + \mathbf{h}.\mathbf{N}\}\, d^3r \quad (5.21)$$

where

$$\mathbf{N} = -\dot{\mathbf{n}}\boldsymbol{\omega} \times \mathbf{n} \quad (5.22)$$

The vector $\mathbf{N}$ represents the rate of change of the director with respect to the background fluid. Eqn (5.21) is the fundamental equation of 'nemato-dynamics.' It displays the two types of dissipation which we announced at the beginning of this section: dissipation by shear flow and dissipation by rotation of the optical axis.

† Eqn (3.110) is written in terms of another tensor $\boldsymbol{\sigma}^d$, but $\boldsymbol{\sigma}^d$ and $\boldsymbol{\sigma}^e$ have the same antisymmetric part [see eqn (3.100)].

It is also interesting to note that the entropy source vanishes for rigid rotations: i.e., if both the molecules and the director rotate around a certain axis, at the same angular velocity, we have $A_{\alpha\beta} = 0$, and $N_\alpha = 0$, and thus by eqn (5.21), $T\dot{S} = 0$. This property is sometimes considered as obvious, and then used as a starting point to derive (5.21). The more detailed (and painful) derivation which was used here is essentially based on eqn (5.15); as explained in Chapter 3 this equation simply expresses that the distortion free energy F_d is a scalar (a function of div $\mathbf{n}$, of $\mathbf{n}.\text{curl } \mathbf{n}$, and of $(\mathbf{n} \times \text{curl } \mathbf{n})^2$): thus the physical content is the same in both derivations.

5.1.4. The laws of friction

5.1.4.1. Definition of fluxes and forces. It is customary, when dealing with irreversible processes, to write each contribution to the entropy source as the product of a 'flux' by the conjugate 'force.' (For a general introduction to these notions, we again refer the reader to the book by de Groot and Mazur, ref. [7].) Here the entropy source is given by eqn (5.21). We choose to take as fluxes the components of the symmetric tensor $A_{\alpha\beta}$ and the components of the vector N_μ.† The tensor **A** has six independent components, and the vector **N** has three components.‡ According to eqn (5.21) we may then say that:

$$\sigma_{xx}^s \text{ is the force conjugate to } A_{xx}$$

$$2\sigma_{xy}^s \text{ is the force conjugate to } A_{xy} \quad \text{etc.} \qquad (5.23)$$

$$h_\alpha \text{ is the force conjugate to } N_\alpha$$

The factor of 2 occurring in the non-diagonal elements expresses the fact that a term such as $\sigma_{xy}^s A_{xy}$ appears twice in the entropy source (5.21).

5.1.4.2. The friction coefficients. We may now proceed to write down a set of phenomenological equations expressing the forces in terms of the flux (or vice versa). At this stage we assume that the fluxes are *weak on the molecular scale*—for instance that the rotation velocities described by **N** are very small in comparison with the rotation frequencies characteristic of one molecule. This limit of slow motions is required for all hydrodynamic theories.

† With this choice, all the fluxes are odd under time reversal, while all forces are even.
‡ The components of **N** are linked by one relation ($N_\mu n_\mu = 0$). However a linear dependence between fluxes does not invalidate the Onsager relations. See de Groot and Mazur (ref. [7] §6.3.)

In the limit of weak fluxes, the forces will be linear functions of the fluxes; we write this in the form

$$\sigma^s_{\alpha\beta} = L_{\alpha\beta\gamma\delta}A_{\gamma\delta} + M_{\alpha\beta\gamma}N_\gamma \tag{5.24}$$

$$h_\gamma = M'_{\alpha\beta\gamma}A_{\alpha\beta} + P_{\gamma\delta}N_\delta. \tag{5.25}$$

All the coefficients L, M, M', P have the dimension of a viscosity, namely mass $\times$ length^{-1} $\times$ time^{-1}; to see this, it is enough to note that σ and $\mathbf{h}$ have the dimensions energy per unit volume, while $\mathbf{A}$ and $\mathbf{N}$ have the dimensions of frequency.

For the diamagnetic materials of interest, the influence of the magnetic field $\mathbf{H}$ on the friction constants is completely negligible; this observation simplifies the situation considerably.

Another simplification is provided by the Onsager theorem [7], which asserts that the matrices M and M' are identical

$$M_{\alpha\beta\gamma} = M'_{\alpha\beta\gamma}. \tag{5.26}$$

To check that the factors of two occurring in the definition (5.23) of the forces are correctly included in eqn (5.26), the reader may consider, for instance, the coefficient of N_γ in the expression (derived from eqn 5.24) for the force $2\sigma^s_{xy}$; this is equal to $2M_{xyy}$. Now the coefficient of A_{xy}, in eqn (5.25) for h_γ, is $2M'_{xyy}$. These two coefficients must be equal by Onsager's theorem; the result is in agreement with (5.26).

Finally, the structure of the matrices L, M, P must be compatible with the local symmetry $(D_{\infty h})$ of the nematic. Thus the only vector which may appear in the definition of L, M, P is the local director $\mathbf{n}$. Also, as we have repeatedly pointed out, all measurable properties must be invariant when we change $\mathbf{n}$ into $-\mathbf{n}$. In such a change $\mathbf{A}$ and σ^s are invariant, while $\mathbf{N}$ and $\mathbf{h}$ are odd. The most general structure for eqns (5.25) and (5.26) compatible with these requirements is

$$\sigma^s_{\alpha\beta} = \rho_1\,\delta_{\alpha\beta}A_{\mu\mu} + \rho_2 n_\alpha n_\beta A_{\mu\mu} + \rho_3\,\delta_{\alpha\beta}n_\gamma n_\mu A_{\gamma\mu} + \alpha_1 n_\alpha n_\beta n_\mu n_\rho A_{\mu\rho} +$$
$$+ \alpha_4 A_{\alpha\beta} + \tfrac{1}{2}(\alpha_5 + \alpha_6)(n_\alpha A_{\mu\beta} + n_\beta A_{\mu\alpha})n_\mu +$$
$$+ \tfrac{1}{2}\gamma_2(n_\alpha N_\beta + n_\beta N_\alpha) \tag{5.27}$$

$$h_\mu = \gamma'_2 n_\alpha A_{\alpha\mu} + \gamma_1 N_\mu. \tag{5.28}$$

Again all the coefficients $(\rho'_s, \alpha'_s, \gamma'_s)$ have the dimension of a viscosity. The Onsager relation (5.26) imposes that

$$\gamma'_2 = \gamma_2. \tag{5.29}$$

Thus, in the general case, the equations of nematodynamics (eqns 5.27 and 5.28) involve eight independent coefficients. However, for the majority of the experiments, one is concerned only with motions which are very slow in comparison with sound waves: it is then permissible to treat the fluid as *incompressible*: we shall now focus our attention on this case.

5.1.4.3. *Incompressible nematics: the Leslie presentation.* If the density of the fluid is constant, the trace of the A tensor vanishes

$$A_{\mu\mu} = \text{div } \mathbf{v} = 0 \qquad (5.30)$$

Thus the (ρ_1) and (ρ_2) terms in eqn (5.27) drop out. The term (ρ_3) does not contribute to the entropy source (5.21) because it reduces to three equal diagonal components; it can be lumped into the scalar pressure p and omitted from the equations (recall that, for an incompressible fluid, p is not given by an equation of state, but is an unknown, to be fixed at the end of the calculation by the condition of constant density).

For the current applications, where the starting point is the acceleration equation (eqn 5.2), it is convenient to transform eqn (5.27) into an equation for the complete viscous stress σ'. The antisymmetric part of σ' is given explicitly in terms of the molecular field $\mathbf{h}$ by eqns (5.8) and (5.17). The field $\mathbf{h}$, in turn, is written in eqn (5.28). Collecting these results, one arrives at the following equations

$$\sigma'_{\alpha\beta} = \alpha_1 n_\alpha n_\beta n_\mu n_\rho A_{\mu\rho} + \alpha_4 A_{\alpha\beta}$$
$$+ \alpha_5 n_\alpha n_\mu A_{\mu\beta} + \alpha_6 n_\beta n_\mu A_{\mu\alpha}$$
$$+ \alpha_2 n_\alpha N_\beta + \alpha_3 n_\beta N_\alpha \qquad (5.31)$$

$$h_\mu = \gamma_1 N_\mu + \gamma_2 n_\alpha A_{\alpha\mu}. \qquad (5.32)$$

Together with the relations

$$\gamma_1 = \alpha_3 - \alpha_2 \qquad (5.33)$$

$$\gamma_2 = \alpha_2 + \alpha_3 = \alpha_6 - \alpha_5. \qquad (5.34)$$

The coefficients α_i are usually called the *Leslie coefficients*: there are six α_is, linked by one relation (5.34) (first derived by Parodi [3]). Thus the dynamics of an incompressible nematic involves *five independent coefficients*, all with the dimension of a viscosity. For the few examples where data are available at the present time, the five coefficients appear to be of comparable magnitude—typically in the range 10^{-2} to 10^{-1} poises. Values for MBBA are given in Table 5.

5.1.4.4. *Another choice of fluxes and forces.* In the preceding paragraphs, starting from the entropy source (5.21) we defined as fluxes the quantities $A_{\alpha\beta}$ and N_α (all of which are odd under time reversal), and as forces the quantities $\sigma^s_{\alpha\beta}$ and h_α (all of which are even under time reversal).

However, it is sometimes convenient to make a different choice, and to take $\sigma^s_{\alpha\beta}$ and N_α as fluxes, while $A_{\alpha\beta}$ and h_α are now the forces. This choice does lead to very compact formulae for the study of *small motions* in a nematic single crystal, which have been introduced first by the Harvard group [4], and which we shall now write down.

Let us immediately restrict our attention to the problem of small amplitude motion. The unperturbed director $\mathbf{n}_0$ is taken along the z-axis. The perturbed director has small components $n_x(\mathbf{r})$, $n_y(\mathbf{r})$. We shall discuss the stresses only to first order in n_x and n_y. As already explained, the component of $\mathbf{h}$ along $\mathbf{n}$ may be chosen arbitrarily. For the present problem, we may then assume that h_z vanishes identically and write the entropy source as

$$T\dot{S} = \sigma^s_{\alpha\beta}A_{\alpha\beta} + h_x N_x + h_y N_y, \tag{5.35}$$

thus displaying explicitly the only independent components of $\mathbf{h}$ and $\mathbf{n}$ (h_x, h_y and N_x, N_y). Let us now write a set of linear relations giving σ^s and N in terms of A and h: for an incompressible fluid, the most general form is

$$\sigma^s_{\alpha\beta} = 2\nu_2 A_{\alpha\beta} + 2(\nu_3 - \nu_2)(A_{\alpha\mu}n^0_\mu n^0_\beta + A_{\beta\mu}n^0_\mu n^0_\alpha)$$
$$+ 2(\nu_1 + \nu_2 - 2\nu_3)n^0_\alpha n^0_\beta n_\mu A_{\mu\rho}n_\rho$$
$$- \tfrac{1}{2}\lambda(n^0_\alpha h_\beta + n^0_\beta h_\alpha); \tag{5.36}$$

$$N_i = h_i/\gamma_1 + \lambda A_{iz} \qquad (i = x, y). \tag{5.37}$$

A number of comments are required to help the reader at this point.

The parameters ν_1 ν_2 ν_3 have the dimensions of viscosity. Why they are chosen as the fundamental parameters in this version of the theory will become more apparent if we write down explicitly the components of $\boldsymbol{\sigma}^s$

$$\begin{aligned}
\sigma^s_{xx} &= 2\nu_2 A_{xx} \\
\sigma^s_{xy} &= 2\nu_2 A_{xy} \\
\sigma^s_{xz} &= 2\nu_3 A_{xz} - \tfrac{1}{2}\lambda h_x \\
\sigma^s_{zz} &= 2\nu_1 A_{zz} \\
\sigma^s_{yz} &= 2\nu_3 A_{yz} - \tfrac{1}{2}\lambda h_y.
\end{aligned} \tag{5.38}$$

The parameter λ is a dimensionless number. For a system of fluxes, one group of which is odd under time reversal (namely $\mathbf{A}$) while the other ($\mathbf{h}$) is even, the Onsager relations demand that the crossed coefficients be *opposite* (see de Groot and Mazur [7]). This is why we have $(-\lambda)$ in eqn (5.36) while we have $(+\lambda)$ in eqn (5.37).

The eqns (5.36) (5.37) must be identical in content with the equations (5.31) (5.32) of the Leslie formulation. Comparing the two sets [and also making use of the identities (5.33) and (5.34)] one arrives at the relations

$$2\nu_2 = \alpha_4$$

$$2\nu_3 = \alpha_4 + \beta \tag{5.39}$$

(where $\beta = \alpha_6 + \alpha_3 \lambda = \alpha_5 + \alpha_2 \lambda$)

$$2\nu_1 = \alpha_1 + \alpha_4 + \alpha_5 + \alpha_6$$

and also

$$\lambda = -\gamma_2 / \gamma_1. \tag{5.40}$$

The entropy source may be expressed in terms of the fluxes A and h in the form

$$T\dot{S} = 2\nu_1 A_{zz}^2 + 4\nu_2(A_{zx}^2 + A_{zy}^2) + 2\nu_3(A_{xx}^2 + A_{yy}^2) + (h_x^2 + h_y^2)/\gamma_1. \tag{5.41}$$

The dissipation does not involve λ: for this reason the parameter λ is classified as a 'reactive parameter' by the Harward group [4]. The condition of increasing entropy ($T\dot{S} > 0$), (evaluated with the constraint $A_{\alpha\alpha} \equiv \operatorname{div} \mathbf{v} = 0$ for an incompressible fluid) imposes that ν_1, ν_2, ν_3, and γ_1 be positive.

Let us now discuss the equation giving the *bulk force* (per unit volume) in a nematic with distortions and flow. This is defined by eqn (5.2), namely

$$\rho \frac{dv_\beta}{dt} = \partial_\alpha \{\sigma_{\alpha\beta}^e + \sigma_{\alpha\beta}^s + \tfrac{1}{2}(n_\beta h_\alpha - n_\alpha h_\beta)\} \tag{5.42}$$

Here $\boldsymbol{\sigma}^e$ is the Ericksen stress, and [[as can be seen from eqn (3.100)] it is of second order in the deviations n_x, n_y: thus, in the small motion approximation, it drops out completely. $\boldsymbol{\sigma}^s$ is defined by eqn (5.36) or eqn (5.38). The last term in eqn (5.42) represents the antisymmetric part of σ', as derived from eqns (5.8) and (5.17). In this term, to first order in n_x, n_y, we replace $n_\beta h_\alpha$ by $n_\beta^0 h_\alpha$, etc. Finally we obtain for the bulk force

$$\rho \frac{dv_\beta}{dt} \cong \partial_\alpha \sigma_{\alpha\beta}^s + \tfrac{1}{2}(n_\beta^0 \, \partial_\alpha h_\alpha - n_\alpha^0 \, \partial_\alpha h_\beta) \tag{5.43}$$

As already pointed out at the beginning of this chapter, the stress tensor leading to this bulk force is not unique. The Harward group has made use of this observation, and has constructed a *symmetric* stress tensor $\sigma_{\alpha\beta}^H$ which gives rise to the same set of bulk forces, provided that $H = 0$ (no external torques in the bulk). The Harward stress $\boldsymbol{\sigma}^H$ is defined by

$$\sigma_{\alpha\beta}^H = \sigma_{\alpha\beta}^s + \tfrac{1}{2}\{-n_\mu^0 \, \partial_\mu \Pi_{\alpha\beta} - n_\mu^0 \, \partial_\mu \Pi_{\beta\alpha} + n_\alpha^0 \, \partial_\mu \Pi_{\beta\mu} + n_\beta^0 \, \partial_\mu \Pi_{\alpha\mu}\}. \tag{5.44}$$

The corresponding force is

$$\partial_\alpha \sigma_{\alpha\beta}^{\rm H} = \partial_\alpha \sigma_{\alpha\beta}^{\rm s} + \tfrac{1}{2}\{-n_\mu^0 \, \partial_\mu h_\beta - n_\mu^0 \, \partial_\mu \, \partial_\alpha \Pi_{\beta\alpha} + n_\alpha^0 \, \partial_\alpha \partial_\mu \Pi_{\beta\mu} + n_\beta^0 \, \partial_\mu h_\mu\} \quad (5.45)$$

where we have made use of the equation for the molecular field $\mathbf{h}$, which reduces here (for $H = 0$) to $h_\alpha = \partial_\mu \Pi_{\mu\alpha}$.† The second and third terms in the bracket of eqn (5.45) cancel out, and (5.45) is seen to be identical to (5.43).

For many applications connected with small distortions of a nematic single crystal, the calculations are somewhat simpler when performed with the Harvard stress tensor—especially because the structure of $\sigma^{\rm s}$ is very simple. On the other hand the ELP approach gives a more detailed insight for the internal torques, etc.

5.1.4.5. Summary of equations and unknowns. Let us now summarize what are the unknowns, and what are the equations, for a study of motions in an incompressible nematic

(1) *The unknowns;*
   (a) the velocity field $\mathbf{v}(\mathbf{r}, t)$;
   (b) the director $\mathbf{n}(\mathbf{r}, t)$;
   (c) the pressure $p(\mathbf{r}, t)$.

Since $\mathbf{n}$ is a unit vector, it involves only two independent parameters. Thus the total number of unknown quantities is $3+2+1 = 6$.

(2) *The equations* (for definiteness we shall refer to the ELP presentation):
   (a) the acceleration equation (5.2) where the stresses are defined by eqn (3.100) for the reversible part $\sigma^{\rm e}$, and by eqn (5.31) for the viscous part;
   (b) the equations (5.32) giving the rate of change of the director in terms of the velocity gradients plus the molecular field $\mathbf{h}$. The latter is given explicitly in terms of the director by eqn (3.89). At first sight we have three equations (5.32), since they involve the components of a vector. However it must be recalled that the molecular field $\mathbf{h}(\mathbf{r}, t)$ is defined only within the transformation

$$\mathbf{h}(\mathbf{r}, t) \to \mathbf{h}(\mathbf{r}, t) + \lambda(\mathbf{r}, t)\mathbf{n}(\mathbf{r}, t)$$

since this transformation leaves the torque $\mathbf{\Gamma}$ (eqn 5.17) unchanged, and it is only $\mathbf{\Gamma}$ which is observable. Because of this freedom we obtain only two conditions from (5.32);
   (c) the incompressibility condition (5.30).

---

† h_α is defined by eqn (3.89). In this equation the term $\partial F_d/\partial n_\alpha$ is of second order in n_x, n_y, and may thus be omitted here.

We see that the number of equations is then $3+2+1 = 6$, just equal to the number of unknowns. In principle we have an adequate tool for the study of macroscopic motions. In practice there are many difficulties:

(1) the equations are quite complicated—their consequences have been explored only for very simple types of macroscopic flows, or for small deviations from a fully aligned state;

(2) the description does not include disclination lines (or points). In practice the lines may play a significant role in certain dissipative flow properties. This implies that all hydrodynamic experiments, devised to check the ELP equations, are meaningful only if all disclinations are carefully eliminated from the sample, and from all upstream regions.

5.2. Experiments measuring the Leslie coefficients

At the time of writing (1972), experimental studies on the flow properties of nematics are rather scarce. However, a number of methods are beginning to be used; we list below only those methods for which the interpretation is reasonably simple.

5.2.1. Laminar flow under a strong orienting field

In this type of experiment, the molecules are firmly aligned in one direction by a constant magnetic field $\mathbf{H}$.† The word 'firmly' corresponds to the following conditions:

1) The lateral walls limiting the flow may tend to impose some preferred direction of alignment, different from the direction of $\mathbf{H}$. To avoid this, the diameter D of the flow must be much larger than the coherence length ξ defined in Chapter 3. Then the misalignments due to wall effects are confined to a small sheet of thickness ξ near the walls, and are thus negligible:

$$\xi \ll D \quad \text{or} \quad \mathrm{H} \gg \sqrt{\left(\frac{K}{\chi_a}\right)}\frac{1}{D} \tag{5.46}$$

2) The flow itself tends to disrupt the alignment; in a velocity gradient A there will be, in general, a hydrodynamic torque acting on the molecules. It is given by eqns (5.17) and (5.32), and is of order γA (per cm³) where γ is some average of γ_1 and γ_2. It must be balanced by the restoring torque due to the magnetic field, which is $\chi_a H^2 \sin\theta$, θ being

† Alignment by an electric field E may also be used (if the material has a positive dielectric anisotropy); however this method requires special care to avoid all possible electrohydrodynamic instabilities (see Section III).

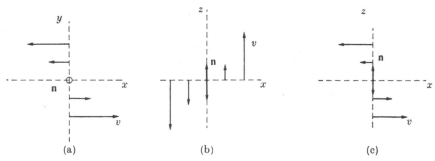

FIG. 5.1. The three fundamental geometries for viscosity measurements in a well-aligned nematic (director **n**).

the angle between **n** and **H**. [For a discussion of this torque, see Chapter 3, eqn (5.48)]. Thus $\sin \theta \sim \gamma A / \chi_a H^2$. The alignment will be essentially unperturbed if $\theta \ll 1$, or

$$A \ll \chi_a H^2 / \gamma \qquad (5.47)$$

Let us now assume that the two conditions (5.46, 5.47) are satisfied, and discuss the possible types of laminar flow: depending on the relative orientations of the field **H**, the velocity **v** and the velocity gradient, we find three typical geometries for *simple shear*. These possibilities are shown and listed under (a), (b), (c) in Fig. 5.1. The corresponding viscosities have been measured by Miesowicz, using a slowly-moving plate to create the shear (Fig. 5.2) [8]. He studied *p*-azoxyanisole and found the following values at 122°C†

$$\eta_a = 3.4 \times 10^{-2} \text{ poise}$$
$$\eta_b = 2.4 \times 10^{-2} \text{ poise}$$
$$\eta_c = 9.2 \times 10^{-2} \text{ poise}$$

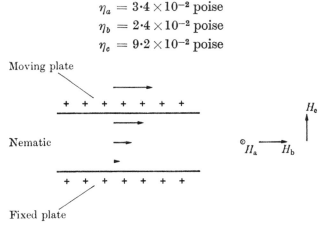

FIG. 5.2. Principle of the Miesowicz experiment. The field **H** aligning the molecule is applied in one of the three fundamental directions a, b, and c. The moving plate oscillates at a very low frequency, and viscosity is derived from the damping of the oscillations.

† The value of η_a is an extrapolation from data at different temperatures.

How are the measured values of η_a, η_b, η_c related to the Leslie coefficients? To find this we return to eqn (5.31) giving the viscous stress in terms of the shear rate tensor $A_{\alpha\beta}$ and of the effective director velocity $\mathbf{N} = d\mathbf{n}/dt - (\boldsymbol{\omega}\times\mathbf{n})$. In the present cases $d\mathbf{n}/dt = 0$ along each flow line and $\boldsymbol{\omega} = \frac{1}{2}\mathrm{curl}\ \mathbf{v}$ is simply expressed in terms of the velocity gradient.

If β is the direction of $\mathbf{v}$, and α the direction of the velocity gradient, we must derive the viscous force along β per unit area in a plane normal to α. With our convention for the ordering of indices (see eqn 5.2) this force is given by the component $\sigma'_{\alpha\beta}$ of the stress tensor. When $\sigma'_{\alpha\beta}$ is known, we can derive an effective viscosity η by the equation

$$\eta = \frac{\sigma'_{\alpha\beta}}{2A_{\alpha\beta}}. \tag{5.48}$$

Let us now consider explicitly the three cases in Fig. 5.1 and label the axes as shown. (The molecules and the field $\mathbf{H}$ are always along the z axis.) We get for these situations

(a)
$$\left.\begin{aligned}
A_{yx} &= \frac{1}{2}\frac{\partial v}{\partial y} \\[2mm]
\sigma'_{yx} &= \alpha_4 A_{yx} \\[2mm]
\eta_a &= \frac{\sigma'_{yx}}{2A_{yx}} = \frac{1}{2}\alpha_4
\end{aligned}\right\} \tag{5.49}$$

(b)
$$\left.\begin{aligned}
A_{zx} &= \frac{1}{2}\frac{\partial v}{\partial x} \\[2mm]
N_x &= -\omega_y = A_{xz} \\[2mm]
\sigma'_{xz} &= \alpha_3 N_x + (\alpha_4 + \alpha_6)A_{xz} \\[2mm]
\eta_b &= \frac{1}{2}(\alpha_3 + \alpha_4 + \alpha_6)
\end{aligned}\right\} \tag{5.50}$$

(c)
$$\left.\begin{aligned}
A_{zx} &= \frac{1}{2}\frac{\partial v}{\partial z} \\[2mm]
N_x &= -\omega_y = -A_{zx} \\[2mm]
\sigma_{zx} &= \alpha_2 N_x + (\alpha_4 + \alpha_5)A_{zx} \\[2mm]
\eta_c &= \frac{1}{2}(-\alpha_2 + \alpha_4 + \alpha_5) = \eta_b - \gamma_2.
\end{aligned}\right\} \tag{5.51}$$

These may also be expressed in terms of the Harvard coefficients $(v_1, v_2, v_3, \gamma_1, \lambda)$ by a suitable application of eqns (5.39) and (5.40):

$$\begin{aligned}
\eta_a &= v_2 \\
\eta_b &= v_3 + \tfrac{1}{4}\gamma_1(1-\lambda)^2 \\
\eta_c &= \eta_b + \lambda\gamma_1.
\end{aligned} \tag{5.52}$$

Thus, from measurements on simple shear flow we can extract three relations between the five friction coefficients.

5.2.2. Attenuation of ultrasonic shear waves

This type of experiment has been set up recently for nematics by Candau and Martinoty [9a]. The principle is shown in Fig. 5.3. An ultrasonic shear wave (of frequency ω) propagates in a solid crystal, penetrates

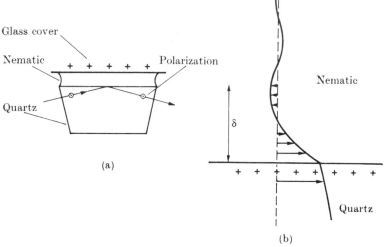

Fɪɢ. 5.3. The Candau–Martinoty experiments on shear wave attenuation in nematics: (a) set up (b) aspect of the wave penetration in the nematic. The penetration thickness δ is in the micron range.

slightly into the nematic, and is reflected back; what is measured is a reflection coefficient. In ref. [9a] the incidence is oblique (for practical reasons). In the present analysis, to avoid some minor complications, we shall restrict our attention to normal incidence Fig. 5.3b.

There is an apparent similarity between the set-up of Fig. 5.3b and the Miesowicz experiment, (Fig. 5.2) where an oscillating plate was moved in the fluid. However, there are some crucial differences:

(1) The frequency range: in the high-frequency ultrasonic experiment shear is induced only in a very small thickness of the nematic, near the vibrating solid. For a conventional fluid of viscosity η, and density ρ, this penetration thickness δ is known to be [10]

$$\delta = \left\{ \frac{\eta}{\rho\omega} \right\}^{\frac{1}{2}} \tag{5.53}$$

As we shall see, this result remains correct for a nematic fluid, η being a certain combination of the Leslie coefficients. Typically, with $\rho = 1$, $\omega/2\pi = 10_y$ and $\eta = 0\cdot1$ poise, we have $\delta = 4~\mu\mathrm{m}$. On the other hand, in the Miesowicz experiment, the plate oscillates very slowly ($\omega \to 0$) the thickness δ is larger than the sample dimensions, and the velocity gradient is the same at all points in the nematic.

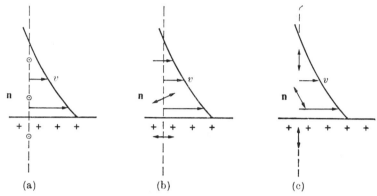

Fig. 5.4. The three fundamental geometries for acoustic measurements of shear viscosities. Note that in cases (b) and (c) the director n is significantly tilted by the flow.

(?) The orienting agent. In the Miesowicz experiment the molecules are aligned by a strong magnetic field **H**. In the ultrasonic experiment, a preferred direction is defined by the boundary conditions at the crystal–nematic interface. (Three typical situations are displayed in Fig. 5.4.)

However, as we shall see, in spite of the directional effects of the wall, the ultrasonic shear does in general impose a non-negligible tilt angle to the optical axis. For this reason the effective viscosities derived from the ultrasonic experiment differ from the Miesowicz viscosities η_a, η_b, η_c.

Let us now show on one example how the effective 'ultrasonic viscosities' $\tilde{\eta}_a$, $\tilde{\eta}_b$, $\tilde{\eta}_c$ can be related to the friction coefficients. We shall not consider case (1) (which is a very particular case with no tilt of the optical axis) but rather focus our attention on case (2), as being more illustrative. We call $0x$ the axis normal to the interface, and $0z$ the easy direction at the wall. The flow velocity $v(x)$ is also parallel to $0z$. The director **n** is nearly parallel to $0z$, but is slightly tilted in the (x, z) plane ($n_z \sim 1$, $n_x \ll 1$, $n_y = 0$). The shear rate tensor $A_{\alpha\beta}$ has one non-vanishing component $A_{xz} = \frac{1}{2}\partial v/\mathrm{d}x$ and the local rotation vector **ω** reduces to $\omega_y = -\frac{1}{2}(\partial v/\partial x)$. We are interested in the component σ_{xz} of the strain, giving the force (along z) per unit area in the (y, z) plane. This is given by Leslie's equation (5.31) which may be written as

$$\sigma_{xz} = \eta_b \frac{\partial v}{\partial x} + \alpha_3 \mathrm{i} \omega n_x.$$ (5.54)

The tilt of the director is derived from eqn (5.32), which reads

$$\gamma_1 \mathrm{i} \omega n_x = h_x - \tfrac{1}{2}(\gamma_1 + \gamma_2) \frac{\partial v}{\partial x}.$$ (5.55)

By eqn (3.23) the molecular field h_x is of order $K \nabla^2 n_x \sim K n_x / \delta^2$ where δ is defined in eqn (5.53). Thus

$$h_x \sim \frac{K \rho \omega}{\eta} n_x$$

$$\frac{h_x}{\gamma_1 \omega n_x} \sim \frac{K \rho}{\eta \gamma_1} \sim \frac{K \rho}{\eta^2}$$ (5.56)

where we do not distinguish between γ_1 and the average viscosity η. The parameter

$$\mu = \frac{K \rho}{\eta^2}$$ (5.57)

is a dimensionless quantity which plays an important rôle in many dynamical problems connected with nematics.† In all known cases μ is *small*. Typically with $K = 10^{-6}$ dynes, $\rho = 1$ gram/cm^3; where $\eta = 10^{-2}$ poise, we have $\mu = 10^{-2}$. Higher values of η would make μ even smaller. The order-of-magnitude estimate (5.56) shows that h_x may be neglected in eqn (5.55). Inserting the resulting form for $\mathrm{i} \omega n_x$ in eqn (5.54) we arrive at $\sigma_{xz} = \bar{\eta}_b (\partial v / \partial x)$ where the effective viscosity $\bar{\eta}_b$ is given by

$$\bar{\eta}_b = \eta_b - \alpha_3 \frac{\gamma_1 + \gamma_2}{2 \gamma_1}.$$ (5.58)

The main interest of this derivation based on the Leslie approach, is to show explicitly that there is a tilt in the optical axis ($n_x \neq 0$). On the other hand, if one is interested only in the final result (5.58), one may reach it more concisely through the Harvard formulation; using eqn 5.38) we have:

$$\sigma_{xz}^s = 2 \nu_3 A_{xz} - \frac{\lambda}{2} h_x.$$

As shown above, h_x is negligible. The differences between the complete Harvard tensor $\boldsymbol{\sigma}^{\mathrm{H}}$ (eqn 5.44) and $\boldsymbol{\sigma}^s$ are also of order h, and may be

---

† μ may be interpreted as the ratio of two diffusion constants: $\mu = D_\mathrm{n}/D_\mathrm{v}$, where D_v describes the diffusion of the *vorticity*, well known for isotropic fluids $D_\mathrm{v} = \eta/\rho$. The other constant D_n describes the diffusion of *orientation*, and is given by $D_\mathrm{n} = K/\eta$. This formula for D_n will be justified later (eqs 5.70–72).

omitted. Thus one arrives at

$$\tilde{\eta}_b = \nu_3, \qquad (5.58')$$

which is in fact identical to (5.58), but much simpler in notation. Repeating this argument for the three cases (a) (b) (c) one obtains

$$\tilde{\eta}_a = \eta_a = \nu_2$$
$$\tilde{\eta}_b = \tilde{\eta}_c = \nu_3. \qquad (5.59)$$

The equality between $\tilde{\eta}_b$ and $\tilde{\eta}_c$ had been noticed first by A. Rapini by careful scrutiny of the equations (such as 5.58) in the Leslie formulation, which had been derived by Candau and Martinoty [10].† In the Harvard formulation the equality is obvious (as first pointed out by P. Martin): clearly this formulation is more suitable for the ultrasonic problem, since the variables are **A** and **h**, and h may be neglected. On the other hand, in the Miesowicz problem, the Leslie choice of variables (**A** and $\dot{\mathbf{n}}$) is more suitable, because **n** is fixed by the magnetic field.

Finally, it may be shown that, for all relevant mechanical parameters (acoustic impedances, wave reflection coefficients etc) the nematic should behave exactly like an ordinary fluid, of viscosity $\tilde{\eta}_a$ (or $\tilde{\eta}_b$ or $\tilde{\eta}_c$) depending on the geometrical conditions. A careful reader may find this simple result somewhat surprising—his point being that the tilt $n_x(x)$ derived from eqn (5.55), with $h_x \simeq 0$, does not satisfy the correct boundary condition for strong anchoring ($n_x(0) = 0$). However, this defect is not serious when μ (as defined by eqn 5.57) is small. One can show that there is a thin layer (thickness $\mu^{\frac{1}{2}}\delta$) near the crystal surface, where h_x cannot be neglected and where n_x adjusts to the boundary condition. But this layer is so thin that it plays a negligible role in the mechanical properties of the fluid.

Measurements of the reflected intensity at the interface between quartz and MBBA have been undertaken by Candau and Martinoty [10a] for the geometries (a) and (b). It is very much to be hoped that case (c) will also be feasible, and that the Rapini equality ($\tilde{\eta}_b = \tilde{\eta}_c$) will be compared to experimental data.

At first sight, one might also be tempted to repeat the ultrasonic experiments under a strong magnetic field **H**, parallel to the easy axis on the walls; then the tilt of the optical axis would be negligible and the effective viscosities would be given by the Miesowicz set η_a, η_b, η_c.

---

† The experts in classical mechanics (Ericksen, Truesdell, etc) consider that the use of the Onsager relations in hydrodynamics requires special caution. The Rapini equality, which does depend on the validity of these relations, provides a direct experimental check on this point.

However, this is not feasible in practice. To understand why, let us consider for instance case (b) and insert a magnetic contribution into eqn (5.55). In the limit of small μ we now have

$$\left(i\omega + \frac{\chi_a H^2}{\gamma_1}\right) n_x = -\frac{\gamma_1 + \gamma_2}{2\gamma_1} \frac{\partial v}{\partial x}.$$

To reduce n_x significantly we should achieve $\chi_a H^2/\gamma_1 > \omega$ or $H > (\gamma_1 \omega / \chi_a)^{\frac{1}{2}}$. Taking $\gamma_1 = 0.1$ poise, $\frac{1}{2}\omega/\pi = 10^7$ and $\chi_a = 10^{-7}$ c.g.s. units, this would correspond to fields of order 10^8 gauss!

Another method for generating shear waves in nematics is based on the *capillary* waves which propagate at a nematic–air interface. These waves could be generated by mechanical means. In practice, it is convenient to observe the *spontaneous* thermal fluctuations of the surface by inelastic light scattering [9b].

At first sight, one might expect two types of novel effects connected with these waves in nematics:

(1) when the surface undulates (with a certain wave vector $\mathbf{q}$) the molecular arrangement below it is distorted (within a thickness $\sim q^{-1}$). Thus the effective surface tension $\tilde{A}$ might be modified. In fact, we have qualitatively:

$$\tilde{A} - A = Kq^2 \frac{1}{q} = Kq$$

where A is the natural surface tension and K is a Frank constant. The relative shift $(\tilde{A} - A)/A$ is then very small (of order qa) for the long wavelengths ($\gtrsim 10 \ \mu$m) of interest, and the effect is negligible.

(2) The damping of the capillary waves depends on the Leslie coefficients; in particular, for tangential (or conical) boundary conditions at the interface, the damping depends on the angle between the direction of wave propagation ($\mathbf{q}$) and the nematic axis ($\mathbf{n}$).

These friction effects have been studied with great care, on MBBA, by the Kastler group [9b]. They do give certain combinations of the Leslie coefficients (see Table 5.3).

5.2.3. Laminar flow in the absence of external fields

With conventional isotropic fluids, studies on laminar flow in a capillary, or between rotating cylinders ('Couette flow' represent the most direct means of measuring the viscosity. In nematics, measurements of this type have also been carried out, mainly by Porter, Johnson, and coworkers [11]. However, as already mentioned in the introduction to this chapter, the interpretation of these early experiments is delicate, because:

(1) The boundary conditions at the walls were not controlled.

(2) The possible role of disclination lines was not ascertained.

Recently, more precise experiments have been carried out by Fisher and Wahl [12a] in a situation of simple shear flow between two parallel plates, with homeotropic boundary conditions on both plates. What is measured in these experiments is *not* a mechanical property. Rather, by optical methods, one probes the distortions of the molecular alignment due to the flow. The aspect of these distortions depends critically on the parameter $\lambda = -\gamma_2/\gamma_1$.

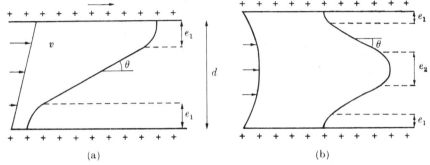

(a)   (b)

FIG. 5.5. Distortions induced in a nematic plate by laminar flows. Strong normal anchoring is assumed on both walls; $|\lambda| > 1$. (a) Simple shear flow: at high enough shear rates, the nematic molecules tend to make a fixed angle θ with the flow lines, except in two boundary layers, of thickness $e_1 \cong (K/\eta_s)^{\frac{1}{2}}$. (b) Poiseuille flow: we now have two regions of alignment at the angle θ separated by an adjustment sheet of thickness $e_2 = d^{\frac{1}{2}}e_1^{\frac{1}{2}}$. The difference between e_2 and e_1 stems from the fact that the local shear $\partial v/\partial z$ is small in the central region.

(1) If $|\lambda| > 1$ there is a certain critical angle θ between **n** and **v** (defined by $\cos 2\theta = 1/\lambda$) for which the hydrodynamic torque $\mathbf{\Gamma}$ (given by equations 5.17 and 5.32) vanishes. Then, far from the walls, the molecules tend to lie precisely at this angle. In the vicinity of the walls, since the molecules must adjust to a prescribed boundary condition, their orientation changes progressively; this takes place in a certain 'transition layer' of thickness $e \sim \{K/\eta s\}^{\frac{1}{2}}$ K being an elastic constant, η an average viscosity, and s the shear rate. This behaviour is shown on Fig. (5.5a) for simple shear flow. The slightly more complicated case of laminar flow between two fixed walls is displayed in Fig. (5.5b).

(2) If $|\lambda| > 1$ the hydrodynamic torque $\mathbf{\Gamma}$ is non-zero for all orientations of the molecules.† This implies that the nematic structure

---

† This is true except for the configuration in Fig. 5.1a, where the hydrodynamic torque vanishes for symmetry reasons.

is very strongly deformed. For the simplest case of shear flow between two plates, the director, as seen by an observer moving from one plate to the other rotates by many turns: the pitch of this "cycloidal structure" is of order e^2/D where D is the distance between the plates† (Fig. 5.6‡.)

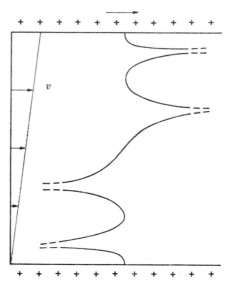

FIG. 5.6. Distortions induced in a nematic plate by simple shear flow, when $|\lambda| \equiv |\gamma_2/\gamma_1| < 1$. When moving from one glass plate to the center of the slab, one finds that the director rotates by a large angle (of the order of $7\pi/4$ in the example shown).

In practice, one often finds that $|\lambda|$ is slightly larger than unity; in a thick nematic sample, the molecules tend to become aligned along the direction of flow (θ close to 0). For instance, in MBBA at 22°C, Wahl and Fisher find [12]

$$\theta = 8° \qquad |\lambda| = +1{\cdot}04.$$

If $\theta \sim 0$, we are automatically in case (b) of Fig. 5.1. Provided that λ is close to 1, we expect that a bulk viscosity measurement in a thick capillary give η_b as the effective viscosity: this has indeed been approximately verified in a few cases.

Detailed theoretical calculations on the distortions induced by flow are listed under [13a]. As regards the mechanical properties, the most

† This estimate applies when $1-|\lambda|$ is of order unity. If $1-|\lambda|$ becomes small, the pitch increases.

‡ The cycloidal structure should also be obtained with $|\lambda| > 1$ if a suitable magnetic field is applied at 45° from the flow lines: see P.G. de Gennes, *Physics Letters* 41A, 479 (1972).

important consequence of these distortions is that the apparent viscosity η_{app} becomes a *function of the shear rate s*. Ericksen has shown, by a purely dimensional argument [13b] that

$$\frac{\eta_{app}(s)}{\eta_{app}(0)} = f\left(\frac{e_1}{d}\right) = f\left[\left(\frac{K}{\gamma_1 s}\right)^{\frac{1}{2}}\frac{1}{d}\right]$$

In this formula e_1 is the thickness of the boundary layer defined for instance in Fig. 5.5a, and d is the flow diameter. The function f is dimensionless and depends only on:

(1) The ratio between various Leslie coefficients such as λ.

(2) The particular laminar flow under study.†

The Ericksen law holds for all (disclination free) laminar flows with strong anchoring boundary conditions and agrees very well with the existing data [13c]. To appreciate the importance of this law, it may be useful to compare it with the scaling law which holds for dilute polymer solutions

$$\frac{\eta(s)}{\eta(0)} = f(s\tau)$$

where τ is the relaxation time of one macromolecule, and is independent of the flow size d.

Finally, it must be emphasized that all the preceding discussion was restricted to *stable flows*. In fact, inside the laminar domain (i.e. keeping the Reynolds number low) we do sometimes find remarkable instabilities due to coupling between orientation and flow. For instance, the simple shear flow of case (a) in Fig. 5.2 becomes unstable above a certain critical shear rate (P. Pieranski and E. Guyon, *Sol. State Commun.* **13**, 435, (1973).

5.2.4. Variable external fields

It is possible to induce motions in a nematic fluid by suitable time-dependent external fields. The motions may concern the director (rotation of the optical axis) or the molecular centres of gravity (hydrodynamic flow) or both. The fields may be electric or magnetic. However, in most practical situations, the coupling between a nematic and an electric field involves very special charge transport processes. For this reason all electric effects will be discussed separately, later in this chapter (Section 5.3.). For the moment we restrict our attention to the comparatively more simple case of a magnetic field $\mathbf{H}(t)$. We also assume that $\mathbf{H}$

---

† In Fig. (5b) we find two characteristics lengths e_1 and e_2. But their ratio is only a function of e_1/d and the scaling property is maintained.

is spatially uniform. These restrictions still allow for many possible set-ups, which can be interesting either for a determination of certain Leslie coefficients, or to induce some remarkable magneto-optic effects. Here, we shall discuss briefly a few typical examples.

5.2.4.1. Oscillating fields. Consider for instance the nematic slab with one free surface shown in Fig. 5.7. At the bottom of the slab the

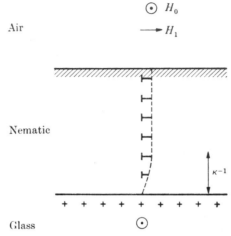

FIG. 5.7. A nematic droplet (with tangential boundary conditions) under a large static field H_0 (normal to the sheet) plus a small oscillating field H. We use here the Friedel–Kleman graphical notation to show tilted molecules: each 'nail' represents a molecule pointing partly out of the sheet, the tip of the nail being directed towards the observer.

molecules are strongly anchored (along the x direction) by a polished glass surface. Furthermore, a static field H_0 may also be added in the same direction. At the free surface, the boundary condition is assumed to be tangential.

We apply on this system a small a.c. field H_1 (of angular frequency ω) along the y axis (i.e. in the plane of the slab, but normal to H_0). The molecules then assume a slightly twisted arrangment; at all points the director is still in the (x, y) plane, but it makes a small angle $\phi(z, t)$ with the unperturbed direction (x). There are at least two methods to detect the distortion ϕ:

(1) *Magnetic detection.*† The magnetization density in the liquid crystal is given by eqn (3.45).

$$\mathbf{M} = \chi_\perp \mathbf{H} + \chi_a (\mathbf{H} \cdot \mathbf{n}) \mathbf{n},$$

† This method has been first described to the author in a private communication by Dr. D. Johnson (Kent State University).

where $\mathbf{H} = \mathbf{H_0} + \mathbf{H_1}$. The transverse component of interest (modulated at the frequency ω) is

$$M_y = \chi_\perp \mathbf{H_1} + \chi_a \mathbf{H_0} \phi \qquad (5.60)$$

The two terms may be separated through their frequency dependence, since, as we shall see, the ϕ term drops out at high frequencies. What is measured finally is the average of ϕ over the sample volume. Since M_y is a small (diamagnetic) response, it can be measured only on rather large samples.

(2) *Optical detection.* A plane wave, moving vertically upwards, enters the nematic slab from below. The initial polarization is along $0x$. On distances of order λ (the optical wavelength) the changes in twist angle ϕ are small; thus, inside the nematic slab, the polarization of the light wave follows adiabatically the direction of the optical axis. When the wave emerges at the free surface, it's plane of polarization has been rotated by an angle $\phi(z_s)$ (z_s being the surface level). Thus in this method, one probes the molecular rotation at the free surface.

From a theoretical point of view, this situation of oscillating twist is particularly simple; it can be shown from the Leslie equations (5.31) and (5.32) that there is no hydrodynamic flow ($v \equiv 0$). Then the torque equation—deduced from eqns (5.17) and (5.32)—reduces to

$$\Gamma_z = \gamma_1 \frac{\partial \phi}{\partial t} = K_2 \frac{\partial^2 \phi}{\partial z^2} + \chi_a H_0 (H_1 - H_0 \phi) \qquad (5.61)$$

In eqn (5.61) we have retained only terms of first order in H_1 or ϕ. The boundary conditions to be imposed on ϕ are

$$\phi = 0 \text{ at the glass–nematic interface} \qquad (5.62)$$

$$\frac{\partial \phi}{\partial z} = 0 \text{ at the free surface.}$$

(The latter condition is a condition of zero surface torque.) Replacing $\partial/\partial t$ by $i\omega$ in eqn (5.61) and solving, one finds

$$\phi = \frac{H_1}{H_0} \frac{1}{1 + i\omega\theta} \left\{ 1 - \frac{\cosh(\kappa z)}{\cosh(\kappa d)} \right\} \qquad (5.63)$$

In eqn (5.63), θ is a characteristic time, defined by

$$\theta = \frac{\gamma_1}{\chi_a H_0^2} . \qquad (5.64)$$

Typically with $\gamma_1 = 10^{-1}$ poise, $\chi_a = 10^{-7}$ and $H = 10^4$ gauss, $\theta \sim 10^{-2}$ s. The origin of ordinates (z) in eqn (5.63) has been taken at the free surface; d is the thickness of the nematic slab, and κ^{-1} is a (complex) length, defined by

$$\kappa^2 = \frac{\chi_a H_0^2 + i\omega\gamma_1}{K_2} = \frac{1}{\xi_2^2}(1 + i\omega\theta), \qquad (5.65)$$

ξ_2 being the magnetic coherence length.

In practice the frequencies of interest are $\omega \sim 1/\theta$ and thus κ^{-1} is comparable to ξ.

For $\kappa d > 1$, eqn (5.63) shows that the response ϕ is essentially equal to

$$\phi_{\text{bulk}} = \frac{H_1}{H_0}\frac{1}{1 + i\omega\theta} \qquad (5.66)$$

except for a thin layer of thickness κ^{-1} near the anchored wall. Using eqn (5.66), and measuring the amplitude and phase angle of the response ϕ at various frequencies ω, one can determine θ. Finally if the anisotropy of the susceptibility χ_a is known, one can derive the friction constant γ_1 through eqn (5.64).

Similar considerations should apply to more complicated geometries; if we have a large sample, submitted to a static field H_0 plus a crossed oscillating field H_1, we expect to measure in the bulk a tilt of the optical axis which is still given by eqn (5.66). Again, we should find no hydrodynamic flow (except possibly in a thin sheet near the outer surface of the sample) for the following general reason: in a region where the field **H** and the magnetization **M** are both spatially uniform, there is no bulk magnetic force tending to induce a flow. Thus we can have driving forces only near the sample surface. But, at finite frequencies ω, such forces will induce flow only in a thin shell below the surface (see eqn 5.53).

5.2.4.2. *Rotating fields.* Early experiments on the torque received by a nematic sample subjected to a large, rotating, magnetic field were carried out by Tsvetkov [14]. But it has progressively become clear that the effects of the walls of the container may be quite complex in a situation of this kind. Let us illustrate this by returning to the nematic slab of Fig. 5.6 with one anchored surface and one free surface. Instead of the fields $\mathbf{H_0} + \mathbf{H_1}$, we now apply a rotating field in the plane of the slab†

$$H_x = H \cos \omega t$$
$$H_y = H \sin \omega t$$

† Experimentally, it may sometimes be more convenient to rotate the sample in a fixed field. Both situations are equivalent.

If ω is low ($\omega\theta \ll 1$, where θ is always defined by eqn 5.64) and if the slab thickness d is much larger than the magnetic coherence length ξ, the molecules in the bulk of the sample would like to rotate essentially in phase with H. However, this cannot give a steady-state regime; because the molecules at the glass surface are anchored, it would in fact correspond to a twist between both surfaces which would increase linearly with time.

The nematic must invent a process to relax the twist. There are many ways for him to do this; we shall quote two of them:

(1) Relaxation of twist by *disclination loops*. This is probably the most common procedure. The loop separates a region of low twist from a region of higher twist, and expands.

How is the loop created? As explained in Chapter 4, nucleation of a loop in the bulk is difficult. In practice, it is observed to take place on a 'source' (a speck of dust, or some irregularity on the glass surface). The loop then grows and detaches from the source.

(2) Relaxation of twist by *emission of* 180° *walls*. The walls are emitted by the glass surface, and move towards the free surface.†
This process has been observed in flat samples, where disclination loops cannot nucleate easily [15].

These two examples lead us to think that the dissipation measured in the Tsvetkov experiment [14] may be influenced by the nucleation, migration, and annihilation, of orientational defects in the nematic structure—lines or walls. The experiment is thus meaningful only if the anchoring conditions at the interfaces and the type of defect involved, are well under control.

However, in a *bulk* sample at low ω the boundary effects are found experimentally to be rather unimportant: this probably means that process (1) is dominant and that the fraction of the sample volume where the orientation is perturbed by the lines is small.

In this regime, the optical axis rotates and follows the field **H** with a certain phase lag ϕ. The value of ϕ is such that the friction torque and the magnetic torque $\frac{1}{2}\chi_a H^2 \sin 2\phi$ balance. It may be shown that there is no hydrodynamic flow ($\mathbf{v} \equiv 0$) and thus the friction torque reduces to $\gamma_1\omega$. The equation

$$\gamma_1\omega = \tfrac{1}{2}\chi_a H^2 \sin 2\phi$$

---

† If, instead of a sample of thickness d with one free surface, we had a sample of thickness $2d$ between two polished glass surfaces, the process would be very similar; walls (of opposite twist) would be emitted by both glass surfaces, then migrate towards the centre and annihilate by pairs.

has solutions ϕ provided that $\omega\theta < \frac{1}{2}$. Measuring the mechanical torque, on the sample, and knowing ω, one derives γ_1: this Tsvetkov method has been revived recently by Gasparoux and Prost (Physics Letters, *36A*, 245, 1971).† Another approach has been used by Luckhurst and co-workers [14c]; they probe the angle ϕ by an e.s.r. technique.

Finally we should discuss the case of higher frequencies ($\omega\theta > \frac{1}{2}$). If one again neglects all boundary effects, one finds theoretically that the director should still rotate, (although not with constant speed). The average rotation velocity $\bar{\omega}$ is then smaller than ω. However, in this regime, it is more difficult to obtain reproducible data; the role of the boundaries is probably more important.

5.2.4.3. Pulsed fields. Measurements on a physical system using pulsed fields or oscillating fields are usually equivalent in content, provided that the physical system under study responds linearly to the applied field. On the other hand, the two approaches become non-equivalent, if the response is strongly non-linear. For liquid crystals, this will occur in the vicinity of a phase-transition induced by the field, the typical case for nematics being the Frederiks transition between two orienting glass plates (see Chapter 3). For each of the three typical 'Frederiks geometries' (listed as 1, 2, 3 in Fig. 3.13) there is a well-defined critical field H_c. Two simple ways of conducting the experiment are then the following: to raise the field abruptly from 0 to a certain value H_0 (larger than H_c) and study the subsequent distortions in the nematic; or to decrease the field from H_0 to 0, starting from an equilibrium distorted state in the field H_0, and following its relaxation towards the unperturbed state.

The instantaneous state of distortion may be monitored by optical methods, or by various types of transport studies. For instance, we mentioned thermal conductance measurements as a probe of static distortions. It turns out that they also provide an adequate means of following the dynamics of the Frederiks transition, for the following reasons: thermal lags in the thermocouples can be made very small by using evaporated metallic films, and the intrinsic time lags associated with heat transport in the nematic film are of order $d^2/\pi^2 D_t$, where d is the thickness and D_t is the thermal diffusivity (the ratio thermal conductance: specific heat). On the other hand, as we shall see, the time constants associated with the orientational effects of interest are of order $d^2\eta/\pi^2 K$ where η is an average viscosity and K a Frank elastic

† See also: *Journal de Physique*, **32**, 953, (1971).

constant. The thermal diffusivity D_t turns out to be at least ten times larger than the orientational diffusivity K/η: thus thermal lags are negligible. Dynamical experiments of this type have been carried out recently by Guyon and Pieranski, and interpreted theoretically by F. Brochard [16].

To explain the main dynamical features, we shall restrict our attention here to the Frederiks transition of type 2, involving pure twist (see Fig. 3.13). This case is in fact not suitable for transport measurements across the slab, but it is pedagogically convenient because (as already mentioned) there is no hydrodynamic flow: the molecules rotate without any translational motion, and this simplifies considerably the analysis based on the Leslie equations. We shall also assume that the maximum field H_0 is only slightly larger than the critical field H_c. Then, as explained in Fig. 3.14, the tilt angle ϕ of the molecules is small, and it's value at equilibrium $\phi_0(z)$ is, to a good approximation, given by a simple sine wave:

$$\phi_0(z) = a_0 \sin\left(\frac{\pi z}{d}\right), \tag{5.67}$$

(where the glass surfaces are located at $z = 0$ and $z = d$). The amplitude a_0 may be derived from a variational calculation, and is given by

$$a_0 \simeq 2\left\{\frac{H_0 - H_c}{H_c}\right\}^{\frac{1}{2}}. \tag{5.68}$$

The dynamical equation for $\phi(z, t)$ (deduced from eqns (5.17) and (5.32) with $v \equiv 0$) has the form

$$\Gamma_z = \gamma_1 \frac{\partial\phi}{\partial t} = K_2 \frac{\partial^2\phi}{\partial z^2} + \chi_a H^2 \sin\phi\cos\phi \tag{5.69}$$

(together with the boundary conditions $\phi(0) = \phi(d) = 0$).

Let us consider first the case where H is *decreased* abruptly from H_0 to 0 at time $t = 0$. Equation (5.69) then reduces to

$$\gamma_1 \frac{\partial\phi}{\partial t} = K_2 \frac{\partial^2\phi}{\partial z^2}, \tag{5.70}$$

and the solution—coinciding with $\phi_0(z)$ at $t = 0$—is

$$\phi(z, t) = \phi_0(z)e^{-t/\theta}, \tag{5.71}$$

with a relaxation time

$$\theta = \frac{\gamma_1 d^2}{\pi^2 K_2}. \tag{5.72}$$

With $d = 10 \ \mu\text{m}$, $\gamma_1 = 10^{-1}$ poises, $K_2 = 10^{-6}$ dynes we have $\theta = 10^{-2}$ s. Equation (5.72) is quite typical of nematic relaxation processes in zero field, which are very important for many technical applications of nematics.

Let us now turn to the case where H is abruptly *increased* (from 0 to H_0) at time $t = 0$. Since, even in the final state described by eqn (5.67), ϕ will still be small, we may expand the dynamical eqn (5.69) in powers of ϕ. However, to reach the correct equilibrium condition at large times, we must include non-linear terms up to order ϕ^3. This leads to

$$\gamma_1 \frac{\partial \phi}{\partial t} = K_2 \frac{\partial^2 \phi}{\partial z^2} + \chi_a H_0^2 \phi (1 - \tfrac{2}{3}\phi^2) + \cdots \qquad (5.73)$$

To a first approximation $\phi(zt)$ is always a simple sine wave in z:

$$\phi(z, t) = a_0 u(t) \sin\left(\frac{\pi z}{d}\right) \qquad (5.74)$$

where $u(0)$ and $u(\infty) = 1$. Multiplying both sides in eqn (5.73) by $\sin(\pi z/d)$ and integrating over the thickness d, one arrives at

$$\theta' \frac{\mathrm{d}u}{\mathrm{d}t} = u - u^3 \qquad (5.75)$$

with

$$\theta' \cong \frac{H_c}{2(H_0 - H_c)} \quad (\gg \theta)$$

Equation (5.75) is readily integrated to give

$$u(t) = m(t)\{1 + m^2(t)\}^{-\tfrac{1}{2}}$$

$$m(t) = \text{const.} \times \exp(t/\theta') \qquad (5.76)$$

Equation (5.76) describes first the exponential growth of a small fluctuation (regime $m \ll 1$)—followed by saturation (regime $m \gg 1$, $u \to 1 - \tfrac{1}{2}m^{-2}$). The time constant for each of these steps is of order θ'. On the whole, we see that these studies on dynamical twist provide a means of measuring the friction constant γ_1. For the other types of Frederiks transitions (cases 1 and 3 of Fig. 3.13) the situation is more complicated, but also more interesting; various combinations of the Leslie coefficients are involved and can be measured. For a more thorough experimental and theoretical discussion the reader is referred to ref. [16].

5.2.5. *Inelastic scattering of light*

We have seen (Section 3.4) that, in a nematic single crystal, the long wavelength fluctuations of the optical axis give rise to a large scattering of light. It must be realized that these fluctuations are not static; in fact their dynamical character was already displayed in the early observations on 'flicker effects' by Friedel, Grandjean, and Mauguin [17]. If, in a region of space, the director $\mathbf{n}$ differs from its average orientation $\mathbf{n}_0$ the fluctuation $\delta \mathbf{n} = \mathbf{n} - \mathbf{n}_0$ will relax towards zero in a certain time. For long wave-length fluctuations, this relaxation process may be predicted by the macroscopic equations of nematodynamics. Experimentally, we can probe these time dependent fluctuations of $\mathbf{n}$ because they result in a frequency modulation of the scattered light. The corresponding frequency broadenings are small (in the kilocycle range), but measurable with the present laser sources and photon beat techniques [18]. This type of study has been carried out in some detail on p-azoxyanisole by the Orsay Group [19]. As we shall see, these experiments give rather detailed information on the Leslie coefficients.

The scattered amplitude, for a given scattering wave vector $\mathbf{q}$, depends on the two independent Fourier components $n_1(\mathbf{q})$ and $n_2(\mathbf{q})$ of the fluctuation $\delta \mathbf{n}$. These components have been defined (and studied from a static point of view) in Chapter 3 (see in particular Fig. 3.20). For a general $\mathbf{q}$, the component n_1 describes a mixed deformation involving splay and bend, while n_2 involves twist and bend. Each of them may be analysed separately by a suitable choice of polarizations in the light scattering experiment. When this separation is performed, the main experimental results appear to be the following:

(1) for each mode the power spectrum (or frequency distribution) has the form of *one single Lorentzian*, centred on the incident beam frequency. This means that the behavior of the fluctuations is purely viscous (no oscillations);

(2) when q retains a fixed orientation, but varies in magnitude, the line widths $\Delta\omega_1(\mathbf{q})$, $\Delta\omega_2(\mathbf{q})$ are essentially *proportional to* q^2. (in the absence of any external magnetic field);

(3) the widths $\Delta\omega_\alpha(\mathbf{q})$ ($\alpha = 1, 2$) do depend somewhat on the orientation of $\mathbf{q}$.

An interpretation of these facts can be given in terms of the Leslie equations, suitably linearized for small deviations from equilibrium [20]. The conclusions may be summarized as follows.

(1) A fluctuation $n_\alpha(\mathbf{q})$ sees a restoring 'force' in the Onsager sense,

13

which is simply the corresponding component of the molecular field

$$h_\alpha = -(K_\alpha q_\perp^2 + K_3 q_z^2)n_\alpha = -K_\alpha(\mathbf{q})n_\alpha \qquad (5.77)$$

$$(\alpha = 1, 2).$$

Equation (5.77) is a direct consequence of eqn (3.74). (Here for simplicity we have also assumed that no magnetic field is applied.)

(2) If the parameter μ (defined by eqn 5.57) is much smaller than unity (as it always appears to be) the Leslie equations predict a purely viscous type of relaxation; i.e., they can be reduced to the form

$$\frac{\partial}{\partial t} n_\alpha(\mathbf{q}) = -\frac{1}{\tau_\alpha(\mathbf{q})} n_\alpha(\mathbf{q}) \qquad (5.78)$$

(the time τ being real). In terms of a power spectrum, this gives a single Lorentzian of width $\Delta\omega_\alpha(\mathbf{q}) = 1/\tau_\alpha(\mathbf{q})$, and is in agreement with the experiments on PAA.

(3) The relaxation rate (i.e. the right hand side of eqn (5.78) is proportional to the restoring force (5.77), and inversely proportional to a certain effective viscosity $\eta_\alpha(\mathbf{q})$:

$$\frac{1}{\tau_\alpha(\mathbf{q})} = \frac{K_\alpha(\mathbf{q})}{\eta_\alpha(\mathbf{q})} \qquad (5.79)$$

(note the dimensions: $K_\alpha(\mathbf{q}) \sim K q^2$ is an energy/cm³ and the viscosity η has the dimensions $ML^{-1}T^{-1}$). The viscosity $\eta_\alpha(\mathbf{q})$ depends only on the orientation of $\mathbf{q}$, and is given explicitly by [18b]

$$\eta_1(\mathbf{q}) = \gamma_1 - \frac{(q_\perp^2 \alpha_3 - q_z^2 \alpha_2)^2}{q_\perp^4 \eta_b + q_\perp^2 q_z^2(\alpha_1 + \alpha_3 + \alpha_4 + \alpha_5) + q_z^4 \eta_c} \qquad (5.80)$$

$$\eta_2(\mathbf{q}) = \gamma_1 - \frac{\alpha_2^2 q_z^2}{q_\perp^2 \eta_a + q_z^2 \eta_c}$$

In eqn (5.80) η_a, η_b, η_c are the Miesowicz viscosities defined by eqns (5.49–5.52). Some limiting cases of eqn (5.80) deserve special mention:

(a) Omitting first all the angular factors, we find $1/\tau = D_0 q^2$ where $D_0 = K/\eta \sim 10^{-5}$ cm²/sec may be interpreted as the diffusion coefficient for orientation.

(b) If $\mathbf{q}$ is along the nematic axis (along $0z$) the deformation for both modes reduces to a pure bend, and the effective viscosity is

$$\eta_{\text{bend}} = \gamma_1 - \frac{\alpha_2^2}{\eta_c}$$

(c) If **q** is normal to the nematic axis, mode (2) becomes a pure twist deformation, of viscosity $\eta_{\text{twist}} = \gamma_1$
Mode (1) is then a case of pure splay, and $\eta_{\text{splay}} = \gamma_1 - \alpha_3^2/\eta_b$.

The data on mode (2) in PAA at 125°C can be fitted with $\gamma_1 = 5.9 \times 10^{-2}$, $\alpha_2^2/\eta_a = 0.1$, $\alpha_2^2/\eta_c = 0.05$ (all viscosities being measured in poise). For technical reasons, the data on mode (1) are more difficult to obtain. All that has been possible to extract from them at present is a rough estimate of the splay viscosity

$$4.8 \times 10^{-2} < \eta_{\text{splay}} < 6.5 \times 10^{-2}.$$

Thus, at the present stage, the light scattering experiments provide us with three relations between the five unknown Leslie coefficients, plus another approximate relation (for η_{splay}). An interesting comparison between these results and the Miesowicz data [8] on η_a, η_b, η_c has been carried out by the Harvard Group [4]. Using as a starting point η_a, η_b, η_c, plus the Orsay value for γ_1, they derive

$$\lambda = -\frac{\gamma_2}{\gamma_1} = 1.15$$

$$\nu_3 = 2.4 \times 10^{-2} \text{ poise}$$

Then they show that the three other Orsay data (α_2^2/η_a, α_2^2/η_c, and η_{splay}) are reproduced with reasonable accuracy. This does confirm to some extent the validity of the general nematodynamic equations, involving five independent parameters. In the future, the light-scattering experiments will undoubtedly prove more and more useful to check these equations. In particular, recent improvements in detection methods should soon allow for a complete study of mode (1).

In Table 5.1 we give a list of data for the friction coefficients of MBBA. The coefficient γ, which is particularly important, has been compiled for eight nematic liquids by C. K. Yun (*Phys. Lett.* **43A**, 369, 1973).

5.3. Convective instabilities under electric fields

The alignment in a nematic single crystal may often be destroyed by a rather small voltage (of the order of 10 volts) applied between two points

TABLE 5.1
Viscosities of MBBA (units = centipoises)

| Leslie coefficients (eqs 31, 34) | | | | | | | | Miesowicz (eqs 49, 51) | | |
α_1	α_2	α_3	α_4	α_5	$\dfrac{\alpha_3+\alpha_2+\alpha_5}{\alpha_6}$	$\dfrac{\alpha_3-\alpha_2}{\gamma_1}$	$\dfrac{\alpha_3+\alpha_2}{\gamma_2}$	$\dfrac{\alpha_4}{2}$ η_a	$\dfrac{-\alpha_2+\alpha_4+\alpha_5}{2}$ η_b	$\dfrac{\alpha_3+\alpha_4+\alpha_6}{2}$ η_c
6·5±4	−77·5±1·6	−1·2±0·1	83·2±1·4	46·3±4·5	−34·4±2·2	76·3±1·7	−78·7±1·7	23·8±0·3	103·5±1·5	41·6±0·7
						155		42		
						110		25·2		
						125				
						80				
						130±5				
						86±2				

$$\gamma_1/\gamma_2 = 0.959 \pm 0.05$$

Light scattering (eq. 80)			Ultrasound (eq. 58)		Capillary waves	Dynamical Fredericks transition		
$\dfrac{\alpha_3^2}{\gamma_1-\eta_b}$	$\dfrac{\alpha_2^2}{\gamma_1-\eta_c}$	γ_1	$\eta_1-\dfrac{\alpha_3^2}{\gamma_1}$	$\dfrac{\alpha_4}{2}$	$\gamma_1+\dfrac{\gamma_2^2}{\gamma_1}$	$\dfrac{\alpha_2^2}{\gamma_1+6\eta_3}$	$t°C$	
(bend)	(splay)	(twist)	$\tilde\eta_b=\tilde\eta_c$	$\tilde\eta_a$		γ^*		
							room temp.	Gahwiller, Phys. Letters, 36A, 311 (1971).
							20°C	Prost-Gasparoux, Phys. Lett. 36A, 245 (1971).†
							24°C	Martinoty-Candau, Mol. Cryst. 14, 243 (1971).†
			27	42			25°C	Langevin, J. de Phys. 33, 249 (1971).†
			16·3	25·2	16·1		$t-t_c = 3°C$	Martinand-Durand, Phys. Rev. Lett. 29, 562 (1972).
19±3							24°C	Wahl-Fischer, Optics Comm. 5, 341 (1972).
21±2	126						22°C	Haller, J. Chemical Phys. 57, 1400 (1972).
							25°C	Haller-Litster, Mol. Cryst. 12, 277 (1971).
					107		22°C	Brochard-Pieranski-Guyon, C.R.A.S. 273, 486 (1971) et J. de Phys.
							room temp.	Cladis, Phys. Rev. Letters, 28, 1629 (1972).
					110±5		23°C	Leger, Solid St. Comm. 11, 1499 (1972).†
							25°C	Solid St. Comm. 10, 697 (1972).

† This reference also gives data as a function of temperature.

in the sample. This type of electro-optic effect was discovered independently by a number of experimentalists but its practical importance has been first realized by the RCA group under G. Heilmeier [20]. It may lead to very interesting applications connected with display systems. The fundamental process underlying the effect was at first mysterious, but W. Helfrich [21] proposed an explanation—based on a combination of charge transport and convection effects—which seems to account for the most salient facts. Because convection is involved, the problem essentially belongs to nematodynamics, and we shall try to summarize here the main experimental results together with their interpretation. The effect is rather complex; for this reason, we shall present only a very special selection of experiments, with no reference to historical order, and often omitting contributions which were important in their time, but which did not point to (what we now believe to be) the crucial phenomena.

5.3.1. Basic electrical parameters

We shall be concerned here with experiments which involve only d.c. or low-frequency a.c. electric fields (typical frequency range $0-10^3$ Hz), for which the most important parameters are the static dielectric constant and the static conductivity.

5.3.1.1 Dielectric constants. In Chapter 3 we discussed briefly the static dielectric constants $\epsilon_{\parallel}$ (measured along the optical axis) and $\epsilon_{\perp}$ (normal to the axis). By suitable insertion of permanent dipoles in the chemical formula of the material, one can often achieve either $\epsilon_{\parallel} > \epsilon_{\perp}$ (if the dipole is parallel to the long axis of the molecule) or $\epsilon_{\parallel} < \epsilon_{\perp}$ (if the dipole is normal). Typical values for MBBA at room temperature are $\epsilon_{\parallel} = 4 \cdot 7$, $\epsilon_{\perp} = 5 \cdot 4$.

5.3.1.2. Conductivities. Another important electrical property is the static conductance; it is usually small (typically in the range 10^{-9} to 10^{-8} ohm^{-1} cm^{-1}) and anisotropic: The anisotropy of σ was measured very early by Svedberg *Ann. Phys* **44** (1914) and **46** (1916). In most of the examples studied up to now the parallel conductance $\sigma_{\parallel}$ is somewhat larger than the perpendicular conductance $\sigma_{\perp}$. With usual samples of MBBA, $\sigma_{\parallel}/\sigma_{\perp} \sim 1 \cdot 5$.

It must be emphasized, however, that all conductance results are strongly dependent on the amount, and on the chemical nature, of the impurities present in the sample; it is helpful to have charge carriers; they are required for the electro-optic effects to be described below.

But it would be a great advance if the carriers could be controlled. (a) Starting with a pure material, and doping it with ions which are (slightly) soluble in it—such as some substituted ammonium derivatives—one can in principle vary at will the magnitude of $\sigma_\parallel$ and $\sigma_\perp$, keeping a constant ratio $\sigma_\parallel/\sigma_\perp$. (b) By changing the stereochemical shape of the dissolved ions by suitable chemical substitutions, one should be able to change the ratio $\sigma_\parallel/\sigma_\perp$. (c) Recently, values of $\sigma_\parallel/\sigma_\perp$ smaller than unity have been achieved in the vicinity of a smectic–nematic transition (F. Rondelez, *Sol. State Commun.* **12**, 1675, (1972).

For notational purposes, it may sometimes be helpful to divide nematics into four classes $(++)$ $(+-)$ $(-+)$ $(--)$ where the first symbol gives the sign of $\epsilon_\parallel - \epsilon_\perp$, and the second symbol the sign of $\sigma_\parallel - \sigma_\perp$. Spectacular electro-optic effects have been found and studied principally in the $(-+)$ class: we shall mainly focus our attention on this case.

5.3.1.3. Electrode effects. One important complication connected with electrical currents in organic materials such as nematics, is brought in by electrode effects. All d.c. studies require metallic electrodes in direct contact with the fluid. Chemical reactions take place at these electrodes, with the following consequences.

(1) *Injection* of supplementary carriers. These injection effects have been studied in isotropic organic fluids of high purity, and they can, by themselves, lead to certain convective instabilities [22]. Unfortunately, the injection process is often chemically complex. It can be controlled accurately only for very special electrodes (semipermeable membranes).

(2) *Chemical degradation* of the nematic material (this may be a serious nuisance for technical applications).

5.3.1.4. Elimination of electrode effects. Clearly, from the point of view of fundamental studies at least, it is preferable to eliminate the complications due to specific electrode effects. At present, the best method which has been found to achieve this in nematics amounts to the use of *low frequency a.c. fields* rather than d.c. fields.

Using a.c. fields, it is possible to insert thin insulating foils (e.g. teflon) between the electrodes and the nematic slab; there does remain a field (or a voltage drop V) between the foils because, at finite frequencies, the electrical carriers in the nematic are not able to screen out entirely the charges Q on the electrodes; for a conventional lossy dielectric the relation between V and Q reads

$$V = \frac{Q}{C}\frac{i\omega\tau}{1+i\omega\tau}$$

(where C is the capacitance in the absence of losses, and $\tau = 4\pi\sigma/\epsilon$ is the dielectric relaxation time, ω being the angular frequency).

Even using a.c. fields, the effects observed with or without the teflon shields are often identical! This probably means that, in the latter case, the injected carriers are confined to a thin sheet near the electrode, and are not able to generate extra instabilities.

5.3.2. Experimental observations at low frequencies

A nematic material of the $(-+)$ class is placed between two semi-transparent electrodes (typical slab thickness 30 μm). A d.c. or (preferably) low-frequency a.c. voltage V is applied across the slab. For increasing values of the r.m.s. amplitude V we then observe the following sequence of regimes.

5.3.2.1. Single crystal regime.
At low voltages V (typically in the range $V \sim 1$ V) the molecules are alined normal to the electric field E (along a certain axis x) as expected for a dielectric material with $\epsilon_\parallel < \epsilon_\perp$. In this regime we have a nematic single crystal; in particular any light beam incident on the slab is reflected (or transmitted) specularly.

5.3.2.2. Williams domains.
When V reaches a critical threshold V_c (of the order of 5 V) a periodic distortion of the nematic alignment is observed. In many cases, it is a simple one-dimensional type of distortion, which was first observed by Williams [23] and studied in particular by the groups at IBM, Orsay and Ford [24] [25]. The aspect of the distortion is shown in Fig. 5.8(a, b, c) (see between pp 148–9). It can be detected by various optical means.

Consider a light wave propagating along z and polarized along x. At a point such as P_1 the polarization is parallel to the optical axis: the refraction index is then the 'extraordinary' index n_e. On the other hand, at point Q_1, the polarization makes a certain angle with the optical axis and the effective index n_{eff} is a certain combination of n_e and of the ordinary index n_0 ($< n_e$). Thus at point Q_1, $n_{eff} < n_e$. We conclude that, for a polarization along x, the slab behaves as a periodic array of cylindrical lenses $P_1, P_2 \ldots$. An incident plane wave is then focused at a series of lines F_1, F_2,.... . A photograph showing these focusing effects is shown on Fig. 5.8a. Note that the focusing effects disappear completely if the light is polarized along (y); this proves that the molecules do remain in the (x, z) plane.

With the (x) polarization, the sample may be used as a periodic grating, and the periodicity may be derived from a study of the selective reflections on this grating; the repeat period along $x(\lambda_x)$ is observed to be a linear function of the sample thickness d.

At low frequencies ω, the distortions in the molecular alignment are found to be *static*; the pattern stays the same when the electric field reverses.

The distortions are also accompanied by a certain amount of *cellular flow* in the nematic liquid; this motion may be displayed by following the motion of a dust particle floating in the slab. The flow lines are observed to have the same periodicity than the distortions; the geometrical relations between the two effects are shown on Fig. 5.8c. This flow pattern is somewhat reminiscent of what is observed when a slab of isotropic fluid is heated from below (the Benard phenomenon [27]). The analogy strongly suggests that a convective instability is involved.

The voltage threshold V_c is essentially *independent of the sample thickness*; at first sight one might be tempted to interpret this property in terms of an electrochemical process at a metal–nematic interface. However, this cannot be correct, since the same voltage V_c is maintained with teflon shields in a.c. regimes. We shall see that a constant V_c is a natural consequence of the Helfrich model.

5.3.2.3. *Dynamic scattering.* If the voltage V is increased above V_c, the distortion amplitudes and the associated flow velocities increase. Finally, at some higher voltage V_t, a new regime is reached:
  (1) The Williams domains become disordered and mobile
  (2) The flow is now turbulent
  (3) The long-range nematic alignment is completely upset.
Optically the new, fluctuating, disordered state can be observed without any special specification on the light polarization; this shows that the molecules are not confined any more to orientations in the (x, z) plane. Observation under the microscope also shows a number of disclination loops nucleating near the limiting surface.

On a macroscopic scale, the practical consequence of (3) is a strong, diffuse, scattering of light. This 'dynamic scattering'—as it has been called by the RCA group—is technologically interesting because it involves low voltages, low power dissipation, and small sample size, and also because it operates by reflection of any type of light (e.g. sunlight).

On the other hand, the processes involved in the transition towards turbulence (particularly the role of disclination lines) are very poorly understood at present: most fundamental studies have been directed towards the interpretation of the lower threshold V_c; in the following discussion, we shall restrict our attention to V_c.

5.3.3. *The Helfrich interpretation*

The effects which we have listed above are observed mainly with materials of the $(-+)$ *class* prepared in the *planar* texture.

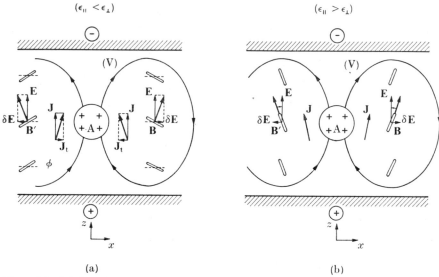

FIG. 5.9. (a) Carr–Helfrich effect for a material with negative dielectric anisotropy and positive conductance anisotropy $(-+)$ (boundary conditions: tangential). (b) Carr–Helfrich effect for a $(++)$ material (boundary conditions: normal). The molecules are represented by small rods.

They can be explained rather simply in terms of convective instabilities. The general idea was first invoked by Tsvetkov and Carr [28] but the detailed discussion of small motion instabilities is due to W. Helfrich [21].

To understand the basic concepts, let us consider a slab of a $(-+)$ material under electric fields E as shown in Fig. 5.9a. In the unperturbed situation the molecules are in the planar texture (say along x).

The perturbed state has a small periodic distortion of the bending type. Of course the Frank elastic energy is increased by the distortion and gives rise to a restoring force. On the other hand, if $\sigma_{\parallel} > \sigma_{\perp}$ there is a component J_t of the current, along x, which tends to pile up a positive charge density q in the region around point A. This charge accumulation has two main effects.

(1) The field at point B is shifted from E to $E + \delta E$. The molecules at B tend to remain orthogonal to the overall field; as seen in Fig. 5.9a, this *electrostatic torque* tends to increase the initial distortion.

(2) The fluid around A is subjected to a bulk force qE: this gives rise to a certain flow pattern, shown qualitatively on the figure. The result at point B is a strong *hydrodynamic torque* which also tends to increase the distortion.

If ϕ is the angular amplitude of the distortion, and k its wave-vector along x, the restoring torque due to elastic distortions is $-K_3 k^2 \phi$. On the other hand, both the electrostatic and the hydrodynamic torques are proportional to $E^2 \phi$. We also know, from the experiments described above, that the wavelength of the distortion (along x) is comparable to the sample thickness d (i.e. $k \sim 1/d$).

Thus we conclude that there is a threshold in field E_c defined by

$$E_c^2 = \text{const. } K/d^2, \tag{5.81}$$

or that $V_c = E_c d$ is independent of sample thickness, as is indeed observed. The exact value of the constant in eqn (5.81) would have to be derived from a rather complex two-dimensional nematodynamic calculation. An approximate value has been derived by Helfrich, using a one-dimensional approximation; all quantities such as the tilt angle ϕ are assumed to depend only on the transverse coordinate x, and not on z: the boundary conditions at both sides of the plate are omitted, except that the wave vector k is taken to be equal to C/d where C is a fixed numerical constant (or order unity). The result is then comparatively easy to reach, and reads

$$V_c^2 = \frac{V_0^2}{\zeta^2 - 1} \tag{5.82}$$

where

$$V_0^2 = 4\pi C^2 \frac{K_3 \epsilon_{\parallel}}{\epsilon_{\perp}(\epsilon_{\perp} - \epsilon_{\parallel})} \tag{5.83}$$

and ζ^2 is a dimensionless parameter, which must be larger than unity to obtain an instability

$$\zeta^2 = \left(1 - \frac{\sigma_{\perp}}{\sigma_{\parallel}} \frac{\epsilon_{\parallel}}{\epsilon_{\perp}}\right)\left(1 + \frac{\alpha_2 \epsilon_{\parallel}}{\eta_c \epsilon_a}\right) \tag{5.84}$$

For MBBA, (assuming that the ratios between Leslie coefficient are close to those estimated for PAA) one finds $\zeta^2 \sim 3 \cdot 2$.

Helfrich has shown that, with plausible values for K_3 and a constant C of order π (i.e. a half-wavelength equal to the sample thickness) the estimated values of V_c are of the order of a few volts—quite acceptable in view of the many unknowns involved. Improved calculations of the boundary effects [26] essentially confirm these results.

Let us now discuss the instabilities which may occur with the other nematic classes. For simplicity, we shall assume that for all cases, in low fields E, the easy direction imposed by the walls coincides with that imposed by E; this avoids further complications due to Frederiks transitions in the field E. Consider for instance a sample of the $(++)$ class (Fig. 5.9b); here the unperturbed state has the molecules normal to the slab, and the fluctuation which may induce local charge accumulation is a splay. However it is seen that the electric torque at point B now tends to stabilize the structure. To study the hydrodynamic torque, we note that at point B the flow lines are nearly parallel to the molecules. As explained in Section 5.2 the hydrodynamic torque is then very weak (in other words, the parameter $\lambda = -\gamma_2/\gamma_1$ is close to unity, and the torque is proportional to $\lambda - 1$). For points above or below B, we do have some hydrodynamic torques but their sign depends on the precise location studied. Finally, in this case, we find only two strong torques (elastic and electric) and both tend to stabilize the structure: no instability is expected.

A similar conclusion can be derived for the $(--)$ class. On the other hand, for the $(+-)$ class an instability is predicted (created mainly by the electric torques).

The limiting case $\epsilon_a = 0$ is delicate. Consider for instance a $(0+)$ sample in the homeotropic texture of Fig. 5.9b. If $\lambda > 1$ and $\epsilon_a = 0$, the one-dimensional argument (based on the torque at point B) predicts instability. If we go to an improved, two-dimensional, calculation, and choose $k \sim \pi/d$ ('cylindrical' rolls), we find that the system is stable: the torques near the top and bottom surfaces are opposed in sign to the torques at B. But if we choose $k \gg \pi/d$ ('thin' rolls), the one-dimensional argument *must* become correct: an instability should occur for thin rolls.

5.3.4. Extension to higher frequencies

5.3.4.1. Williams domains and chevrons. The experimental variation of the threshold V_c with frequency $(\omega/2\pi)$ is displayed on Fig. 5.10. Note that, to obtain this type of curve, the sample must not be too thin (typically, we require $d > 10$ μm). Then one observes two very different regimes, respectively below and above a certain threshold frequency $(\omega_c/2\pi)$.

For $\omega < \omega_c$ the threshold V_c is rather low, and also independent of sample thickness. For $V \geqslant V_c$ we have Williams domains, with a spatial periodicity comparable to the thickness d.

For $\omega > \omega_c$ the threshold is much higher, and V_c is proportional to d; the real threshold parameter in this regime is the field $E_c = V_c/d$, where

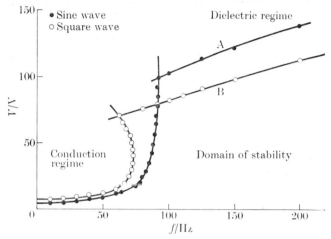

Fig. 5.10. Typical threshold voltage versus frequency for commercial MBBA (courtesy G. Durand).

E_c varies like $\omega^{\frac{1}{2}}$. The onset of instability at V_c is manifested optically by parallel striations (shown on Fig. 5.11; see between pp. 148–9). The distance between striations in now much smaller than d, and depends on ω (or on E_c). It is proportional to $1/E_c \sim \omega^{-\frac{1}{2}}$.

This second type of instability has been detected first by Heilmeier and Helfrich [29] and studied in some detail by the Orsay Group [30]— who coined the name 'chevrons' for the pattern of Fig. 5.11. A very useful 16 mm film, showing the Williams domains, the chevrons, and also the turbulent structures observed at voltages $V \gg V_c$, has been taken by R. Kashnow [31].

The cut-off frequency $\omega_c/2\pi$ is typically of the order of 100 Hz and is found to increase linearly with the sample conductivity [30]. Thus with very pure materials of low conductance, the AC effects should always correspond to the chevron regime.

5.3.4.2. Interpretation. All the features displayed above may be interpreted in terms of the Carr–Helfrich model, suitably extended to cover time dependent phenomena [32]. The geometrical conditions are still those of Fig. 5.7b or 5.9a. In the distorted state the molecules are deflected by a small angle ϕ in the (x, z) plane. The most important parameter from the point of view of charge accumulation is not exactly ϕ, however, but rather the curvature $\psi = \partial\phi/\partial x$ of the molecular pattern. We use ψ and the charge density q as our fundamental variables. Again, as in the original Helfrich calculation for static regimes [21], we consider only a one-dimensional problem (ϕ and q depend only on x).

Positive charges pile up in the regions of negative curvature (when E is along $+z$) as is shown on Fig. 5.9a: the charge source is proportional to $-\psi E$. This leads to an equation for the charge of the form

$$\dot{q}+\frac{q}{\tau}+\sigma_{\mathrm{H}}\psi E \to 0, \tag{5.85}$$

where τ describes dielectric relaxation, and σ_{H} is associated with the Carr process. Using Ohm's law and Poisson's equation in an anisotropic medium, one obtains [32]:

$$\frac{1}{\tau}=\frac{4\pi\sigma_{\|}}{\epsilon_{\|}} \tag{5.86}$$

$$\sigma_{\mathrm{H}}=\sigma_{\|}\left(\frac{\epsilon_{\perp}}{\epsilon_{\|}}-\frac{\sigma_{\perp}}{\sigma_{\|}}\right) \qquad (\sigma_{\mathrm{H}}>0) \tag{5.87}$$

(for MBBA $\sigma_{\mathrm{H}} \sim \frac{1}{2}\sigma_{\|}$).

We must also write down an equation for the curvature ψ. One origin for ψ is the bulk electrostatic force qE, giving rise to cellular flow and hydrodynamic torques. From Fig. 5.9a we see that a negative qE will tend to give a positive increase in ψ. This suggests an equation of the form

$$\dot{\psi}+\frac{\psi}{T}+\frac{qE}{\eta} = 0. \tag{5.88}$$

where $1/T$ is a relaxation rate for molecular orientation, and η has the dimensions of a viscosity. In fact, the complete calculation, including both hydrodynamic and electric torques, leads to eqn (5.88) with the following coefficients

$$\frac{1}{T}=\frac{\eta_{\mathrm{c}}}{\gamma_1\tilde{\eta}_{\mathrm{c}}}\left[\frac{(\epsilon_{\perp}-\epsilon_{\|})\epsilon_{\perp}}{4\pi\epsilon_{\|}}E^2+K_3k^2\right] \tag{5.89}$$

$$\frac{1}{\eta}=\frac{\eta_{\mathrm{c}}}{\gamma_1\tilde{\eta}_{\mathrm{c}}}\left[\frac{\epsilon_{\perp}-\epsilon_{\|}}{\epsilon_{\|}}-\frac{\alpha_2}{\eta_{\mathrm{c}}}\right] \tag{5.90}$$

In these formulae, η_c and $\tilde{\eta}_c$ are the effective viscosities defined by eqns (5.51) and (5.58), while k is the wave vector of the bend deformation.†

Equations (5.85) and (5.88) must then be solved, in the presence of a given a.c. field $E = E_m \cos \omega t$. The general solutions have the form

$$q(t) = q_p(t)\, e^{st}$$
$$\psi(t) = \psi_p(t)\, e^{st},$$

where q_p and ψ_p are periodic functions (of period $2\pi/\omega$), and s is a parameter depending on the field amplitude E_m. The threshold is obtained when the real part of s vanishes.‡

The complete discussion is rather complex—especially since the relaxation rate for orientation $1/T$ (eqn 5.89) does depend on the instantaneous value of the field $E(t)$. Here we shall give only a simplified discussion of the two regimes.

(1) *At low frequencies* ω, the fields E near threshold are rather small, and T is rather large. Thus $\omega T > 1$, and the only important Fourier component of $\psi(t)$ is at zero frequency

$$\psi(t) \to \bar{\psi}$$

Then the charge source, proportional to $-\psi E$ (eqn 5.85) is a simple sinusoidal wave $-\sigma_H \bar{\psi} E_m \cos \omega t$. The corresponding charge is given by

$$q(t) = q' \cos \omega t + q'' \sin \omega t \qquad (5.90)$$

$$q' = -\sigma_H \tau E_m \bar{\psi}\, \frac{1}{1+\omega^2\tau^2}$$

Let us now turn to the equation for the average curvature $\bar{\psi}$, obtained by averaging eqn (5.88) over one period; it may be shown to reduce to

$$\frac{\bar{\psi}}{\bar{T}} + \tfrac{1}{2}q' E_m = 0, \qquad (5.91)$$

where $1/T$ is the time-average of the relaxation rate (5.89). This gives a threshold condition

$$1 = \frac{E_m^2}{2} \cdot \frac{\sigma_H \tau \bar{T}}{\eta} \cdot \frac{1}{1+\omega^2\tau^2},$$

which, after using (5.89), may be cast in the form

$$\bar{E}^2 = \tfrac{1}{2}E_m^2 = \frac{4\pi\epsilon_\|}{(\epsilon_\perp - \epsilon_\|)\epsilon_\perp}\, K_3 k^2\, \frac{1+\omega^2\tau^2}{\zeta^2 - (1+\omega^2\tau^2)} \qquad (5.92)$$

† I am very much indebted to A. Rapini, who noticed that the simplified forms (5.89) and (5.90) could be used instead of the heavier expressions found in ref. [32].
‡ For the cases at hand s appears to be real at threshold.

where ζ^2 has been defined in eqn (5.84). The threshold field will correspond to the minimum allowable wave vector k: following the Helfrich assumption, we may assume that this is of the form C/d. This gives a voltage threshold (r.m.s.):

$$V_c(\omega) = V_c(0)\left(\frac{1+\omega^2\tau^2}{\zeta^2-(1+\omega^2\tau^2)}\right)^{\frac{1}{2}} \tag{5.93}$$

Equation (5.93) shows that $V_c(\omega)$ increases with ω and finally becomes very large when ω approaches a cut-off frequency

$$\omega_c = \frac{1}{\tau}\sqrt{(\zeta^2-1)}. \tag{5.94}$$

The theoretical curve (5.93) is in rather good agreement with the data on MBBA [30]. Also, according to eqn (5.86), $1/\tau$ (and thus ω_c) is a linear function of the conductivity σ—in agreement with experiment.

Thus, at the threshold, for all frequencies $\omega < \omega_c$, the instability pattern corresponds to a static distortion ($\bar{\psi} \neq 0$) and to oscillating charges; for this reason it is often referred to as the *conducting regime*.

(ii) The above calculation breaks down when ω reaches ω_c, because in this region the fields become large, ωT reaches values of the order of unity, and the curvature ψ becomes time dependent. The situation is comparatively simple if we consider only the domain $\omega \gg \omega_c$. This implies $\omega\tau > 1$. Then the charge q cannot follow the excitation:

$$q \to \bar{q}.$$

The bulk force $qE \to \bar{q}E_m\cos\omega t$ is sinusoidal, and the curvature response $\psi(t)$, given by eqn (5.88), is a more or less complicated periodic function of time. Qualitatively we may guess that, to get a sizeable response, the phase lag of ψ must not be too large; the threshold condition corresponds to

$$\omega T = \text{constant}.$$

Thus, in this regime, T must be short. The system achieves this by two means which can be understood from eqn (5.89). One is to have high field amplitudes. The other (first pointed out by O. Parodi and E. Dubois Violette) is to use a high value of the wave vector k. At threshold, both effects contribute roughly equally, and we have

$$\frac{\epsilon E_c^2}{4\pi} \sim K_3 k^2 \sim \eta\omega$$

where ϵ and η are suitable combinations of dielectric and viscosity coefficients. These rules do explain the experimental observation which we quoted before:

threshold field $E_c \sim \omega^{\frac{1}{2}}$

E_c independent of sample thickness

spatial period of the striations $\pi/k \sim \pi/k \sim \omega^{-\frac{1}{2}}$

In the high-frequency regime (for $\omega \gg \omega_c$) the molecular pattern oscillates while the charges are static; for this reason this is often called the *dielectric regime*. Another name is the *fast turn-off mode* [29]; if the a.c. voltage is turned off from a value slightly above V_c to 0, the striation pattern disappears rapidly. A plausible explanation for this feature is that the relaxation rate in zero field (deduced from eqn 5.89) is

$$1/T \simeq K_3 k^2/\eta$$

and is still high because the wave vector k of the striations is large.

On the whole, the Helfrich model [21] appears well substantiated by these a.c. studies near the threshold V_c, and some information on the Leslie coefficients will probably be extracted from the data.†

5.4. Molecular motions

Very little is known—and even less is understood—concerning the dynamics of nematic fluids on the molecular scale. Here we shall present only a selected list of experiments which appear to be relevant.

5.4.1. Dielectric relaxation

For a few typical nematics, the dielectric constants $\epsilon_\parallel(\omega)$ and $\epsilon_\perp(\omega)$ have been measured on oriented samples, and with a wide frequency range (radiofrequencies and microwaves). The main results appear to be the following:

(1) Both $\epsilon_\parallel$ and $\epsilon_\perp$ usually show a normal (Debye) type of relaxation at microwave frequencies ($\omega/2\pi \sim 10^{10}$) [33]. Similar relaxation times are also found in the isotropic phase.

(2) If the molecule under study has a non-zero component of electric dipole along its long axis, there is an additional relaxation process at much lower frequencies, which concerns only the parallel dielectric constant $\epsilon_\parallel$. In materials where the nematic range falls around 100°C (such as PAA) this occurs in the radiofrequency range. The effect was discovered by Maier and Meier [34], and interpreted in some detail by Meier and Saupe [35] in terms of a 180° rotation of the

† The instabilities under a field E stem from a coupling between charge transport and molecular orientation. Similar instabilities are generated by *thermal* transport—see Pieranski P., Guyon E., and Dubois Violette, E. *Phys. Rev. Lett.* **30,** 736, (1973).

14

molecule around one of the *short* axes of the molecule; for a long
molecule, such a rotation is clearly difficult in the nematic phase, and
the resulting relaxation rate is correspondingly slow—typically 10^3
times slower than for rotations around the long axis. It is tempting to
relate this slowing down to the (small) probability for one molecule to
be orthogonal to the nematic axis. However, these considerations
cannot be made very precise, because of short range order effects two
neighbouring molecules, whose dipoles are accidentally parallel may
rotate synchronously and cross the perpendicular conformation more
easily than a single one, etc.

5.4.2. Nuclear spin–lattice relaxation

The spin lattice relaxation rate $1/T_1$ of a nucleus inside the
nematic liquid gives some information on the motion of its
immediate surroundings, on a time scale *ca.* 10^{-6} seconds. Some of the
most salient results for the nematic phase are reviewed in refs (36–8).

In practice, we have two slightly different types of nuclear probes,
as exemplified by the following (current) cases: protons (spin $\frac{1}{2}$)
where the main relaxing agent is the dipole–dipole interaction
between nuclear spins; and deuterons (^{2}H) or nitrogen nuclei (^{14}N)
[39] where the agent is a local electric field gradient.

In isotropic liquids, the molecular motions which dominate the
relaxation process can usually be described in terms of one single
correlation time τ_c; as a result, the dependence of the relaxation rate on
the nuclear frequency ω_n is simple [40]. In nematic liquids the situation
is very different: the frequency dependence of $1/T_1$ is more complex and
cannot be discussed in terms of one correlation time τ_c.

In fact, it is clear that many processes can contribute to $1/T_1$. Let us
first restrict ourselves to an ideal situation, with rigid molecules and
relaxation by intramolecular couplings only.† Then we can (at least)
think of contributions due to:

(1) Rotations around the long molecular axis.

(2) large angle rotations around a short molecular axis (such as
those involved in the low frequency relaxation of $\epsilon_\parallel$).

(3) small amplitude oscillations of the long axis around its average
orientation.

Process (1) has a short correlation time $\tau_1 \sim 10^{-10}$ s ($\omega_n \tau_1 \ll 1$). It

---

† This means, for instance, that we neglect all dipole–dipole interactions between
nuclei belonging to different molecules.

should give a small, frequency independent, contribution to the relaxation rate $1/T_1$.† Process (2) is rare, as we have seen but it has a long correlation time, which enhances its efficiency for nuclear relaxation. It can contribute to the frequency dependence of $1/T_1$. Process (3) cannot be characterized by one correlation time: we know this, because the lower end of the frequency spectrum corresponds to the slow motions of long wave length fluctuations, and such motions can be analysed by the Leslie equations. Using eqn (5.79) we see that a fluctuation of wave vector q has a correlation time of order η/Kq^2 (where η is a viscosity and K a Frank constant): thus different Fourier components have different correlation times. In fact the most important fluctuations (for nuclear relaxation) are those for which the correlation time is comparable to the nuclear period $\eta/Kq^2 \sim \omega_n^{-1}$. This corresponds to wavelengths

$$\lambda = \frac{2\pi}{q} = 2\pi \left(\frac{K}{\eta\omega_n}\right)^{\frac{1}{2}}.$$

Taking $\omega_n = 10^7$, $K = 10^{-6}$ and $\eta = 10^{-1}$ we find $\lambda = 600$ Å. λ is thus significantly larger than the molecular length a, and a discussion in terms of the continuum theory is not unreasonable. This has been carried out by Pincus [41] with subsequent improvements by Doane and Johnson [42] and by Lubensky [43]. Qualitatively, the resulting contribution to the nuclear spin–lattice relaxation may be written as

$$\frac{1}{T_1} \sim (\gamma H_L)^2 \frac{k_B T S^2}{K} \left(\frac{K}{\eta} + D\right)^{-\frac{1}{2}} \omega_n^{-\frac{1}{2}}, \qquad (5.95)$$

where H_L is the local field describing the relaxing agent, γ the nuclear gyromagnetic factor, S is the nematic order parameter, and D is a translational self-diffusion coefficient. To understand why D plays a role, consider the limiting case where the fluctuations relax very slowly $(K/\eta \to 0)$: the nematic is still distorted by thermal agitation, but the distortions are frozen. The molecule carrying the nuclear spin under study, moves in this structure by brownian motion and has a variable position $\mathbf{r}(t)$. It is plausible to assume that the molecule constantly adjusts its long axis parallel to the local (distorted) director $\mathbf{n}\{\mathbf{r}(r)\}$; thus, when $\mathbf{r}(t)$ changes the local field seen by the nuclear spin is modulated.

Experimental relaxation rates in PAA can be fitted (between 5

---

† Of course this type of rotation is efficient only if the vector linking the two nuclei involved in the dipole–dipole coupling is not parallel to the long axis.

and 10 megacycles) by a law of the form:

$$\frac{1}{T_1} = A + B\omega_n^{-\frac{1}{2}}$$

This would appear as the superposition of two relaxation processes: one of them (A) due to local motions with a short correlation time, the other ($B\omega_n^{-\frac{1}{2}}$) being of the Pincus type. However, this agreement is not entirely significant, for various reasons: (1) as emphasized by Vilfan *et al.* [37] it is known that, in certain isotropic liquids, the intermolecular spin interactions (modulated by the relative motion of the molecules) can lead to a relaxation rate of the form:

$$\frac{1}{T_1} = C - D\omega_n^{+\frac{1}{2}}$$

This is not easily distinguished from the 'augmented Pincus form' quoted above. In fact, the latter form gives better agreement for MBBA, and possibly for all nematogens with rather long terminal chains [37]. (2) the slow motions listed under (*b*) above may play an important role (3) in materials like PAA, the aromatic protons of interest are strongly perturbed by dipolar coupling with the methyl protons at both ends of the molecule: this has been proven by selective deuteration of the methyl protons [38].

On the whole, the relaxation processes appear rather complex, but suitable tricks (such as deuteration) will help to unravel them.

5.4.3. Acoustic relaxation

Acoustic waves provide a tool to study relaxation processes in liquids, if these processes are comparatively slow ($\sim 10^{-7}$ seconds). The case of nematics is particularly complex, and the results are only partly understood. Nevertheless, the technique is interesting and the existing data deserve a short description.

5.4.3.1. Low-frequency limit. At very low frequencies (say, well below 1 MHz), longitudinal waves propagate in a nematic with a velocity c_0 which is *independent of direction*. This is one characteristic feature of a liquid, where c_0 depends only on the bulk rigidity coefficient

$$E_0 = -V \left(\frac{\partial p}{\partial V} \right)_{\text{adiabatic}}$$

$$c_0 = [E/\rho]^{\frac{1}{2}} \qquad (\rho = \text{density})$$

Experimentally, c_0 is (as usual) mainly a decreasing function of temperature. But it shows a dip near the nematic isotropic transition point T_c [44, 45]. This dip can be qualitatively understood by a thermodynamic argument based on the Maier–Saupe free energy [46].

5.4.3.2. *Dispersion of the sound velocity: $c(\omega)$.* At finite frequencies ω, the sound velocity $c(\omega)$ is slightly modified. This has been studied in MBBA by the groups at Rutgers [47] and M.I.T. [48]. In particular, the dip which we mentioned above disappears above $\omega/2\pi \sim 10$ MHz. At these high frequencies, the order parameter S of the nematic phase is not able to follow the density fluctuations, and the sound velocity c measures a rigidity *at constant S*, which is not singular near T_c.

5.4.3.3. *Anisotropy of the velocity: $c(\theta)$.* At finite frequencies, the sound velocity c becomes slightly dependent on the angle between the optical axis and the direction of propagation. This effect has been shown by the Rutgers group, using a refined method of phase detection, in MBBA samples aligned by a magnetic field H. The result (at one given frequency) is of the form

$$c(\theta) = c(0)[1 - \Delta \sin^2\theta]$$

where Δ is typically of order 10^{-3}, and is positive (maximum velocity along the optical axis).

At temperatures well below T_c, Δ increases with frequency (more or less linearly). Near T_c, Δ becomes essentially independent of ω (in the range $2M_c < \omega/2M < 10M_c$)—presumably because the relaxation time of the order parameter becomes longer than the period in this temperature interval.

5.4.3.4. *Attenuation as a function of frequency and temperature: $\alpha(\omega)$.* A detailed study of the attenuation (for frequencies in the range 0·3 to 23 MHz$_c$) has been carried out recently by the M.I.T. group on MBBA [48]. Unfortunately, the data were taken on unoriented samples where many disclinations are probably present and may influence the damping. However, the main results are probably unaffected by this complication. Far from T_c, the attenuation α is well described in terms of a single relaxation time τ. Closer to T_c, α increases considerably, as expected from the coupling to the order parameter S: this is a general feature of order disorder transitions, emphasized in particular by Landau and Khalatnikov [49] in connection with superfluid helium, and refined in subsequent theories.

The relaxation rates involved are found to be in the range 10^6–10^7 s^{-1}, as could be expected from the results on $c(\omega)$ near T_c. But a picture using a single exponential relaxation is not sufficient to explain the data.

5.4.3.5. Anisotropy of the attenuation. For usual frequencies (in the megacycle range) the attenuation α is much more angular dependent than the velocity c. This has been shown by various groups [50, 52]. As a function of the angle θ between the optical axis and the direction of propagation, one can usually fit the attenuation by the form

$$\alpha(\theta) = \alpha(0)[1 - \delta \sin^2\theta]$$

with $\delta > 0$ and $\delta \sim 0.1$.

It is of some interest to compare this form with the predictions of a purely hydrodynamical theory, including the friction coefficients $\alpha_1 \ldots \alpha_5$ for an incompressible fluid plus two 'bulk viscosities'. The relevant formulae have been written down by the Harvard group [4a]. They have the form

$$\alpha(\theta) = a + b \sin^2\theta + c \sin^2 2\theta$$

The coefficient b involves the bulk viscosities, while the coefficient c depends only on $\alpha_1 \ldots \alpha_5$. The experimental results lead us to conclude that the bulk viscosities dominate the attenuation ($b \gg c$).

There is of course a correlation between the attenuation $\alpha(\omega)$ and the velocities $c(\omega)$, which are related respectively to the imaginary and to the real part of certain elastic constants. In an isotropic liquid this correlation is of limited use, because there are too many unknowns; for instance, two velocities at least must be known, $c(\omega = 0)$ and $c(\omega \to \infty)$. If one studies the anisotropies, however, the correlation is more useful, as first noted by Jahnig [53a], because Δc ($\omega = 0$) vanishes identically in a liquid. From his analysis, it appears that at least two different relaxation processes control the frequency dependence of the bulk viscosities in MBBA: (a) a specific mode of the butyl terminal chain is probably important [53b]; (b) near the clearing point T_c, the order parameter S relaxes rather slowly (as expected for a nearly second order transition) and is strongly coupled to the density fluctuations. But its effect cannot be described by a simple exponential relaxation.

5.4.4. Translational motions

5.4.4.1. Self-diffusion. A self-diffusion coefficient D has been qualitatively introduced in the last paragraph. More accurately, one must define two diffusion coefficients $D_{\parallel}$ and $D_{\perp}$. In principle, they can be measured by various techniques, using:

(1) Radioactive tracers [54]. For PAA the results at 125°C are $D = 4 \times 10^{-6}$ cm² s⁻¹, $D = 3 \cdot 3 \times 10^{-6}$ cm² s⁻¹.

(2) Inelastic scattering of neutrons; this makes use of the large, incoherent scattering due to the protons of the molecule. With a monoenergetic ingoing beam, and a scattering wave vector $\mathbf{q}$, the energy width of the outgoing beam $\Delta\omega_{\mathbf{q}}$ is given by

$$\Delta\omega_{\mathbf{q}} = D_{\parallel}q_z^2 + D_{\perp}q_{\perp}^2 \tag{5.96}$$

However, eqn (5.96) applies only when $qa \ll 1$; this corresponds to small widths $\Delta\omega_{\mathbf{q}}$, which are rather hard to measure accurately.

Early experiments with low flux reactors were open to some doubt [55, 56]. More recent measurements with high energy resolution and low q values have recently been performed at Julich, and seem to give diffusion coefficients which are in better agreement with values from other sources.

(3) Studies on nuclear spin precession in a magnetic field gradient: the principles of this method can be found for instance in Abragam's treatise [40]. The application to nematic fluids is difficult, because the spin–spin relaxation time T_2 is rather short. However, an interesting progress has been achieved recently: a special sequence of spin echoes can remove most of the dipolar interactions which are responsible for T_2; this elaborate technique has been set up recently by the Ljubliana group [57], and has allowed for a good measurement of $D_{\parallel}$ in MBBA. Measurements of $D_{\perp}$ require a small technical adjustment and should be available soon.

Theoretical estimates of the self diffusion constants have been given by Franklin [58].

5.4.4.2. Diffusion of a solute. It is often easier to measure $D_{\parallel}$ and $D_{\perp}$ for a solute in the nematic phase than for the nematic molecules themselves. For instance, if the solute is a dye, injected at time $t = 0$ in one small region of the sample, a study of the spatial spread in coloration at later times t is enough to measure D. This technique was used as early as 1917 (to study the diffusion anisotropy of nitrophenol dissolved in azoxyphenetol), by Svedberg [59]; his sample was aligned by a field $H \sim 3000$ G, and he could measure diffusion with concentration gradients either parallel or normal to $\mathbf{H}$. For this particular example he found $D_\parallel/D_\perp = 1\cdot41$.

These diffusion studies can give some useful informations of solvent–solute interactions. They could also be used to detect a macroscopic conformational change in a nematic; one example is discussed in the following problem.

Problem: a nematic slab is twisted by a field H larger than the Frederiks threshold H_c. How does the twist react on diffusion in the plane of the slab?

Solution: the geometrical conditions are shown on Fig. 5.12. The slab is in the (xy) plane, with the walls at $z = 0$ and $z = d$. The concentration gradient is taken along x. The easy axis of the walls $(\mathbf{n_0})$, is at a finite angle ψ from x in the

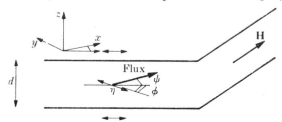

Fɪɢ. 5.12. Lateral diffusion in a nematic twisted by a magnetic field $\mathbf{H}$.

xy-plane. The field H is applied in the plane of the slab, normal to $\mathbf{n_0}$. Below the threshold field H_c we have a single crystal with optical axis along $\mathbf{n_0}$. Above threshold, the molecules at point (x,y,z) make an angle $\psi + \theta(z)$ with the x-axis. For instance, if H is slightly larger than H_c, we have (see Section 3.2):

$$\theta \sim a_0 \sin\left(\frac{\pi z}{d}\right)$$

$$a_0 = 2\left\{\frac{H - H_c}{H_c}\right\}^{\frac{1}{2}}.$$

The diffusion coefficient, for a concentration gradient along x, is

$$D(z) = D_\perp + (D_\parallel - D_\perp)\cos^2\{\psi + \theta(z)\}.$$

We shall restrict our attention to diffusion times $t \gg d^2/D$. In this limit the dye concentration becomes independent of z, and depends only on x. The effective diffusion coefficient is then the average of $D(z)$ over the sample thickness

$$\overline{D} = D_\perp + (D_\parallel - D_\perp)\overline{\cos^2(\psi + \theta)}.$$

In particular, for H slightly larger than H_c, this becomes

$$\overline{D} = D_\psi - (D_\parallel - D_\perp)\sin(2\psi)\bar{\theta}$$
$$D_\psi = D_\perp + (D_\parallel - D_\perp)\cos^2\psi$$

$$\bar{\theta} = a_0\frac{1}{d}\int\limits_0^d \sin\left(\frac{\pi z}{d}\right)\,\mathrm{d}z = \frac{2}{\pi}a_0$$

Clearly the optimum situation to detect a threshold amounts to take $\psi = \pi/4$.

5.4.4.3. Mobility of the charge carriers. In an organic semiconductor, the mobility μ of the charge carriers is never easy to measure. In an organic fluid, the situation is still worse; convective motions induced by the electric field give rise to an apparent mobility which is much larger than the intrinsic μ. However, some information on the carrier mobilities in materials such as MBBA can be obtained by (at least) two methods:

(1) Studies on transient regime in ultrapure specimens with electrodes suppressing all injection [60].

(2) Studies on A.C. instabilities in the limit of very small 'chevrons' [61]. The analysis of instabilities in Section 5.2 was based on Ohm's law $\mathbf{J} = \sigma \mathbf{E}$.† However, when spatial variation become rapid, we must also include a diffusion term:

$$\mathbf{J} = \sigma \mathbf{E} - D_\mathrm{c} \nabla q, \tag{5.97}$$

where the diffusion constant of the carriers D_c is related to σ by Einstein's relation

$$D_\mathrm{c} = \frac{k_\mathrm{B} T \sigma}{n e^2} = \frac{k_\mathrm{B} T}{e} \mu \tag{5.98}$$

(e = charge of the carriers, n = number of carriers per cm³).
If we take the divergence of $\mathbf{J}$ and use Poisson's equation to eliminate E we find

$$\mathrm{div}\,\mathbf{J} = \frac{4\pi\sigma}{\epsilon}\, q - D_\mathrm{c}\nabla^2 q - \left(\frac{4\pi\sigma}{\epsilon} + D_\mathrm{c}k^2\right)q. \tag{5.99}$$

where k is the wave vector of the perturbation. This may also be written as

$$\mathrm{div}\,\mathbf{J} = \frac{4\pi\sigma}{\epsilon}\, q(1 + k^2 r_\mathrm{D}^2) \tag{5.100}$$

where the length r_D is defined by

$$r_\mathrm{D} = \left\{\frac{D_\mathrm{c}\epsilon}{4\pi\sigma}\right\}^{\frac{1}{2}} = \left\{\frac{\epsilon k_\mathrm{B} T}{4\pi n e^2}\right\}^{\frac{1}{2}}, \tag{5.101}$$

where r_D is nothing else but the Debye–Hückel screening radius, associated with the mobile carriers, and familiar from the theory of electrolytes [62]. Equation (5.100) shows that diffusion begins to play an important role in the charge balance when $k r_\mathrm{d} \sim 1$; when the size of the parallel striations decreases down to r_d, the formulae of Section 5.3 break down. This observation, due to Dubois Violette and Parodi [61], allows for a measurement of r_d (which turns out to be typically of order

---

† For simplicity, we neglect here the anisotropy of σ, μ, etc.

1 μm). Knowing r_d, one can derive the carrier density n from (5.101), and from n plus the measured conductivity σ, one finally reaches the mobility μ (eqn 5.98).

Both methods for the measurement of μ have been put into practice only recently, and the results are not yet firmly established. But the existing data (for MBBA at room temperature) give rather low values of μ

$$\mu \sim 10^{-6} \text{--} 10^{-5} \text{ cm}^2 \text{ s}^{-1} \text{ V}^{-1}.$$

These numbers are significantly smaller than might be expected from viscous friction on a Stokes sphere of molecular size. Various interpretations of this effect may be suggested:

(1) The mobile ions may be of unusually large size.

(2) A charged ion deforms the nematic alignment (through its electric field) up to rather large distances. When the ion moves, the distortion must adjust accordingly; this provides an additional friction.

(3) Apart from any distortion, when the ion moves, there is a certain lag in the electric polarization cloud surrounding it. As explained in Section 5.4.1, the component of the dielectric response due to longitudinal dipoles has a slow relaxation; this tends to increase the lag and the resulting friction [63]. In practice, as pointed out by R. B. Meyer, when the individual dipoles in the nematic flip very slowly, the main relaxation channel for the polarization cloud is by spatial diffusion of the dipoles around the ion. A rough calculation based on this idea gives an enhancement of friction

$$\frac{\Delta f}{f} \sim 0 \cdot 1 \frac{\epsilon_\infty (\epsilon_0 - \epsilon_\infty)}{\epsilon_0} \frac{c}{a}$$

where a is the ionic radius, ϵ_0 and ϵ_∞ are the low and high frequency dielectric constants, and c is a characteristic length defined by

$$c \sim \frac{e^2}{\epsilon_0 k_B T}$$

Typically c is of order 100 Å and $\Delta f/f \sim 2$. Clearly, to see if these processes are indeed important, detailed studies with nematics of different dielectric properties will be required.

5.4.5. *Temperature variation of the friction coefficients*

The Leslie coefficients introduced in Section 5.1 (eqn 5.27) are related to certain local correlations in the fluid. From their temperature dependence two simple features emerge:

(1) The friction coefficients show an activation energy behaviour (as is found in most fluids far below their gas–liquid critical point).

(2) however, in the vicinity of the clearing point ($T \lesssim T_c$) there is a difference in behaviour between the various coefficients: α_4, which does not involve the alignment properties, is a rather smooth function of temperature; but all the other αs describe couplings between orientation and flow, and are thus affected by a decrease in the nematic order.

For instance, near T_c, the α_1 term of eqn (5.27)

$$\alpha_1 n_\alpha n_\beta n_\mu n_\rho A_{\mu\rho},$$

is more usefully written in the form:

$$\tilde{\alpha}_1 Q_{\alpha\beta} Q_{\mu\rho} A_{\mu\rho},$$

where $Q_{\alpha\beta}$ is the tensor order parameter introduced in Chapter 2, and $\tilde{\alpha}_1$ is a new coefficient which remains finite for $Q \to 0$. This means that α_1, should then be proportional to Q^2 (or S^2 in the Maier Saupe notation). A similar argument shows that α_2, α_3, α_5, and α_6, should be linear in S. This observation is due to Imura and Okano (Japanese Journal of Applied Physics, *11*, 440, 1972). It leads to rather good agreement with the existing data on the temperature dependence of the various viscosities.

5.4.6. Semi-slow motions above T_c

In Chapter 2, we discussed some short-range order effects occurring above the nematic–isotropic transition temperature T_c. We concluded that, just above T_c, the fluid contained correlated regions of size $\xi(T_2) \sim 200$ Å. Let us consider the dynamical properties of such a structure in more detail.

One molecule in the fluid rotates with a very short time constant (10^{-11} s). But, for an object of size $\xi(T_c)$ any overall rotation or deformation is less rapid (time constant of the order of 10^{-7} s. We shall say that such a process is *semi-slow*. Our information on these motions comes from the following types of measurements (some of which have been announced in Chapter 2):

flow birefringence [64];

inelastic scattering of light [65];

nuclear spin lattice relaxation [39];

ultrasonic attenuation of shear waves [66], and also (less directly) of longitudinal waves.

Time constant of the electric birefringence (Kerr effect) [68] [69].†

Since these phenomena involve rather large regions and semi-slow motions in the fluid, they may be described macroscopically in terms of rate equations for the order parameters $Q_{\alpha\beta}$ (defined as in Chapter 2) [67]. However, just as below T_c, we have a certain coupling between orientation and flow. To describe it, we must again introduce two sets of fluxes.

The first is the rate of change of the order parameter

$$\frac{\delta Q_{\alpha\beta}}{\delta t} = R_{\alpha\beta}. \tag{5.101}$$

Note that in principle, the derivative $\delta/\delta t$ should represent the variation of Q, along one flow line, and with respect to the background fluid; $\delta Q/\delta t$ is the analogue, above T_c, of the vector $\mathbf{N} = \dot{\mathbf{n}} - \boldsymbol{\omega} \times \mathbf{n}$ of Section 5.1. However, for the motions of interest above T_c, we may usually treat $Q_{\alpha\beta}$ and the flow velocities v_α as infinitesimal quantities of first order; then, the difference between $\delta Q/\delta t$ and the partial derivative $\partial Q/\partial t$ is negligible.

The other set of fluxes remains, as in Section 5.1, the shear rate tensor $A_{\alpha\beta}$ defined by eqn (5.18). We also restrict our attention to incompressible flows ($A_{\alpha\alpha} \equiv 0$): this will be justified for the applications listed above, because the semi-slow motions of interest are at much lower frequencies than sound waves.

We now write the entropy source

$$T\dot{S} = \dot{Q}_{\alpha\beta} R_{\alpha\beta} + \sigma'_{\alpha\beta} A_{\alpha\beta}, \tag{5.102}$$

where $\sigma'_{\alpha\beta}$ is again the viscous stress. The force $R_{\alpha\beta}$ is the restoring force derived from the free energy F: (eqn 38) of Chapter 2.

$$R_{\alpha\beta} = -\frac{\partial F}{\partial Q_{\alpha\beta}} = -A(T) Q_{\alpha\beta} \tag{5.103}$$

In eqn (5.103) we have kept only the term which was linear in Q, and we have assumed $\mathbf{H} = 0$. We can now construct a set of phenomenological equations relating the forces to the fluxes—all of them being symmetric traceless tensors of rank 2. The most general form for these equations, compatible with rotational invariance and with the Onsager symmetry relation, is then [67]

$$\sigma'_{\alpha\beta} = 2\eta A_{\alpha\beta} + 2\mu R_{\alpha\beta} \tag{5.104}$$

$$\dot{Q}_{\alpha\beta} = 2\mu A_{\alpha\beta} + \nu R_{\alpha\beta}. \tag{5.105}$$

† This last information is not quite clear-cut, because the *static* Kerr effect has an unexplained temperature dependence.

If Q is dimensionless (and normalized so that $Q_{zz} = 1$ in a completely ordered phase) the three coefficients η, μ, γ have the dimension and the magnitude of a viscosity. Nematodynamics above T_c involves only three parameters, while below T_c, five parameters were required (in an incompressible fluid). This simplification comes from the fact that $Q_{\alpha\beta}$ is small above T_c. Writing that $T\dot{S}$ is positive definite for all motions, and inserting into eqn (5.102), one obtains the conditions

$$\eta > 0, \qquad 2\mu^2 < \nu\eta, \qquad \nu > 0 \tag{5.106}$$

Let us now show briefly how the three parameters η, μ, ν, can be related to experimental data.

5.4.4.1. Flow birefringence. Consider for instance a shear flow with v along x, and a velocity gradient along z ($A_{xz} \neq 0$). This is a steady state, with $R_{\alpha\beta} = 0$. Inserting this into eqn (5.105) we get, after making use of eqn (5.103):

$$Q_{xz} = -\frac{2\mu}{A(T)} A_{xz} = -\frac{\mu}{A(T)} \frac{\partial v}{\partial z}, \tag{5.107}$$

while all other components of $Q_{\alpha\beta}$ are equal to zero. The tensor $Q_{\alpha\beta}$ has then two principal axes (labelled 1 and 2) which are the bisectors of x and z.

$$Q_{11} = -Q_{22}, \qquad Q_{33} = 0$$

$$|Q_{11} - Q_{22}| = \frac{2\mu}{A(T)} \left|\frac{\partial v}{\partial z}\right|. \tag{5.108}$$

Thus the material is birefringent, and the magnitude of the birefringence is large, since $A(T)$ is small near T_c.

5.4.4.2. Inelastic scattering of light. At small scattering wave vectors $\mathbf{q}$ the fluctuations of the hydrodynamic velocity are very slow (relaxation rates $\eta q^2/\rho$) while those of Q are only semi-slow; then the coupling between orientation and flow becomes ineffective and, in eqn (5.105), the shear rate component $A_{\alpha\beta}$ is negligible. This gives a simple exponential relaxation for $Q_{\alpha\beta}$, with a rate (deduced from eqns 5.103 and 5.105):

$$\Gamma(T) = \frac{A(T)}{\nu}.$$

Because of the small factor $A(T)$, $\Gamma(T)$ is reduced: this is the origin of the semi-slow motions. We may say that the restoring force (5.103) is small near T_c.

The width in frequency of the light scattered by orientation fluctuations is equal to $\Gamma(T)$: it has been measured on MBBA using refined

Fabry Perot Techniques [65]. The values of Γ are in the range 10^7 sec^{-1} and $\Gamma(T)$ increases rapidly when T increases above T_c. This increase is due mainly to the factor $A(T)$, which can be derived from the data on magnetic birefringence (see Chapter 2). By comparing $\Gamma(T)$ and $A(T)$, it appears that the friction coefficient varies with temperature very much like the average viscosity η (Fig. 5.13).

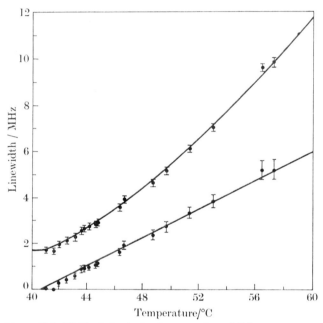

FIG. 5.13. Line-width $\Gamma(T)$ of the inelastic scattering of light: upper curve: raw data for MBBA; lower curve: $\Gamma\eta(T_c)/\eta(T)$ (the temperature dependence of the friction coefficients is removed). After ref. [55].

The above discussion was restricted to very small scattering wave vectors $\mathbf{q}$. At larger q values,† the admixture between orientational and hydrodynamic modes complicates the situation slightly (see ref. [57]) but the orders of magnitude remain the same.

5.4.4.3. Shear wave attenuation. From eqns (104) and (105) after elimination of the internal variables $Q_{\alpha\beta}$, it is easy to derive an effective, frequency dependent, viscosity $\eta(\omega)$

$$\eta(\omega) = \eta - \frac{2\mu^2}{\nu}\frac{i\omega}{\Gamma+i\omega} \qquad (5.110)$$

† i.e. when $\eta q^2/\rho \sim \Gamma$.

Eqn (5.110) shows a dispersion anomaly at $\omega \simeq \Gamma$: above this frequency, the molecular alignment is not able to adjust to the shear. In the transient Kerr effect [68, 69] it is probably also the constant Γ which gives the relaxation rate.

5.4.4.4. Nuclear spin–lattice relaxation. This is also a probe for the dynamical fluctuations of $Q_{\alpha\beta}$. In this case, the analysis requires more sophistication: the spatial derivatives of Q must be included in the free energy and in the friction equations.

Qualitatively, one may understand the relaxation line T_1 from the following argument. The relaxation rate contains a component, due to the fluctuations of Q, of the form:

$$\frac{1}{T_1} \simeq (\gamma H_L)^2 \langle Q^2 \rangle \tau_c \qquad (5.111)$$

where $H_L Q$ measures the local field seen by one nucleus (we omit all component indices). γ is the nuclear gyromagnetic ratio, and τ_c a correlation time for the fluctuations of Q†: here τ_c will be essentially equal to Γ^{-1}. The average $\langle Q^2 \rangle$ is not exactly to be taken at one point, but must be taken as a smeared out average $\langle Q(0)Q(\mathbf{R}) \rangle$ with $R \sim \xi$.‡ To a good approximation, the correlation function $\langle Q(0)Q(R) \rangle$ has the Ornstein Zernike form:

$$\langle Q(0)Q(R) \rangle = \frac{a}{R} \exp{-R/\xi_c)} \qquad (5.112)$$

where a is a molecular length. Eqns (5.111) and (5.112) lead to a relaxation rate

$$\frac{1}{T_1} \simeq (\gamma H_L)^2 \frac{a}{\xi} \tau \qquad (5.113)$$

We have seen that τ_c varies essentially as ξ^2. Thus we expect $1/T_1$ to be proportional to ξ, i.e. to $(T - T^*)^{-\frac{1}{2}}$ in the Landau approximation. A power law of this type has indeed been observed with the relaxation of ^{14}N (controlled by modulations of a quadrupolar field) in PAA [39].

---

† Eqn (5.111) holds only for nuclear resonance frequencies ω_n which are not too high ($\omega_n \tau_c < 1$). For an introduction to the concepts leading to eqn (5.111), see for instance ref. (40).

‡ Among other reasons for this smearing is the fact that, during the time τ_c, the molecule diffuses through the swarm.

REFERENCES

CHAPTER 5

[1] ERICKSEN, J. L. *Archs. ration. Mech. Analysis* **4**, 231 (1960); *Physics Fluids* **9**, 1205 (1966).

[2] LESLIE, F. M. *Quart. J. Mech. appl. Math.* **19**, 357 (1966); *Archs. ration. Mech. Analysis* **28**, 265 (1968).

[3] PARODI, O. *J. Phys. (Fr.)* **31**, 581 (1970).

[4] FORSTER, D., LUBENSKY, T., MARTIN, P., SWIFT, J., and PERSHAN, P. *Phys. Rev. Lett.* **26**, 1016 (1971); MARTIN, P. C., PARODI, O., and PERSHAN, P. J. *Phys. Rev.* A **6**, 2401 (1972); see also STEPHEN, M. J. *Phys. Rev.* A **2**, 1558 (1970) and JAHNIG, F. and SCHMIDT, M. *Ann. Phys.* **71**, 129 (1971).

[5] MARTIN, P. C., PERSHAN, P. J., and SWIFT, J. *Phys. Rev. Lett.* **25**, 844 (1970).

[6] HALLER, I. and LITSTER, J. D. *Phys. Rev. Lett.* **25**, 1550 (1970).

[7] DE GROOT, S. and MAZUR, P. *Non-equilibrium thermodynamics*, North-Holland, Amsterdam (1962).

[8] MIESOWICZ, M. *Nature* **17**, 261 (1935); *Bull. Acad. pol. Sci.*, A **228**, (1936); *Nature* **158**, 27 (1946).

[9] (a) MARTINOTY, P. and CANDAU, S. *Mol. Cryst. liquid Cryst.* **14**, 243 (1971); MARTINOTY, P. Thèse, Strasbourg University (1972); (b) LANGEVIN, D. *J. Phys. (Fr.)* **33**, 249 (1971).

[10] See, for instance, LANDAU, L. D. and LIFSHITZ, E. M. *Fluid mechanics*, § 24. Pergamon, London (1959).

[11] PORTER, R. S. and JOHNSON, J. F. *J. phys. Chem.* **66**, 1826 (1962); *J. chem. Phys.* **45**, 1452 (1966); see also *J. appl. Phys.* **34**, 51, 55 (1963).

[12] FISHER, J. and WAHL, J. *Mol. Cryst. liquid Cryst.* (to be published); *Opt. Commun.* **5**, 341 (1972); see also GAHWILLER, J. *Phys. Rev. Lett.* **28**, 1554 (1972); MEIBOOM, S. and HEWITT, R. C. *Phys. Rev. Lett.* **30**, 261 (1973).

[13] (a) LESLIE, F. M. *Archs. ration. Mech. Analysis* **28**, 265 (1968); ATKIN, R. J. and LESLIE, F. M. *Quart. J. Mech. appl. Math.* **23**, 3 (1970); CURRIE, P. K. *Archs. ration. Mech. Analysis* **37**, 222 (1970); (b) ERICKSEN, J. L. *Trans. Soc. Rheol.* **13**, 9 (1969); (c) FISHER, J. and FREDERICKSON, A. G. *Mol. Cryst. liquid Cryst.* **8**, 267–84 (1969).

[14] (a) TSVETKOV, V. *Zh. eksp. teor. Fiz.* **9**, 603 (1935) TSVETKOV, V. and SOSNOVSKI, A., *Acta Physicochemica USSR* **18**, 358 (1943); (b) GASPAROUX, H. and PROST, J. *Phys. Lett.* A**36**, 255 (1971); (c) LESLIE, F. M., LUCKHURST, G. R. and SMITH, H. J. *Chem. Phys. Lett.* **13**, 368 (1972).

[15] DE GENNES, P. G. *J. Phys. (Fr.)* **32**, 789 (1971); LEGER, L. *Solid State Commun.* **10**, 697 (1972).

[16] BROCHARD, F., GUYON, E., and PIERANSKI, P. *Phys. Rev. Lett.* **28**, 1681 (1972).

[17] See, for instance, FRIEDEL, G. *Annls. Phys.* **18**, 273 (1922) (in particular, p. 359).

[18] (a) ORSAY GROUP on liquid crystals. *J. Chem. Phys.* **51**, 816 (1969). (b) See, for instance, the review by BENEDEK, G. in *Polarisation, matière, rayonnement*, Presses Universitaires de France, Paris (1969).

[19] ORSAY GROUP on liquid crystals, *Mol. Cryst. liquid Cryst.* **13**, 187 (1971); LEGER, L. Thèse 3e cycle, Orsay (1971).

[20] HEILMEYER, G., ZANONI, L. A., and BARTON, L. *Proc. Inst. elect. electron. Engrs.* **56**, 1162 (1968).

[21] HELFRICH, W. *J. chem. Phys.* **51**, 4092 (1969).

[22] See the review by FELICI, N. *Revue gén. elect.* **78**, 77 (1969).

[23] WILLIAMS, R. *J. chem. Phys.* **39**, 384 (1963).

[24] TEANEY, D. and MIGLIORI, A. *J. appl. Phys.* **41**, 998 (1970); ORSAY GROUP on Liquid crystals, *C.r. hebd. Séanc. Acad. Sci., Paris* **270**, 97 (1970); PENZ, P. A. *Phys. Rev. Lett.* **24**, 1405 (1970).

[25] For a recent review see HELFRICH, W. *Mol. Cryst. liquid Cryst.* (to be published).

[26] PENZ, P. A. and FORD, G. W. *Phys. Rev.* **6**, 414 (1972). *Phys. Rev.* **6**, 1676 (1972).

[27] See, for instance, CHANDRASEKHAR, S. *Theory of hydrodynamic stability*, Oxford University Press (1961).

[28] CARR, E. F. in *Ordered fluids and liquid crystals*, p. 76. Adv. Chem. Series, Am. chem. Soc. Pub. (1967).

[29] HEILMEIER, G. and HELFRICH, W. *Appl. Phys. Lett.* **16**, 1955 (1970).

[30] RONDELEZ, F. Thèse 3e cycle, Orsay (1970); ORSAY GROUP on liquid crystals, *Mol. Cryst. liquid Cryst.* **12**, 251 (1071).

[31] Film available from the General Electric Company.

[32] DUBOIS VIOLETTE, E., DE GENNES, P. G., and PARODI, O. *J. Phys. (Fr.)* **32**, 305 (1971).

[33] See, in particular, CARR, E. F. *J. chem. Phys.* **38**, 1536 (1963); **39**, 1979 (1963); **42**, 738 (1965); **43**, 3905 (1965).

[34] MAIER, W. and MEIER, G. *Z. Naturf.* A **16**, 262 (1961).

[35] MEIER, G. and SAUPE, A. *Mol. Cryst.* **1**, 515 (1966). MARTIN A. J., MEIER G., SAUPE A., *Symp. Faraday Soc.* **5**, 119 (1971).

[36] VISINTAINER J., DOANE J. W. FISHEL D., *Molecular Cryst. liquid cryst.* **13**, 69, (1971).

[37] VILFAN, M., BLINC, R., and DOANE, J. W. (to be published).

[38] (a) AYANT, Y. *Mol. Cryst. liquid Cryst.* (to be published); (b) FARINHA MARTINS, A. *Phys. Rev. Lett.* **28**, 289 (1972).

[39] CABANE, B. and CLARKE, G. *Phys. Rev. Lett.* **25**, 91 (1970). GOSH, S., TETTAMANTI, E., INDOVINA, E., *Phys. Rev. Lett.* **29**, 638 (1973).

[40] See, for instance, ABRAGAM, A. *Principles of nuclear magnetism*, Chapter 8. Oxford University Press (1961).

[41] PINCUS, P. *Solid State Commun.* **7**, 415 (1969). See also BLINC, R., HOGENBOOM, D., O' REILLY, D. PETERSON, E. *Phys. Rev. Lett.* **23**, 969 (1969).

[42] DOANE, W. and JOHNSON, D. L. *Chem. Phys. Lett.* **6**, 291 (1970).

[43] LUBENSKY, T. *Phys. Rev.* A **2**, 2497 (1970).

[44] W. HOYER and A. W. NOLLE, *J. chem. Phys.* **24**, 803 (1956).

[45] A. KAPUSTIN and G. EVEREVA, *Kristallographia* **10**, 723 (1965).

[46] A. KAPUSTIN and L. MARTIANOVA, *Kristallographia*, **16,** 649 (1970).

[47] M. MULLEN, B. LÜTHI, and M. J. STEPHEN, *Phys. Rev. Lett.* **28,** 799 (1972).

[48] D. EDEN, C. GARLAND, and R. WILLIAMSON, *J. chem. Phys.* **58,** 1861 (1973).

[49] L. LANDAU and I. KHALATNIKOV, *Dokl. Acad. Nauk. SSSR.* **96,** 469 (1954).

[50] A. E. LORD and M. LABES, *Phys. Rev. Lett.* **25,** 570 (1970).

[51] E. LIEBERMAN, J. LEE, and F. MOON, *Appl. Phys. Lett.* **18,** 280 (1971).

[52] K. KEMP and S. LETCHER, *Phys. Rev. Lett.* **27,** 1634 (1971).

[53] (a) F. JAHNIG, *Z. Phys.* **258,** 199 (1973); (b) F. JAHNIG, to be published.

[54] YUN, G. and FREDERIKSON, G. *Mol. Cryst. liquid Cryst.* **12,** 73 (1970).

[55] BLINC, R., DIMIC, V., PIRŠ, J., VILFAN, M., and ZUPANČIČ, I., *Mol. Cryst. liquid Cryst.* **14,** 97 (1971); MURPHY, J. A. and DOANE, J. W. *Mol. Cryst. liquid Cryst.* **13,** 93 (1971).

[56] JANIK, J. A., JANIK, J. M., OTNES, K., and RISTE, T. *Mol. Cryst. liquid Cryst.* **15,** 189 (1971).

[57] BLINC, R., PIRŠ, J. ZUPANČIČ, I., *Phys. Rev. Lett.* **30,** 546 (1973).

[58] FRANKLIN, W. *Mol. Cryst. liquid Cryst.* **14,** 227 (1971).

[59] SVEDBERG, T. *Kolloidzeitscrift.* **22,** 68 (1918).

[60] (a) HEILMEIER, G., RANONI, L., BARTON, L., *I.E.E.E. Trans. on electron devices* **ED17,** 22 (1970); KOELMANS H., VAN BOXTEL A., *Mol. Cryst.* **7,** 395 (1969); (b) BRIERE, G., GASPARD, F., and HERINO, R. *Chem. Phys. Lett.* **9,** 285 (1971); BRIERE, G. HERINO, R., MONDON F. *Mol. Cryst.* **19,** 157, (1972).

[61] See GALERNE, J., DURAND, G., and VEYSSIE, M. *Phys. Rev. liquid Cryst.* A **6,** 484 (1972).

[62] See, for instance, LANDAU, L. D. and LIFSCHITZ, E. M. *Statistical physics* (§ 91), Pergamon, London (1958).

[63] DE GENNES, P. G. *Comments in Solid St. Phys.* **3,** 148 (1971).

[64] ZVETKOV, V. *Acta Phys.-chim. URSS* **19,** 86 (1944).

[65] LITSTER, J. D. and STINSON, T. W. *J. appl. Phys.* **41,** 996 (1970); *Phys. Rev. Lett.* **25,** 503 (1970); LITSTER, J. D. in *Critical phenomena* (R. E. Mills, ed.) p. 394, McGraw-Hill (1971).

[66] MARTINOTY, P., DE BEAUVAIS, F., and CANDAU, S. *Phys. Rev. Lett.* **27,** 1123 (1971).

[67] DE GENNES, P. G. *Phys. Lett.* A **30,** 454 (1969); *Mol. Cryst. liquid Cryst.* **12,** 193 (1971).

[68] PROST J., LALANNE B., *Phys. Rev.* A (to be published).

[69] WONG W., SHEN Y., *Phys. Rev. Lett.* **30,** 895 (1973).

CHOLESTERICS

'*Donde escono quei vortici di foco pien d'orror?*'

DON GIOVANNI

6.1. Optical properties of an ideal helix

6.1.1. *The planar texture*

THE helical arrangement characteristic of the cholesteric phase has been presented in Chapter 1. In this ideal state the director $\mathbf{n}(\mathbf{r})$ varies in space according to the law

$$\left.\begin{aligned}
n_x &= \cos\theta \\
n_y &= \sin\theta \\
n_z &= 0
\end{aligned}\right\} \tag{6.1}$$

$$\theta = q_0 z + \text{constant}, \tag{6.2}$$

where we have taken the helical axis along z. A cholesteric single crystal, described by eqns (6.1) and (6.2), can often be obtained in thin slabs (thickness d in the 100 μm)—provided that the boundary con ditions on both sides of the slab are *tangential*. The configuration is represented on Fig. 6.1; it is usually called the 'planar texture' or 'Grandjean texture.' Typical examples are the following:

(1) Between one polished glass surface and a free surface. At the glass surface ($z = 0$) the angle $\theta(0)$ is fixed by the polishing direction. At the free surface ($z = d$), the angle $\theta(d)$ is free to adjust.

(2) Between two polished glass surfaces, or in a cleavage gap between two mica sheets. Here both angles $\theta(0)$ and $\theta(d)$ are fixed. The helix must in general adjust its pitch slightly in order to comply to these conditions: the wave vector q' in eqn (6.2) is now different from q_0:

$$q'd = \theta(d) - \theta(0) + n\pi$$

$$(n = \text{integer}),$$

the value of n being chosen to minimize the distortion energy—i.e. to minimize $|q' - q_0|$.

Let us now recall briefly a point of notation: the nominal pitch P of the helical structure is equal to $2\pi/q_0$; however, since the states $(\mathbf{n})$ and

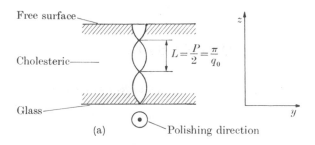

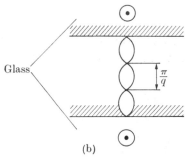

FIG. 6.1. The planar texture of cholesterics. The director is horizontal everywhere. In case (a) the helix has the optimum pitch. In case (b) the pitch is enforced by the boundary conditions.

$(-\mathbf{n})$ are indistinguishable, the periodicity interval along z is $L = P/2 = \pi/q_0$.

We must also remember the question of sign: if the (xyz) frame used in eqn (6.1) is a right-handed frame, and if the wave vector q_0 is positive, we have a right-handed helix; this is found for instance with cholesterol chloride. If q_0 is negative, we have a left-handed helix: most of the aliphatic esters of cholesterol belong to this class.

6.1.2. Bragg reflexions

A light beam, of angular frequency ω, is sent parallel to the helical axis (z). In a zero-order approximation, we may think of the cholesteric as of a nearly isotropic medium, with a certain average index of refraction $\bar{n}$. The optical wavelength in the medium is then

$$\lambda = \frac{2\pi c}{\bar{n}\omega}. \tag{6.3}$$

We may also define the beam by the wave vector $\mathbf{k}_0$, directed along z, and of magnitude $\omega \bar{n}/c$: both notations will be useful.

In the next approximation, we note that the medium is not exactly isotropic; the optical properties are modulated with a spatial period

$L = \pi/q_0$. This may in principle give rise to Bragg reflections, provided that

$$2L = m\lambda \qquad (m = \text{integer}). \qquad (6.4)$$

Experimentally, one does observe *one* Bragg reflection ($m = 1$). The higher order reflections ($m = 2, 3, ...$) are forbidden for normal incidence. The polarization features of the waves are also remarkable [1]:

(1) The reflected light is circularly polarized: at any instant t the electric field pattern in the reflected wave is a helix, identical in shape to the cholesteric helix (see Fig. 6.2a).

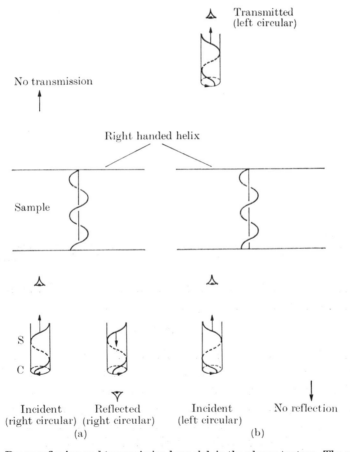

F ig. 6.2. Bragg reflexion and transmission by a slab in the planar texture. The cylinders with helices S represent 'snapshots' of the electric field **E** associated with one wave. The vertical arrows give the direction of propagation. The circles C show the rotation of **E** as seen by an observer at one fixed point in space. The reflected wave emitted by the sample is an image of the cholesteric helix, translated downwards. Examination of the corresponding projected path C shows that it is *right circular*.

(2) If we analyse the incident wave into two components of opposite circular polarizations, we find that only one component is strongly reflected—i.e. the component for which the instantaneous electric pattern is again identical in shape to the cholesteric helix. The other component is transmitted without any significant reflexion through the slab. These features are also displayed in Fig. 6.2.

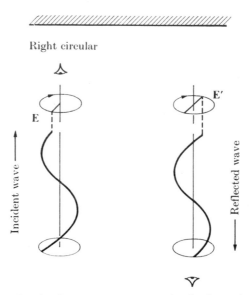

Right circular

Incident wave →

E

Reflected wave →

E′

FIG. 6.3. Reflexion of a circular wave on a conventional mirror. At the mirror surface the reflected field **E′** is just opposite to the incident field **E**. Compare the reflected wave with the reflected wave of Fig. 6.2: they are of opposite circular polarizations.

All these properties can be explained in terms of the scattering amplitude α introduced in Chapter 3 (eqn 3.81):

$$\alpha = \mathbf{f}.\boldsymbol{\varepsilon}(\mathbf{q}).\mathbf{i} \qquad (6.5)$$

where $\mathbf{f}$ and $\mathbf{i}$ represent the polarizations of the reflected wave (of wave vector $\mathbf{k}_1$) and of the incident wave (wave vector $\mathbf{k}_0$). $\mathbf{q} = \mathbf{k}_0 - \mathbf{k}_1$ is the scattering wave vector, and $\boldsymbol{\varepsilon}(\mathbf{q})$ the Fourier transform of the dielectric tensor. In the present case, the three vectors $\mathbf{k}_0$, $\mathbf{k}_1$, and $\mathbf{q}$ will be parallel to (z). As explained in Chapter 1, at any point $\mathbf{r}$ a cholesteric behaves locally like a uniaxial material: the dielectric tensor may thus be written

$$\epsilon_{\alpha\beta}(\mathbf{r}) = \epsilon_\perp \delta_{\alpha\beta} + (\epsilon_\parallel - \epsilon_\perp) n_\alpha(\mathbf{r}) n_\beta(\mathbf{r}). \qquad (6.6)$$

Using eqn (6.1) for $\mathbf{n}(\mathbf{r})$ we can compute $\boldsymbol{\varepsilon}(\mathbf{r})$ and then $\boldsymbol{\varepsilon}(\mathbf{q})$. For $q \neq 0$ a constant term like $\epsilon_\perp$ does not contribute. To illustrate the calculation let us discuss the (xx) component of the dielectric tensor

$$\epsilon_{xx}(\mathbf{q}) = \epsilon_a \int d\mathbf{r} \cos^2(q_0 z) \, e^{iqz} \tag{6.7}$$

$$(\epsilon_a = \epsilon_\parallel - \epsilon_\perp).$$

Writing
$$\cos^2(q_0 z) = \tfrac{1}{2} + \tfrac{1}{4}(e^{2iq_0 z} + e^{-2iq_0 z})$$

and eliminating again the constant term, we see that the integral (6.7) vanishes except when $q = \pm 2q_0$. Let us take $q_0 > 0$ (right handed helix). Then the case $q = -2q_0$ corresponds to k_1 larger than k_0, and is forbidden, since the frequency of the scattered wave must coincide with ω.

The case of interest is $q = 2q_0$, corresponding to $k_0 = -k_1 = q_0$. The reader will check that this condition is identical to eqn (6.4) with $m = 1$. Then

$$\epsilon_{xx}(2q_0) = \tfrac{1}{4}\epsilon_a V,$$

where V is the sample volume.

Similar manipulations give the other components of $\boldsymbol{\varepsilon}(2q_0)$: the only non-vanishing components correspond to polarizations in the xy plane, and the resultant 2×2 matrix is

$$\hat{\varepsilon}(2q_0) = \tfrac{1}{4}\epsilon_a V \hat{M}$$

$$\hat{M} = \begin{pmatrix} 1 & i \\ i & -1 \end{pmatrix} \tag{6.8}$$

Omitting the constant factor the polarization of the reflected wave $\mathbf{f}$ is related to the incident polarization $\mathbf{i}$ by:

$$\begin{pmatrix} f_x \\ f_y \end{pmatrix} = \hat{M} \begin{pmatrix} i_x \\ i_y \end{pmatrix}$$

or explicitly
$$f_x = i_x + i i_y$$
$$f_y = i i_x - i_y. \tag{6.9}$$

We see that $f_y = +i f_x$: the reflected light is circularly polarized. In one case we do *not* get a reflected wave: namely when $i_y = i i_x$; this defines the transmitted wave of Fig. 6.2.

We may also explain why the higher order Bragg reflections are forbidden, in terms of the matrix amplitude $\hat{M}$:

Let us look for instance at the reflection $m = 2$, which would correspond to $q = 4q_0$. This can be obtained by scattering from the initial state k_0 to a virtual photon state $k_0 - 2q_0$, followed by a second scattering from

$k_0 - 2q_0$ to $k_0 - 4q_0$. The matrix amplitude for this process is proportional to $\hat{M}^2$. But, from eqn (6.8) it is easily verified that $\hat{M}^2 = 0$. The proof can be extended to higher order processes involving more than one virtual photon, and also to larger values of m; all higher Bragg reflections are forbidden for normal incidence.

Let us now discuss briefly the case of oblique incidences. Here, we observe reflections, with the geometrical condition:

$$2L \cos r = m\lambda, \tag{6.10}$$

where r is the angle of the refracted beam in the slab (related to the angle of incidence i by $\sin i = \bar{n} \sin r$). The main differences from the case of normal incidence are that [2]:

(1) All orders ($m = 1, 2, 3...$) are observed;

(2) The polarizations are elliptical.

A detailed discussion of oblique propagation and reflections requires heavy numerical calculations [2–4]. We may summarize the explanation of point (i) as follows, taking as an example the second reflection $m = 2$. The intermediate photon (of wave vector $k_0 - 2q_0$) is oblique: the corresponding polarization vectors are not both in the xy plane; it is only their projection in this plane which contributes to the scattering amplitude; this complication is described mathematically by a certain projection operator $\hat{P}$. The second order amplitude is then proportional to $\hat{M}\hat{P}\hat{M}$, and this does not vanish in general.

6.1.3. Transmission properties at arbitrary wavelengths (normal incidence)

6.1.3.1. The Mauguin limit. Cholesterics of large, adjustable pitch P are easily obtained by dilution of an optically active material in a

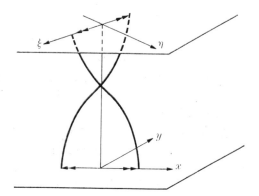

F I G. 6.4. Definition of local axes in a twisted nematic.

nematic matrix: with such dilute solutions, P is typically in the range 10–50 μm and is much larger than the optical wavelength λ. In this limit the optical properties are comparatively simple: they have been discussed first in 1911 by Mauguin [5] (in connection with mechanical twist of nematics).

A thin sheet (corresponding to an interval z, $z+dz$ in the slab), of thickness small compared with P, but large compared to λ, behaves as a uniaxial slab, of optical axis $\mathbf{n}$: in the sheet the eigenmodes of vibration are linearly polarized: one mode along $\mathbf{n}$, with the extraordinary index $n_e = (\epsilon_\parallel)^{\frac{1}{2}}$, and one mode normal to $\mathbf{n}$, with the index $n_0 = (\epsilon_\perp)^{\frac{1}{2}}$. The structure of the two modes may be written down explicitly in the form:

$$\text{Mode 'e'} \quad \begin{aligned} E_x &= \cos\theta\,\cos[\omega\{(zn_e/c)-t\}] \\ E_y &= \sin\theta\,\cos[\omega\{(zn_e/c)-t\}] \end{aligned}$$

$$\text{Mode 'o'} \quad \begin{aligned} E_x &= \sin\theta\,\cos[\omega\{(zn_0/c)-t\}] \\ E_y &= -\cos\theta\,\cos[\omega\{(zn_0/c)-t\}], \end{aligned} \quad (6.11)$$

where $\mathbf{n} = (\cos\theta,\ \sin\theta,\ 0)$.

Let us now consider the whole sample; the successive sheets (dz) do not have the same optical axes. However, in the limit $\lambda \ll P(n_e - n_0)$ it may be shown† that the eqns (6.11) still correctly define the eigenmodes of vibration.

Knowing this mode structure, we may then deduce the transmission properties of the slab. Let us take our x-axis along the easy direction of the bottom plate [$\theta(0) = 0$]. Let us also define two auxiliary axes (ξ, η) in the slab plane, coinciding with the optical axes at the upper end of the slab. If we send a wave polarized along x, entering through the bottom plate, it will come out at $z = d$, as a linear wave, polarized along ξ, and with a phase lag $\phi_e = \omega n_e d/c$. Similarly, an incident wave polarized along y, comes out along η, with a phase lag $\phi_0 = \omega n_0 d/c$. On the other hand, an incident wave of arbitrary linear polarization:

$$\begin{aligned} E_x(z=0) &= \cos\alpha\,\cos(\omega t) \\ E_y(z=0) &= \sin\alpha\,\cos(\omega t), \end{aligned}$$

will come out in the form

$$\begin{aligned} E_\xi(d) &= \cos\alpha\,\cos\{\omega t - \phi_e\} \\ E_\eta(d) &= \sin\alpha\,\cos\{\omega t - \phi_0\} \end{aligned}$$

and, because of the phase difference $\phi_e - \phi_0$, it will be elliptical in general.

† The proof will be described in Section 6.1.4.

The simple 'wave guide' effect obtained with ingoing waves polarized along x or y provides a convenient optical method to measure the total twist angle $\theta(d) - \theta(0)$; this is sometimes of interest with a planar texture between one polished glass plate and a free surface, and leads to a measurement of the pitch P.

A remark about signs; if the cholesteric helix is right-handed, and if $\theta(d) - \theta(0)$ is small, an observer, measuring the rotation of the plane of polarization for the extraordinary wave (the angle between x and ξ), will call the material levogyric† (or more shortly *levo*), according to the usual convention in optics. The geometry is shown in Fig. 6.3.

6.1.3.2.   Rotatory power.   Let us assume now that

$$\lambda > P(n_e - n_0) \tag{6.12}$$

(excluding however, for the moment, the case of exact Bragg reflection $\lambda = P$). Then the experimental observations show that, with normal incidence, the planar texture behaves like an 'optically active' medium; this means that, for a given frequency ω, the two eigenmodes of vibration propagating along $+z$ are circular waves, with different indices $n_1 n_2$.

If we illuminate the slab (at $z = 0$) with a wave of linear polarization

$$\begin{aligned} E_x &= \cos \alpha \cos \omega t \\ E_y &= \sin \alpha \cos \omega t \end{aligned} \tag{6.13}$$

and if we analyse it in terms of the eigenmodes, we find that the corresponding outgoing wave (at $z = d$) is also linear but rotated by an angle

$$\psi = \frac{d\omega}{2c} \{n_1 - n_2\}. \tag{6.14}$$

This is geometrically similar to what is observed in an isotropic, optically-active liquid (IOAL). But the orders of magnitude are strikingly different:

IOAL $\dfrac{\psi}{d} \sim 1$ deg cm^{-1}

cholesteric $\dfrac{\psi}{d} \sim 10^4$ deg cm^{-1}.

The dependence of the optical rotation ψ/d on the wavelength λ of the light is shown in Fig. 6.8. Note in particular the singularity at the Bragg wavelength. A review on optical rotation data can be found in the Ph.D. work of Cano [6].

† Levogyric means: turning to the left.

6.1.4. Interpretation

6.1.4.1. Assumptions on the local dielectric properties. The huge optical rotations observed in the cholesteric phase are clearly not due to an intrinsic spectroscopic property of the constituent molecules (since they do not persist in the isotropic phase). They must reflect the properties of light waves propagating in a twisted anisotropic medium. This has been studied by many authors, the most clear, rigorous, and accessible reference being that of de Vries [7].

The starting point is an assumption on the form of the local dielectric tensor $\boldsymbol{\varepsilon}(\mathbf{r})$ at any point $\mathbf{r}$ in the cholesteric fluid: neglecting the weak intrinsic rotation which persists in the isotropic phase, we may write that the electric displacement $\mathbf{D}(\mathbf{r})$ is a linear functional of the electric field $\mathbf{E}(\mathbf{r})$, taken at the same point $\mathbf{r}$, and use eqn (6.6).

$$\mathbf{D} = \boldsymbol{\varepsilon}\mathbf{E} = \epsilon_\perp \mathbf{E} + \epsilon_a \mathbf{n}(\mathbf{n}.\mathbf{E}) \qquad (6.15)$$

$$\epsilon_a = \epsilon_\parallel - \epsilon_\perp$$

For a wave propagating along the helical axis z, $\mathbf{D}$ and $\mathbf{E}$ are restricted to the xy plane and eqn (6.15), involving two parameters ($\epsilon_\parallel$ and $\epsilon_\perp$), is the most general local form. On the other hand, for oblique propagation, eqn (6.15) represents an approximation; at each point $\mathbf{r}$ the medium is described as *uniaxial* (with axis $\mathbf{n}$), while by symmetry it might be biaxial (with dielectric constants $\epsilon_\perp$, $\epsilon_\parallel$ in the xy plane, and with another constant ϵ_{zz} along the helical axis). However, as already mentioned in Chapter 1, the deviations from uniaxiality ($\epsilon_\perp - \epsilon_{zz}$) are expected to be of order $(q_0 a)^2 \sim 10^{-4}$, and the approximation is probably excellent in all cases.

6.1.4.2. Equation of propagation. Let us now consider specifically the propagation, along z, of an electromagnetic wave of frequency ω. The non-zero field components are

$$E_x(zt) = \text{Re}\{E_x(z)e^{-i\omega t}\}$$
$$E_y(zt) = \text{Re}\{E_y(z)e^{-i\omega t}\} \qquad (6.16)$$

(Re = real part of),
and the Maxwell equations reduce to

$$-\frac{\mathrm{d}^2}{\mathrm{d}z^2}\begin{pmatrix} E_x \\ E_y \end{pmatrix} = \left(\frac{\omega}{c}\right)^2 \hat{\epsilon}(z) \begin{pmatrix} E_x \\ E_y \end{pmatrix}. \qquad (6.17)$$

The explicit form of the matrix $\hat{\epsilon}$ deduced from eqns (6.15) and (6.1) is

$$\hat{\epsilon}(z) = \frac{\epsilon_\parallel + \epsilon_\perp}{2}\begin{pmatrix} 1 & 0 \\ 0 & 1 \end{pmatrix} + \frac{\epsilon_a}{2}\begin{pmatrix} \cos 2q_0 z & \sin 2q_0 z \\ \sin 2q_0 z & -\cos 2q_0 z \end{pmatrix}. \qquad (6.18)$$

Equation (6.17) does not have exactly the structure of an eigenvalue problem. However there are some simple features. The operators on both sides are unchanged by a translation of length L along (z). This implies a Bloch–Floquet theorem [8]; a complete set of solutions can be found, such that, for each of them

$$\begin{pmatrix} E_x \\ E_y \end{pmatrix}_{z+L} = \text{const.} \begin{pmatrix} E_x \\ E_y \end{pmatrix}_z. \tag{6.19}$$

Here we shall find it convenient to write the constant in the form $-e^{ilL}$ (the minus sign is chosen because the helical pitch is $P = 2L$). The wave vector l defines the mode under consideration; note that l may be real (propagating wave) or complex (evanescent wave).

6.1.4.3. Dispersion relation. To derive the solutions explicitly, it is convenient to analyse all fields in terms of circular (rather than linear) waves. This amounts to choose as new variables the quantities

$$E = E_x \pm iE_y.$$

Equation (6.17) then becomes

$$-\frac{\mathrm{d}^2 E^+}{\mathrm{d}z^2} = k_0^2 E^+ + k_1^2 \exp(2iq_0 z)E^-$$

$$-\frac{\mathrm{d}^2 E^-}{\mathrm{d}z^2} = k_1^2 \exp(-2iq_0 z)E^+ + k_0^2 E^-, \tag{6.20}$$

where we have put

$$k_0^2 = \left(\frac{\omega}{c}\right)^2 \frac{\epsilon_\parallel + \epsilon_\perp}{2}$$

$$k_1^2 = \left(\frac{\omega}{c}\right)^2 \frac{\epsilon_a}{2}. \tag{6.21}$$

On eqn (6.20) we can immediately find the form of the modes

$$E^+ = a \exp\{i(l+q_0)z\}$$

$$E^- = b \exp\{i(l-q_0)z\} \tag{6.22}$$

where a and b are two constants, linked by the following relations

$$\{(l+q_0)^2 - k_0^2\}a - k_1^2 b = 0$$

$$-k_1^2 a + \{(l-q_0)^2 - k_0^2\}b = 0. \tag{6.23}$$

The reader may check that eqn (6.22) is compatible with the Bloch–Floquet theorem (eqn 6.19).

The two eqns (6.23) have a non-trivial solution only if the corresponding determinant vanishes:

$$(-k_0^2+l^2+q_0^2)^2-4q_0^2l^2-k_1^4 = 0. \qquad (6.24)$$

For a given frequency ω, k_0, and k_1 are fixed, and eqn (6.24) gives four possible values of l (real or complex). The relation between ω and l for real l is called the *dispersion relation;* it is shown in Fig. 6.5.

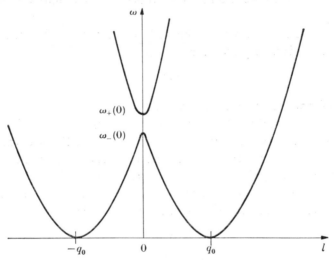

FIG. 6.5. Relation between frequency and wave vector l for propagation of electromagnetic modes in a cholesteric spiral. The mode l, defined by eqn (6.22), is a coherent superposition of two plane waves with wave vectors $l \pm q_0$.

There are two distinct branches, which we call $(+)$ and $(-)$. To locate them, it is useful to consider first the case $l = 0$. This gives

$$k_0^2-q_0^2 = \pm k_1^2.$$

Returning to the definitions of k_0 and k_1 (eqn 6.21) one obtains the following frequencies

$$\omega_+(0) = \frac{cq_0}{n_o} \qquad n_o = \epsilon_\perp^{\frac{1}{2}} = \text{ordinary index,}$$

$$\omega_-(0) = \frac{cq_0}{n_e} \qquad n_e = \epsilon_\parallel^{\frac{1}{2}} = \text{extraordinary index.} \qquad (6.25)$$

$\omega_+(0)$ corresponds to $a = -b$. From eqn (6.22) this describes a linear wave polarized along the direction $\theta(z)+\frac{1}{2}\pi$ (ordinary axis). Similarly $\omega_-(0)$ corresponds to $a = b$, i.e. a linear wave polarized along the local extraordinary axis.

6.1.4.4. Eigen modes for travelling waves. The interval $\omega_-(0) <$
$\omega < \omega_+(0)$ will be called the frequency gap. Let us, for the moment,
choose an ω value outside of this interval. Then as can be seen from
Fig. 6.5 we have four real values of l (grouped in two pairs l_1, $-l_1$, l_2,
$-l_2$). If we are interested in waves which travel along the $(+z)$ direction,
we must retain only the roots which give a positive group velocity
$v_g = \partial\omega/\partial l > 0$. There are *two* such roots (see Fig. 6.5) which we call
l_1 and l_2.†

Each of them defines an eigenmode of vibration. Attached to the
eigenmode (l_i) is an eigenvector

$$\begin{pmatrix} a_i \\ b_i \end{pmatrix}$$

defined only within a multiplicative constant.

In the present regime (l real) eqn (6.23) shows that a and b may both
be taken as real. Then, returning from eqn (6.22) to the real variables of
eqn (6.16), one can see that the electric field associated with the eigenmode is elliptically polarized; furthermore at each point, the axes of the
ellipse coincide with the local optical axes of the cholesteric. A useful
parameter to describe the ellipse is the real number

$$\rho = \frac{-a+b}{a+b}.$$

The axial ratio of the ellipse is $|\rho|$, and the sign of ρ gives the sign of
rotation of the vibration in the (x, y) plane (at fixed z and increasing t).
To calculate ρ in terms of l or ω, we start from the eqns (6.23) written in
the form

$$\frac{a}{b} = \frac{k_1^2}{A} = \frac{B}{k_1^2} \tag{6.26}$$

where

$$A = (l+q_0)^2 - k_0^2$$
$$B = (l-q_0)^2 - k_0^2.$$

Then we can write

$$\rho = \frac{-k_1^2 + A}{k_1^2 + A} = \frac{-B + k_1^2}{B + k_1^2}$$
$$= \frac{-B + A}{B + A + 2k_1^2} = \frac{-2lq_0}{k_0^2 - l^2 - q_0^2 - k_1^2}. \tag{6.27}$$

Equation (6.27) may be transformed, using eqn (6.24). If we define a
positive quantity s^2 by

$$s^2 = +\sqrt{(k_1^4 + 4q_0^2 l^2)}, \tag{6.28}$$

---

† For definiteness we shall call l_1 the larger of the two roots.

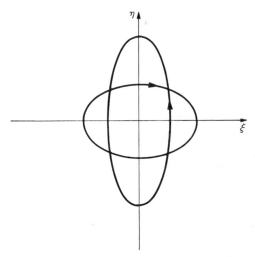

FIG. 6.6. Precession ellipses for the electric field $\mathbf{E}(t)$, at one fixed observation point. The two curves correspond to two eigenmodes with the same ω, but different l. Note that the two ellipses do not have the same axial ratio.

we have

$$\rho = \frac{-2lq_0}{\pm s^2 - k_1^2}.\qquad(6.29)$$

The sign $(\pm)$ in eqn (6.29) depends on the branch used for $\omega(l)$ (Fig. 6.5) The general aspect of the two ellipses associated with one frequency ω is shown in Fig. 6.6. The signs and magnitude of ρ on the various branches associated with travelling waves (real l) are shown in Fig. 6.7.

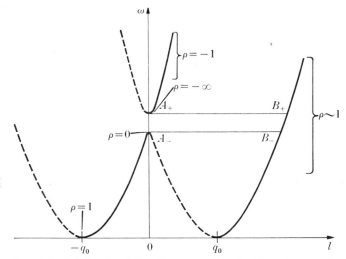

FIG. 6.7. Axial ratio of the ellipses associated with various modes.

Note that the ratio long axis : short axis is *not the same* for the two ellipses associated with one ω value. This is obvious if we compare the points A_+ and B_+ (both corresponding to $\omega = \omega_+(0)$) on Fig. 6.7. At point A_+ we have a linear polarization as explained after eqn (6.25). At point B_+, we can see from eqn (6.28) that $s^2 \sim 2lq_0$ and $\rho \sim +1$, i.e. the polarization is circular.

Mathematically, to have two ellipses of identical shape (differing only by a 90° rotation in the x, y plane) we would require $\rho_1\rho_2 = -1$. This can also be written as

$$a_1a_2 + b_1b_2 = 0.$$

But this relation does *not* hold; the eigenvectors $\begin{pmatrix} a \\ b \end{pmatrix}$ of eqn (6.23) belonging to the same ω and different l are *not* orthogonal. This point has been emphasized in particular by Billard [3] [9].

6.1.4.5. 'Waveguide' regimes. Equations (6.28) and (6.29) show that the mode structure depends critically on the value of the parameter

$$x = \frac{2q_0l}{k_1^2}. \tag{6.30}$$

Let us consider first the regimes where $x \ll 1$. This can be obtained in two ways:

(1) by going to small l values, i.e. in the vicinity of the points A_+ and A_- of Fig. 6.5.

(2) by going to very large l values. Then $l \sim k_0$ and

$$x \sim \frac{2q_0(n_e^2 + n_0^2)}{k_0(n_e^2 - n_0^2)} \sim \frac{2q_0\bar{n}}{k_0(n_e - n_0)}. \tag{6.31}$$

The criterion $x \ll 1$ is the criterion for the *Mauguin limit* $\lambda \ll (n_e - n_0)P$.

In both cases (1) and (2) we find one mode associated with the $(+)$ branch which has $\rho \to \infty$, and represents a linear *ordinary* wave guided by the helix.† The $(-)$ mode has $\rho \to 0$ and represents a linear extraordinary wave, guided in the same way; these conclusions are in agreement with our earlier remarks concerning the gap edges (case 1) and the Mauguin limit (case 2).

6.1.4.6. Circular regimes. In many practical cases the two indices n_e and n_0 are not very different. Then k_1^2, as defined by eqn (6.21), tends to be small and the parameter x defined in eqn (6.30) is *large* for most frequencies of interest. Then $s^2 \to 2q_0 |l|$ and by eqn (6.29) we see that

---

† I.e. where the polarization remains everywhere parallel to the ordinary axis.

$\rho \rightarrow \pm 1$. In this regime the eigenmodes are nearly circular—in agreement with the observations described in paragraph 3.

In fact, all essential properties can be derived directly from the initial eqns (6.22) and (6.23). If we neglect k_1^2 completely (zero order approximation) we see on eqn (6.23) that the l values are

$$l_1 = k_0 + q_0 \text{ associated with } a_1 = 0 \quad b_1 = 1$$

$$l_2 = k_0 - q_0 \text{ associated with } a_2 = 1 \quad b_2 = 0.$$

In the next approximation, we find from eqn (6.24):

$$l_1 = k_0 + q_0 + \frac{k_1^4}{8k_0 q_0 (k_0 + q_0)} + O(k_1^8)$$

$$l_2 = k_0 - q_0 + \frac{k_1^4}{8k_0 q_0 (q_0 - k_0)} + O(k_1^8). \tag{6.32}$$

Since $a_1 \simeq 0$ the mode (1) is, according to eqn (6.22), a circular wave of wave vector $l_1 - q_0$. To make contact with the notation of paragraph 3, we shall represent this wave vector in terms of a refractive index n_1

$$l_1 - q_0 = \frac{\omega}{c} n_1. \tag{6.33}$$

Similarly, for mode (2) we have a circular wave, of opposite sense, and of wave vector

$$l_2 + q_0 = \frac{\omega}{c} n_2. \tag{6.34}$$

The optical rotation per unit length ψ/d is then obtained simply by taking one half of the difference between these two wave vectors (see eqn 6.14)

$$\frac{\psi}{d} = \frac{\omega}{2c} (n_1 - n_2) = \frac{k_1^4}{8q_0(k_0^2 - q_0^2)}. \tag{6.35}$$

This formula is often expressed in terms of a reduced wavelength $\lambda' = \lambda/P = q_0/k_0$. Transforming k_1 by eqn (6.21), one obtains

$$\frac{\psi}{d} = \frac{q_0}{32} \left(\frac{n_e^2 - n_0^2}{n_e^2 + n_0^2}\right)^2 \frac{1}{\lambda'^2(1 - \lambda'^2)}. \tag{6.36}$$

Equation (6.36) is due to de Vries [7]. Note the following features:

(1) Very large magnitude of the rotation: for $\lambda' = 0 \cdot 7$ and $(n_e^2 - n_0^2)/(n_e^2 + n_0^2) = 0 \cdot 1$ we get $\psi/d \sim 10^{-3} q_0$. With a pitch P of 1 μm, corresponding to $q_0 = 6 \times 10^4$, we expect $\psi/d \sim 60$ rad cm^{-1} or 3500 deg cm^{-1}! An example is shown in Fig. 6.8.

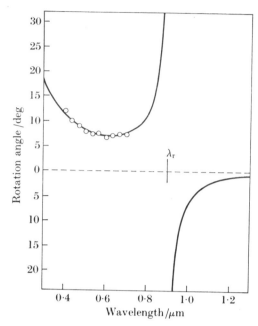

FIG. 6.8. Optical rotation as a function of wavelength for a cholesteric ester. Pitch $P = 600$ nm. Average refractive index $\bar{n} = 1\cdot50$. Sample thickness $25\cdot4\,\mu$m. The theoretical curve was obtained from the de Vries formula (equation 36) by assuming $(n_1 - n_0)/\bar{n} = 3\cdot07 \times 10^{-2}$ (after Baessler et al., Mol. Cryst. **6**, 329, 1970).

(2) 'Dispersion' anomaly (with a change in sign) at the Bragg reflection ($\lambda' = 1$). Of course, when discussing this point, we must remember that very near the Bragg reflection, eqn (6.36) breaks down and the 'waveguide' regime takes over. This compares rather well with the data of Mathieu [1].

(3) Optical rotation proportional to P^3/λ^4 at long wavelengths.

(4) At wavelengths $\lambda \ll P$ (but still too large to be in the Mauguin regime) the rotation becomes proportional to P/λ^2: this regime has been probed in detail by Cano [6] in mixtures of adjustable pitch.

6.1.4.7. Bragg reflection. Let us now choose a frequency ω inside the gap:

$$\frac{cq_0}{n_e} < \omega < \frac{cq_0}{n_0}.$$

For such a case, we see from Fig. 6.5 that eqn (6.24) has only two real roots ($l = \pm l_1$). The other two roots are pure imaginary $l = \pm i\kappa$.

Consider now a thick slab ($d \to \infty$) attacked from below, at normal incidence, by a light beam of polarization $\mathbf{i}(i_x i_y)$. This will in general induce in the slab two waves :[†]

(1) One travelling wave [amplitude proportional to $\exp(il_1 z)$].
According to eqn (6.29) the parameter ρ associated with this wave is
close to $+1$ (for a right-handed cholesteric helix) : the wave is circularly polarized, and the sign of circulation is in agreement with
Fig. 6.3.

(2) One evanescent wave (amplitude $\sim e^{-\kappa z}$).

By a suitable choice of the polarization $\mathbf{i}$, ($\mathbf{i} = \mathbf{i}_R$) it is possible to
extinguish the travelling wave component : this means that a beam of
polarization $\mathbf{i}_R$ will be totally reflected. We conclude that the gap
$[\omega_-(0),\ \omega_+(0)]$ corresponds to the frequency range for possible Bragg
reflections.

It is easy to check that this property agrees with the more elementary
description of Section 6.1.2 in the appropriate limit—namely when
$n_e \rightarrow n_0$. Then the gap shrinks down to one frequency $\omega = cq_0/\bar{n}$; the
corresponding wavelength is $\lambda = 2\pi c/\omega\bar{n} = 2\pi/q_0 = P$, in agreement
with the usual Bragg condition for first order reflections.

6.1.5. Conclusions and generalizations

The optical properties observed on a planar cholesteric texture (Bragg
reflection and optical rotation) are very spectacular : when P corresponds to an optical wavelength in the visible range, the sample shows
some very bright colours in reflection. Also the optical rotations are
huge. All these properties were quite mysterious in the early days of
liquid crystal physics. However, they are explained very accurately by
the dielectric model of Mauguin, Oseen, and de Vries—i.e. by eqn
(6.15)—involving only two dielectric constants $\epsilon_\parallel$ and $\epsilon_\perp$.

We have deliberately skipped the discussion of oblique incidences in
Section 6.1.4; this case is complicated and does not add much to the
insight. It is important however from two points of view :

(1) It is only with oblique waves that one can check the validity of
eqn (6.15) for all electric field directions. A complete comparison between
optical data and theoretical calculations, based on eqn (6.15), for
oblique waves has been carried out on cholesteric mixtures by Berreman
and Scheffer [2]. The agreement found is excellent, and proves unambiguously that a cholesteric fluid is *locally uniaxial*.

(2) Oblique Bragg reflections are often important, in practice especially if we have a polydomain sample rather than a single domain planar

---

[†] The two other waves would correspond to light sources on the upper side of the
slab.

texture. This situation has been studied experimentally by Fergason [10]. A typical geometry is shown on Fig. 6.9. The beam enters the sample with a certain incidence Φ_I, is refracted to Φ_I', then propagates in the sample up to a certain domain, with helical axis $\mathbf{q}_0$, where it suffers a Bragg reflection. We restrict our attention to the case where $\mathbf{q}_0$ is in the plane of incidence. Then *the outgoing beam is also in this plane*, and is characterized by the angles ϕ_R', ϕ_R. Defining an angle r between the beam just before Bragg reflection, and $\mathbf{q}_0$ we can write in the first

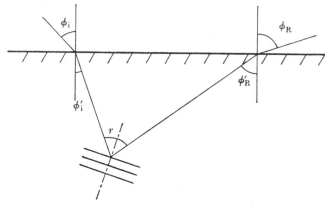

FIG. 6.9. Bragg reflection on a cholesteric polydomain sample. All rays are assumed to be in the plane of the sheet.

approximation of eqn (6.10)

$$P \cos r = m\lambda = m\lambda_v / \bar{n} \tag{6.37}$$

where λ_v is the wavelength in vacuum, $\bar{n}$ the average refraction index, and we assume that $|n_e - n_0| \ll \bar{n}$. Fig. 6.9 gives us the further relations

$$2r = \phi_I' + \phi_R'$$
$$\sin \phi_I = \bar{n} \sin \phi_I' \tag{6.38}$$
$$\sin \phi_R = \bar{n} \sin \phi_R'.$$

Eliminating all internal angles between (6.37) and (6.38) we arrive at [10]

$$\lambda_v = \frac{P\bar{n}}{m} \cos\left\{ \tfrac{1}{2} \sin^{-1}\left(\frac{\sin \phi_I}{\bar{n}}\right) + \tfrac{1}{2} \sin^{-1}\left(\frac{\sin \phi_R}{\bar{n}}\right) \right\} \tag{6.39}$$

This relation allows for a very simple measurement of P in polydomain samples. It is restricted to cases of small birefringence; with large birefringences, multiple reflections become important and the measurements are less accurate.

6.2. Agents influencing the pitch

From the discussions in Section 6.1, we see that the optical properties of a cholesteric material will often depend critically on the value of the pitch P. In the present section we list some typical agents which can be used to change P, and thus lead to various interesting applications.

6.2.1. Physicochemical factors

6.2.1.2. Temperature. In most cholesteric derivatives P is a *decreasing* function of temperature:

$$\frac{\mathrm{d}P(T)}{\mathrm{d}T} < 0.$$

A typical curve for $P(T)$ was shown in Fig. 1.5. The order of magnitude of $\mathrm{d}P/\mathrm{d}T$ is often surprisingly large; as can be seen again on Fig. 1.5, with cholesteryl nonanoate, in a certain temperature range around 74·6°C, one can reach values

$$\frac{1}{P}\left|\frac{\mathrm{d}P}{\mathrm{d}T}\right| \sim 100 \text{ deg}^{-1}.$$

These giant temperature variations have been observed in particular by Fergason and coworkers [10]. Their origin is not yet quite clear. The increase in P when T is decreased may be due to the onset of a short range order of the smectic type; most of the compounds at hand have a smectic phase at lower temperatures, and the smectic stacking in equidistant planes is incompatible with twist [11].

In practice, the changes of P are reflected in colour changes provided that the Bragg reflections can occur in the visible spectrum; both the reflected wavelength and the temperature range of maximum sensitivity can be adjusted with suitable multicomponent mixtures [12]. This gives rise to a number of remarkable applications [13]:

(1) *Measurement of superficial temperatures.* The surface under study is painted with a thin film of cholesteric material; any temperature difference between two points on the surface will show up as a difference in coloration. This is used for medical purposes (detection of tumors, etc) [14] and also for various industrial tests (microelectronic circuits, aircraft wings, etc).

(2) *Conversion of an infrared image into a visible image.* The infrared image is focused on a cholesteric film: in the irradiated areas of the film, some heat is liberated; the temperature rises and the pitch decreases.

The local change of P is probed by a source of visible light, and observed either in reflection or in transmission.

This technique has been useful to visualize light patterns from infrared lasers [15]. It is also applicable to microwave patterns [16].

6.2.1.2. Chemical composition. The overall temperature span of the cholesteric phase, the magnitude, and even the sign of the helical twist q_0, vary greatly from one cholesteric compound to another. (The temperature intervals for aliphatic esters of cholesterol are shown on Fig. 6.10).

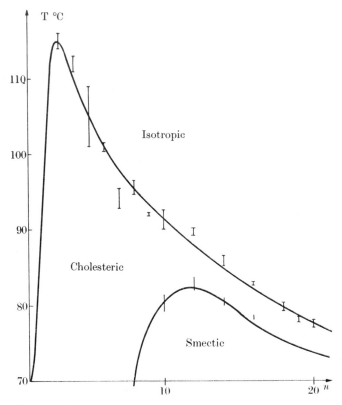

FIG. 6.10. Mesomorphism amongst aliphatic esters of cholesterol: n = number of carbon atoms in the fatty acid component. Data from G. Gray and R. Ennulat, compiled in R. Ennulat, *Mol. Cryst.*, **8**, 247 (1969).

This immediately suggests that the properties of *mixtures* may be both flexible and interesting; a few examples will be discussed below.

(1) *Dilute solutions in a nematic matrix:* when a small concentration c of optically active material is dissolved in a nematic, the result is a cholesteric of large pitch P. In the dilute limit, P is inversely

proportional to c [17] $Pc = \text{constant}$ (6.40)

A microscopic interpretation of this law, in terms of the long range distortions induced in the nematic by a single solute molecule, is given in the next problem. The result is conveniently written in the form

$$q_0 = \frac{2\pi}{P} = +4\pi\beta c \qquad (6.41)$$

where c is the number of solute molecules/cm³, and β is a constant (with the dimension of a surface), dependent on the nature of both solvent and solute. We call β the 'microscopic twisting power' of the solute.

If we choose one particular nematic solvent N, we can measure the magnitude and sign of the microscopic twisting power for all chiral molecules which are soluble in N. In practice, such studies have been carried out for two groups of solvents N:

(a) conventional nematics with two benzene rings such as PAA or MBBA.

(b) special cholesteric mixtures with exact compensation; consider for instance a 1·75:1·00 mixture of cholesterol chloride–cholesterol myristate at 40°C. The pure chloride is right-handed ($q_0 > 0$) while the pure myristate is left-handed ($q_0 < 0$). This particular mixture has $q_0 = 0$ i.e., it is nematic. It has been used as a reference solvent by Baessler and Labes [18]. Typical results are given in Table 6.1 (on a slightly different notation).

TABLE 6.1

Macroscopic twisting power P_t of various solutes in a compensated (cholesterol chloride–cholesterol myristate) mixture at 40°C. P_t is defined by $c_w P P_t = 1$ where c_w is the weight concentration of solute, and P is the resulting pitch (after [18]).

Compound	Number of carbons on side-chain	P_t
Cholesterol	0	+2·0
Cholesterol chloride	0	+2·96
Cholesterol formate	1	−0·67
Cholesterol acetate	2	−1·2
Cholesterol propionate	3	−3·22
Cholesterol butyrate	4	−3·45
Cholesterol valerate	5	−4·45
Cholesterol caproate	6	−4·70
Cholesterol myristate	14	−7·45

The advantage of this mixture is to provide a solvent involving the cholesterol skeleton, which is clearly adequate for a study of twisting powers among cholesterol derivatives. The obvious drawback is that it can be used at one temperature only: if the temperature is changed, we do not have exact compensation any more.

Problem: derive the pitch of a dilute solution of a cholesteric material in a nematic phase, by a superposition of individual distortions.

Solution: our starting point is the discussion of distortions around one floating object in a nematic (problem p. 119, Chapter 3). In the one constant approximation we found that the local rotation $\boldsymbol{\omega}$ at a large distance $\mathbf{r}$ from the object had the form

$$\boldsymbol{\omega} = \boldsymbol{\alpha}\frac{1}{r}+\boldsymbol{\beta}.\nabla\left(\frac{1}{r}\right),$$

where $\boldsymbol{\alpha}$ is a vector, and $\boldsymbol{\beta}$ a dyadic, both dependent on the orientation and shape of the object. As seen in Chapter 3, in the absence of specific torques acting on the object, the vector $\boldsymbol{\alpha}$ vanishes. The dyadic $\boldsymbol{\beta}$ must be an even function of the unperturbed director $\mathbf{n_0}$, and has thus the general form[†]

$$\boldsymbol{\beta} = -\beta+\beta_1\mathbf{n_0}{:}\mathbf{n_0}.$$

However, the second term gives rise only to rotations around $\mathbf{n_0}$, which do not lead to any practical change. Thus we are left with

$$\boldsymbol{\omega} = -\beta\nabla\left(\frac{1}{r}\right) = \beta\frac{\mathbf{r}}{r^3}.$$

Note that $\boldsymbol{\omega}$ is an axial vector, while $\nabla(1/r)$ is a polar vector. Thus β is a pseudo-scalar, and is nonvanishing only if the impurity differs from its mirror image. β has the dimensions of a surface. The general aspect of the distortion described by β is shown on Fig. 6.11.

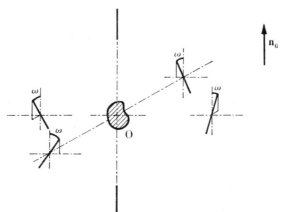

Fɪɢ. 6.11. Static distortions around a chiral object in a nematic matrix.

[†] The choice of a $(-)$ sign in the scalar term will be convenient later: with this definition β will be positive when the solute is cholesterol chloride.

Let us now turn to a dilute solution with c impurities per cm^3, each of them creating a distortion of this type. For small concentrations c, we may assume that the total rotation vector $\omega(\mathbf{r})$ at any point $\mathbf{r}$ is obtained by superposition:

$$\omega(\mathbf{r}) = \sum_p (-\beta)\nabla\frac{1}{r_p}$$

where the sum $\sum_p$ is over all solute molecules. ω is identical to the electrostatic field which would result from charges β located at the various solute positions. Thus by Poisson's equation:

$$\text{div } \omega = +4\pi c\beta$$

and also

$$\text{curl } \omega = 0.$$

The solution corresponds to a helical structure. Putting the helical axis along z, we have:

$$\omega_z = q_0 z$$

$$\omega_x = \omega_y = 0$$

with

$$q_0 = 4\pi c\beta.$$

Thus the helical wave vector q_0 is indeed proportional to the concentration.

(2) *Concentrated solutions*: For most mixtures between cholesterol derivatives, the twist $\bar{q}$ of the mixture is nearly equal to the weight average of the component twists q_i [18]

$$\bar{q}(T) \simeq \sum_t c_i q_i(T) \tag{6.42}$$

where c_i is the (weight) concentration of component i. Note that in eqn (6.42) the various q_is may be positive or negative: in particular, as already mentioned, if we have two components, one with $q_1 > 0$ and the other with $q_1 < 0$, a suitable mixture of the two will be nematic ($\bar{q} = 0$).†

There are, however, some exceptions to the additivity rule (6.42). Mixtures of cholesterol chloride–cholesterol laurate are one such exception; here the plot of $\bar{q}$ versus concentration shows a well defined extremum [19], and is not at all linear on the laurate-rich side. Similar extrema are found with other esters of comparable length (Fig. 6.12). Their microscopic origin is unknown.‡

(3) *Effects of contaminants on the pitch:* certain gases, when absorbed in a cholesteric film, cause a significant change of pitch, and thus of coloration [13]. A device transforming an ultraviolet image into a visible

† Mixtures between two chiral antipodes are essentially ideal and obey eqn (6.42) exactly: see for instance D. Dolphin *et al.*, *J. chem. Phys.*, 1973 (to be published).

‡ One possible interpretation would connect the low $\bar{q}$, observed for the laurate-rich mixtures, with smectic short-range order in the cholesteric phase. X-ray experiments will help to elucidate this point.

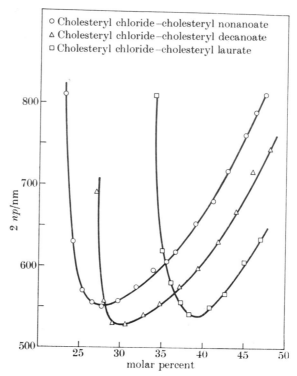

FIG. 6.12. Pitch versus composition for three binary mixtures of cholesterol esters (after J. Adams, W. Haas, and J. Wysocki, *Phys. Rev. Letts.*, **22**, 92 (1969).

image, and based on a similar principle, has been invented by the Xerox group [20]. The u.v. image is focused on a film containing a mixture of cholesteric esters, and in particular some cholesterol iodide; this component is easily decomposed photochemically in the regions irradiated with u.v. light. Thus, in these regions, we have a change in chemical composition and a resulting change in pitch; the latter is then observed under visible light.

6.2.1.3. Pressure. The effects of a hydrostatic pressure on the helical pitch have not been measured up to now: they might be spectacular in the vicinity of the cholesteric–isotropic transition point, or also alternatively near a smectic–cholesteric transition.

6.2.2. External fields

It is possible to distort a cholesteric spiral by a magnetic or an electric field: this gives rise to rather remarkable magneto-optic or electro-optic effects, which we shall now discuss. The starting point is to write a form for the free energy, as a function of the director $\mathbf{n}(\mathbf{r})$, for slow variations of $\mathbf{n}$ in space.

6.2.2.1. Continuum theory for cholesterics. When we discussed the distortion energy F_d for a nematic fluid in Chapter 3, we discarded all terms linear in the gradients of $\mathbf{n}$: these terms are not compatible with an equilibrium conformation where $\mathbf{n} = \text{constant}$. For a cholesteric, however, this argument does not hold, since the equilibrium conformation is twisted. There are two terms which are linear in the spatial derivatives of $\mathbf{n}$, and rotationally invariant: div $\mathbf{n}$ and $\mathbf{n}.\text{curl } \mathbf{n}$. Terms proportional to div $\mathbf{n}$ cannot occur in F_d because the states $\mathbf{n}$ and $-\mathbf{n}$ are indistinguishable. On the other hand, the pseudoscalar quantity $\mathbf{n}.\text{curl } \mathbf{n}$ may appear in F_d, provided that the molecules are different from their mirror images. Adding to this the usual nematic terms (eqn 3.15) we end up with a distortion energy of the form

$$F_d = \tfrac{1}{2}K_1(\text{div } \mathbf{n})^2 + \tfrac{1}{2}K_2(\mathbf{n}.\text{curl } \mathbf{n} + q_0)^2 + \tfrac{1}{2}K_3(\mathbf{n} \times \text{curl } \mathbf{n})^2. \quad (6.43)$$

In eqn (6.43) we do find a term linear in the gradients, namely

$$K_2 q_0 \mathbf{n}.\text{curl } \mathbf{n}$$

and a constant term $\tfrac{1}{2}K_2 q_0^2$. The meaning of eqn (6.43) becomes more transparent if we consider a situation of pure twist

$$n_x = \cos \theta(z) \qquad n_y = \sin \theta(z) \qquad n_z = 0.$$

Then eqn (6.43) reduces to

$$F_d = \tfrac{1}{2}K_2\left(\frac{\partial \theta}{\partial z} - q_0\right)^2. \quad (6.44)$$

We see from this that the equilibrium distortion corresponds to a helix of wave vector $\partial \theta / \partial z = q_0$.

Equation (6.43) is the correct form for the distortion free energy when both $\nabla \mathbf{n}$ and q_0 are small on the molecular scale. If q_0 was large ($q_0 a \sim 1$) the structure of F_d would become more complicated as discussed by Jenkins [21]. However in all practical situations $q_0 a \sim 10^{-3}$ and the Jenkins corrections are not important.

We must add to the distortion F_d some extra terms which describe the coupling to H or E; just as in the nematic case, they have the form

$$F_{\text{mag}} = -\tfrac{1}{2}\chi_a(\mathbf{H}.\mathbf{n})^2 \quad (6.45)$$

$$F_{\text{el}} = -\frac{1}{8\pi} \epsilon_a(\mathbf{E}.\mathbf{n})^2, \quad (6.46)$$

where $\chi_a = \chi_\| - \chi_\perp$ and $\epsilon_a = \epsilon_\| - \epsilon_\perp$. In eqns (6.45) and (6.46), only the **n** dependent terms have been retained. One quantitative point must be mentioned at this stage; for cholesterol esters, where no benzene ring is present, the diamagnetic susceptibilities $\chi_\|$ and $\chi_\perp$ are considerably smaller than on conventional nematics: $|\chi_a|$ is of order 10^{-9} c.g.s. units. Also χ_a is usually *negative* and in a magnetic field H, the director tends to be aligned normal to **H**. To obtain cholesterics with *positive* χ_a, the simplest procedure is to dissolve chiral molecules in a conventional nematic like MBBA.

The dielectric anisotropy $\epsilon_a = \epsilon_\| - \epsilon_\perp$ is positive for cholesterol halides, and negative for the aliphatic esters of cholesterol. It is also negative for dilute solutions of chiral molecules in MBBA.

Finally, it is also possible to use a.c. electric fields in these experiments (ϵ_a is then a function of the a.c. frequency): the a.c. regimes are sometimes useful to eliminate charge transport and electrohydrodynamic instabilities (we come back to this point in Section 6.3).

6.2.2.2. Negative anisotropy. A bulk cholesteric sample, with negative χ_a (or ϵ_a) minimizes its energy by putting the helical axis ($\mathbf{q_0}$) along the field; then the director **n** is normal to the field at all points. No distortion energy is required and the helical pitch is independent of the field.

This very simple effect has been observed with a.c. electric fields (of frequency higher than $\sim$1 kHz) in mixtures of MBBA and cholesterol esters [22]; as explained above, a finite frequency is required to eliminate the convective instabilities usually observed in MBBA (see Chapter 5). This experiment allows to transform a polydomain sample into a well-ordered planar texture, and may be quite useful in practice.

6.2.2.3. Positive anisotropy. Let us start with a rather thick cholesteric sample, so that wall effects may be safely neglected. In low fields H (or E),† the helical structure is undistorted. The susceptibility measured along the helical axis is the perpendicular susceptibility $\chi_\perp$; when measured normal to the helical axis it is the average $\frac{1}{2}(\chi_\perp + \chi_\|)$. If $\chi_\| > \chi_\perp$ this average is larger than $\chi_\perp$. The system will tend to adjust in order to display the maximum susceptibility; the helical axis is thus *normal to the applied field*. This is well confirmed experimentally.

Let us now look for internal distortions of the helical structure. The initial situation for low fields is represented in Fig. 6.13a, and for

† We discuss here the case of a magnetic field, all formulae for electric fields are obtained by the substitution $\chi_a H^2 \to \epsilon_a E^2/4\pi$.

intermediate fields in Fig. 6.13b. In regions such as A, A',... the molecules are favourably aligned along the field. In regions such as B, B',... the molecules have an unfavourable orientation with respect to the field. Thus if the field becomes strong enough region A will expand. Region B, on the other hand, cannot contract very much, since this would require too much twist energy. The overall result is an increase of the pitch P with field.

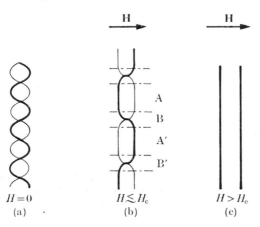

FIG. 6.13. Untwisting of a cholesteric spiral by a field **H**. It is assumed that the molecules tend to lie *along* **H**($\chi_a > 0$).

At somewhat higher fields, this leads to a succession of *180° walls* separating large A regions. Each wall has a finite thickness, of order $2\xi_2(H)$ where ξ_2 is defined as usual:

$$\xi_2 = \left(\frac{K_2}{\chi_a}\right)^{\frac{1}{2}} \frac{1}{H} . \tag{6.47}$$

The distance between walls $L = \frac{1}{2}P(H)$ is now much larger than ξ_2. Finally, at a certain critical field H_c, the walls become infinitely separated ($P \rightarrow \infty$) and we obtain a nematic structure.

This cholesteric–nematic transition has been observed first by Sackmann *et al.* [23] under magnetic fields, and by the Xerox group [24] under electric fields. More detailed studies [24–27] showed that the critical field is inversely proportional to the unperturbed pitch $P(0)$: in practice, to have reasonable critical field values, one always has to work in mixtures of large pitch.

The detailed variation of pitch with field has been measured in a few cases [25] [26] and is shown on Fig. 6.14. At low fields the pitch is

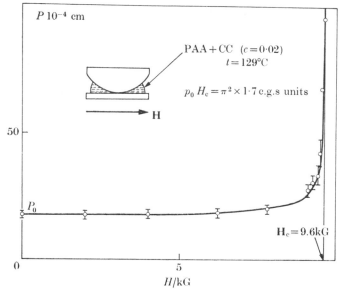

FIG. 6.14. Pitch $P(H)$ as a function of magnetic field for a mixture PAA–cholesterol ester. In this experiment, the pitch was derived from the position of the discontinuities in a Cano wedge (see Fig. 6.20) (after ref. [25]). The continuous curve is theoretical [29].

essentially unperturbed: at higher fields P increases and finally diverges (weakly) at the critical field.

From a theoretical point of view, the transition can be studied rather simply from the continuum free energy (eqns 6.44, 6.45). In particular, the value of the critical field H_c (or E_c) may be derived as follows: for $H \to H_c$ we have well-separated walls, as shown in Fig. (6.13b); the interactions between walls become negligible, and it is enough to study the energy of one single wall in an infinite nematic medium. Assuming always a one dimensional situation of pure twist ($n_x = \cos \theta(z)$, $n_y = \sin \theta(z)$) we find from eqns (6.44) and (6.45) the following equilibrium equation

$$\xi_2^2 \frac{d^2\theta}{dz^2} = \sin \theta \cos \theta.$$

This, as in Chapter 3, has the first integral:

$$\xi_2^2 \left(\frac{d\theta}{dz}\right)^2 = \sin^2\theta, \tag{6.48}$$

ensuring that $d\theta/dz \to 0$ for $\theta \to 0$ or $\theta \to \pi$. The free energy (per unit area) of one wall, compared to the energy the nematic conformation, is then

$$F_{\rm w} = \int \left\{ \tfrac{1}{2} K_2 \left\{ \left(\frac{{\rm d}\theta}{{\rm d}z} - q_0 \right)^2 - q_0^2 \right\} + \tfrac{1}{2} \chi_{\rm a} H^2 \sin^2\theta \right\} {\rm d}z$$

$$\frac{F_{\rm w}}{\chi_{\rm a} H^2} = \int \left\{ \tfrac{1}{2} \xi_2^2 \left(\frac{{\rm d}\theta}{{\rm d}z} \right)^2 - q_0 \xi_2 \frac{{\rm d}\theta}{{\rm d}z} + \tfrac{1}{2} \sin^2\theta \right\}. \tag{6.49}$$

From eqn (6.48) we see that the first and third terms in eqn (6.49) give equal contributions

$$\frac{F_{\rm w}}{\chi_{\rm a} H^2} = \int_0^\pi \xi^2 \left| \frac{{\rm d}\theta}{{\rm d}z} \right| {\rm d}\theta - q_0 \xi_2 \int_0^\pi {\rm d}\theta$$

$$= \int_0^\pi \sin\theta \, {\rm d}\theta - \pi q_0 \xi_2$$

$$= 2 - \pi q_0 \xi_2. \tag{6.50}$$

Thus it becomes unfavourable to have walls when $\xi_2(H) < 2/\pi q_0$. Returning to eqn (6.47) we see that this corresponds to a critical field:

$$H_{\rm c} = \frac{\pi}{2} \left(\frac{K_2}{\chi_{\rm a}} \right)^{\frac{1}{2}} q_0 = \pi^2 \left(\frac{K_2}{\chi_{\rm a}} \right) \frac{1}{P_0}, \tag{6.51}$$

P_0 being the unperturbed pitch. Equation (6.51) was derived independently by Meyer [28] and by the present author [29]. It does show that (for fixed K_2 and $\chi_{\rm a}$) $H_{\rm c}$ is inversely proportional to P_0.

Typically, for a dilute solution of chiral molecules in a conventional nematic, we might choose $K_2 \simeq 10^{-6}$ dynes, $\chi_{\rm a} = 10^{-7}$ c.g.s. units, and $P_0/2 = 10$ μm, corresponding to $H_{\rm c} = 15\ 000$ G.

At fields below critical, the energy of a single wall $F_{\rm w}$ (eqn 6.50) becomes negative; thus walls tend to pile up in the sample, until the (repulsive) interactions between neighbouring walls lead to an equilibrium. These interactions decrease rapidly as a function of the inter-wall distance $L(=P/2)$: they are in fact proportional to $\exp(-L/\xi_2)$. For this reason, as soon as H is below $H_{\rm c}$ we can have a rather large number of walls per unit length: the pitch $P(H)$ is only logarithmically divergent for $H = H_{\rm c}$. As seen on Fig. 6.14, this feature is indeed quite apparent on the experimental data. We shall now list a few experiments giving some information on the distorted state.

(1) *Electron spin resonance line-shape of a solute radical.* The resonance frequency, for a group of spins located in a region where the angle between director and field is θ, is of the form

$$\omega(\theta) = \omega_0 + \omega_1 \cos^2\theta, \tag{6.52}$$

where ω_1 is due to anisotropy in the g-factor, or in the hyperfine interactions of the spin label. From the frequency spread of the absorption line, one can thus derive the distribution law for $\cos^2\theta$ in the distorted structure [30a].

(2) *Optical studies* of the wavelength-dependent transmission for a beam parallel to the helical axis, in the distorted state (Chou, Cheung, and Meyer, *Solid State Comm.*, **2**, 977, 1972).

(3) *Macroscopic magnetic measurements* on the distorted state: in low fields the susceptibility χ is $\frac{1}{2}(\chi_\parallel + \chi_\perp)$. At the critical field the susceptibility is $\chi_\parallel$. The complete $\chi(H)$ curve has been measured by the Bordeaux group [30b].

The transition at $H = H_c$ has some interesting thermodynamic features. It is a second-order transition; the free energy (per cm³) $F(H)$ has a continuous slope at $H = H_c$. However, some hysteresis can be observed in the transition region! For instance, let us start from high fields ($H > H_c$) and assume that we have a good nematic single crystal between two polished glass walls (the easy axis of the wall being parallel to **H**). Then let us decrease progressively; we find that nothing happens at $H = H_c$. For H slightly below H_c, we observe a metastable nematic phase [25]. At a somewhat lower field disclination loops nucleate and the equilibrium pitch is restored. In the metastable regime, as shown by eqn (6.50), it would be favourable to inject some 180° walls in the sample, but the boundary conditions create a barrier opposing wall nucleation.

Our discussion of external field effects with positive anisotropy has been restricted, up to now, to situations of pure twist, with the helical axis normal to the field. As explained at the beginning of this paragraph, this is indeed the usual situation for bulk samples. There are, however, some other possibilities, which we shall quote now.

(1) If the bend constant K_3 happens to be anomalously small ($K_3 < 4K_2/\pi^2$) there should occur, in increasing fields, a transition from the helical conformation (**q** normal to **H**) to a *conical conformation* [28] described by

$$n_y = \cos(qx)\cos\psi$$
$$n_z = \sin(qx)\cos\psi$$
$$n_x = \sin\psi, \qquad\qquad (6.53)$$

where ψ is a function of H only; the 'conical axis' x is parallel to the field. In practice, however, K_3 is larger than K_2 and this conical phase has not been observed.

(2) If the sample is a thin slab, with tangential boundary conditions giving a planer texture, one can apply the field H parallel to the unperturbed helical axis, in spite of the positive anisotropy [31]. In this case, above a certain threshold field a periodic distortion of the cholesteric planes takes place (Fig. 6.17). This possibility was invented by Helfrich, mainly in connection with electric field effects (including charge accumulation) [43]. A more detailed calculation, discussing both magnetic and electric field effects, is due to Hurault [32]. It has been tested by recent experiments of Rondelez *et al.* [31]. For the theory of magnetic distortions, the reader is referred to Section 7.2, where a similar problem is discussed for smectics. The electric case, which is more important in view of practical application, will be discussed later in the present chapter, after the introduction of some useful hydrodynamic concepts.

On the whole, the electro-optic effects, in cholesterics with positive anisotropy, are remarkable. However, they are conveniently observed mainly on materials with large pitch ($P \sim 10\ \mu\text{m}$) and it is not yet clear whether they will find interesting technical applications.

Problem: Construct a coarse-grained version of the continuum theory for cholesterics, applying when the distortions are very gradual in comparison with the pitch. (T. Lubensky and P. G. de Gennes, 1971).

Solution: In the unperturbed state the cholesteric planes are equidistant (interval P_0) and parallel (the unperturbed helical axis will be called the z-axis). In a slightly distorted state, each plane is displaced by an amount $u(\mathbf{r})$ along z; u is a slowly varying function of $\mathbf{r}$. The coarse-grained free energy density F_{cg} must be a function of the gradients of u: the most general form (for small gradients $\nabla u \ll 1$) is

$$F_{\text{cg}} = \tfrac{1}{2}B\left(\frac{\partial u}{\partial z}\right)^2 + \tfrac{1}{2}\tilde{K}\left(\frac{\partial^2 u}{\partial x^2} + \frac{\partial^2 u}{\partial y^2}\right)^2$$
$$+ \tfrac{1}{2}K'\left(\frac{\partial^2 u}{\partial z^2}\right)^2 + K''\frac{\partial^2 u}{\partial z^2}\left(\frac{\partial^2 u}{\partial x^2} + \frac{\partial^2 u}{\partial y^2}\right).$$

There is no term proportional to $(\partial u/\partial x)^2$; such a term would give a non-zero F_{cg} for a uniform rotation around the y-axis. There is also no term of the form

$$\frac{\partial u}{\partial z}\left(\frac{\partial^2 u}{\partial x^2} + \frac{\partial^2 u}{\partial y^2}\right).$$

This term is not compatible with the existence of a twofold axis of symmetry (parallel to the local direction $\mathbf{n}$) in the unperturbed structure. For a Fourier component of wave vector $\mathbf{q}\,(q \ll q_0)$, the K' term if of the order of $\tfrac{1}{2}K'q_z^4 u^2$ and is negligible in comparison with the B term $\tfrac{1}{2}Bq_z^2 u^2$. The K term may also be dropped out for similar reasons. We are then left with *two elastic constants*, E and $\tilde{K}$. F_{cg} may be written in a slightly more general form by introducing a unit vector $\mathbf{d}(\mathbf{r})$

17

normal to the cholesteric planes:

$$F_{cg} = \tfrac{1}{2}B\left(\frac{P}{P_0}-1\right)^2 + \tfrac{1}{2}\tilde{K}(\operatorname{div}\mathbf{d})^2,$$

where P is the local value of the pitch in the distorted structure. The constant E is immediately derived from eqn (6.44) by considering a case where $\partial u/\partial z = $ constant:

$$B = K_2 q_0^2.$$

To obtain $\tilde{K}$ we consider a 'jelly-roll' arrangement of cholesteric planes: in terms of cylindrical coordinates (r, ϕ, z) this corresponds to:

$$n_r = 0$$

$$n_\phi = \cos\theta(r)$$

$$n_z = \sin\theta(r).$$

We impose $\theta(r+P_0) \equiv \theta(r)$ (no change in pitch). The vector $\mathbf{d}$ is parallel to the r axis, and div $\mathbf{d} = 1/r$. Thus

$$F_{cg} = \tfrac{1}{2}\tilde{K}/r^2$$

The local Frank free energy F is derived from eqn (6.43): after some calculation one obtains

$$F = \tfrac{1}{2}K_2\left(\frac{d\theta}{dr} - q_0 - \frac{1}{r}\sin\theta\cos\theta\right)^2 + \tfrac{1}{2}K_3\frac{1}{r^2}\cos^4\theta.$$

We consider F in the limit $q_0 r \gg 1$ (weak distortions). The optimum form of $\theta(r)$ corresponds to:

$$\frac{d\theta}{dr} = q_0 + \frac{1}{r}\sin\theta\cos\theta.$$

In the correction term on the right-hand side we may insert the unperturbed value $\theta = q_0 r + \text{const}$. The terms of order $1/q_0 r$ in θ may then be found by integration, and it may be checked that they are compatible with the periodicity condition $\theta(r+P_0) = \theta(P_0)$.

With this choice of $\theta(r)$ the twist contribution drops out. The bend contribution to the coarse-grained average may be obtained from the unperturbed form of θ; the average over angles of $\cos^4\theta$ is equal to $\tfrac{3}{8}$ and we obtain:

$$\tilde{K} = \tfrac{3}{8}K_3.$$

6.3. Dynamical properties

Cholesterics are liquids, and very similar in their local structure to nematics; here too, we have some remarkable couplings between orientation and flow. The fundamental mechanical equations which describe these couplings have again been discussed by Leslie [33]. For the usual case of very weak twists on the molecular scale, negligible compressibility, and uniform temperature, the equations for cholesterics and nematics are *identical*. The entropy source is always given by eqn

(6.21) of Chapter 5, in terms of the viscous stress $\sigma'_{\alpha\beta}$, the molecular field h_α, the shear rate tensor $A_{\alpha\beta}$ and the relative rotation velocity N_α of the director. The torque balance eqn (5.17) is also maintained, and the relations between fluxes $(A_{\alpha\beta}, N_\alpha)$ and forces $(\sigma'_{\alpha\beta}, h_\alpha)$ retain the form (eqns 5.31, 5.32): they involve five independent coefficients with the dimension of a viscosity. The only difference is that the molecular field **h** must now be derived from the form eqn (6.43) of the free energy.

However, in spite of these formal similarities, the physical flow effects and orientational effects are much more complex on a helical structure: in particular, as we shall see, the *apparent* bulk viscosity of a cholesteric sample may often be 10^5 times larger than the friction coefficients defined in the Leslie equations! Also, from an experimental point of view, it is often difficult to produce a situation where both the flow and the texture properties are adequately controlled; we shall restrict our attention here to a few situations of this class, where experiments have been done, or at least look feasible.

6.3.1. Studies on small motions in a planar texture

6.3.1.1. *Pulsed external fields.* Consider for instance a cholesteric with positive χ_a, subjected to a time-dependent magnetic field $H(t)$ normal to the helical axis (z). The helix will distort with a certain time lag. For the example chosen, the only deformation involved is twist, and it may be shown that there is no backflow. Theoretical calculations for small amplitude motions which are relevant to the present problem, can be found in ref. [34]. Typical relaxation rates are of order $K_2 q_0^2 / \gamma_1$, but the details of the relaxation depend on the average field. Experiments of this type are currently under way [35], and should provide a good measurement of γ_1.

6.3.1.2. *Ultrasonic attenuation of shear waves.* The principle has been discussed in Chapter 5 in connection with nematics. A theoretical analysis for the planar texture of cholesterics has been given by F. Brochard [36a]; the regimes depend critically on the ratio between the acoustic penetration depth δ and the half-pitch $P/2$. To vary δ requires acoustic equipment working on a broad range of frequencies. It is simpler in practice to vary P, using for instance a mixture of MBBA and (dilute) cholesterol esters. The experiment gives two friction coefficients: they are in reasonable agreement with other data for MBBA [36b].

6.3.1.3.   Inelastic scattering of light.   No quantitative data on light
scattering by cholesterics are available up to now, but the situation
should soon improve. Both the intensity [37] and the frequency
spectrum [34] of the scattered light have been analysed theoretically, at
least for one case : namely for scattering wave vectors **k** which are
parallel to the helical axis.

For each **k**, there are two orientational modes which are strongly
coupled to the light. For **k** parallel to $\mathbf{q}_0$, these modes are very simple.
One mode corresponds to *pure twist* : putting the helical axis along z, it
may be described as follows

$$n_x = \cos(q_0 z + u) \simeq n_x^0 - u \sin(q_0 z)$$
$$n_y = \sin(q_0 z + u) \simeq n_y^0 + u \cos(q_0 z) \qquad (6.54)$$
$$n_z = 0$$

with $u = u_0 e^{ilz}$. This type of twist deformation contributes only to one
component of the dielectric tensor

$$\epsilon_{x,y} = \epsilon_a\{n_x^0 \,\delta n_y + n_y^0 \,\delta n_x\} = \epsilon_a \cos(2q_0 z)u_0 e^{ilz}. \qquad (6.55)$$

Thus it is associated with a scattering wave vector

$$k = l \pm 2q_0. \qquad (6.56)$$

The thermal square amplitude of u is derived from eqn (6.44) and has
the form

$$\langle |u_0|^2 \rangle = \frac{k_B T}{K_2 l^2} \qquad (6.57)$$

it diverges for $l \to 0$, or equivalently, for $k \to \pm 2q_0$ (i.e. in the vicinity
of the Bragg peaks). The corresponding relaxation rate is

$$\frac{1}{\tau_{lt}} = \frac{K_2 l^2}{\gamma_1}. \qquad (6.58)$$

The second mode, detected with **k** along q_0, is what we might call the
'umbrella' mode, defined by

$$n_x = \cos(q_0 z)\cos v \sim n_x^0$$
$$n_y = \sin(q_0 z) \cos v \sim n_y^0$$
$$n_z = \sin v \sim v, \qquad (6.59)$$

with $v = v_0 e^{ilz}$. This mode contributes to two components of the
dielectric tensor

$$\epsilon_{x,z} \simeq \epsilon_a n_x^0 \,\delta n_z = \epsilon_a \cos(q_0 z)v_0 \, e^{ilz}$$
$$\epsilon_{y,z} \simeq \epsilon_a n_y^0 \,\delta n_z = \epsilon_a \sin(q_0 z)v_0 \, e^{ilz}.$$

We see that the scattering vector is now

$$k = l \pm q_0. \tag{6.60}$$

The thermal square amplitude of v is found to be

$$\langle |v_0|^2 \rangle = \frac{k_B T}{K_3 q_0^2 + K_1 l^2}. \tag{6.61}$$

It is maximum (but still finite) for $l = 0$ or $k = \pm q_0$, i.e. midway from the Bragg peaks. The relaxation rate for the umbrella mode is [34]

$$\frac{1}{\tau_{lu}} = \frac{(\alpha_2 + \alpha_4 + \alpha_5)(K_3 q_0^2 + K_1 l^2)}{\gamma_1(\alpha_4 + \alpha_5) - \gamma_2 \alpha_2}. \tag{6.62}$$

As is often found, the relaxation rate is minimal at the l value for which the intensity is maximal ($l = 0$). On the whole, we see that studies on the line width for $\mathbf{k}$ parallel to $\mathbf{q_0}$ will give two relations on the Leslie coefficients (eqns 6.58 and 6.62).

6.3.2. Macroscopic flow

6.3.2.1. Apparent viscosities and permeation. A typical viscosity measurement in ordinary fluids make use of a capillary (of radius R), the pressure drop per unit length of the capillary p' is then related to the mass flow Q and to the viscosity η by the Poiseuille law [32]:

$$p' = \frac{8}{\pi} \frac{Q}{\rho R^4} \eta \tag{6.63}$$

(ρ = fluid density).

Measurements of this type (or with slightly more complicated geometries) have also been carried out on (unoriented) samples of cholesterics [39]. Unfortunately the data have also always been taken with one same apparatus of fixed geometrical size (i.e. fixed R). Thus it has never been proven that the Poiseuille law holds for cholesterics—in fact, as we shall see, there are strong reasons to believe that it does *not*. However, it is customary to state the results of each measurement in terms of an *apparent viscosity* η_{app}: for instance, with a capillary, we would put

$$\eta_{app} = \frac{\pi \rho R^4 p'}{8Q}. \tag{6.64}$$

Typical data on η_{app} are shown on Fig. 6.16. Note the following features:
   (1) η_{app} is considerably larger in the cholesteric phase than in the isotropic phase (the ratio may go up to 10^6!)

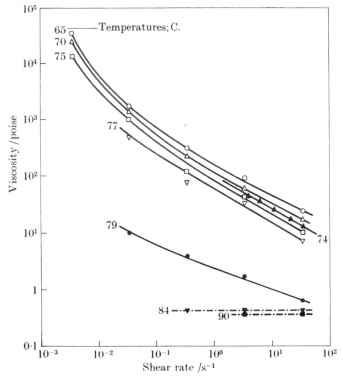

FIG. 6.15. Apparent viscosity of cholesterol myristate as a function of shear rate (after K. Sakamoto and R. S. Porter, *Mol. Cryst.*, **8**, 443 (1969)).

(2) η_{app} is very sensitive to the shear rate: the viscosity is 'non-newtonian'.

These properties have no counterpart in oriented nematic preparations; clearly they must be associated with specific properties of the helical phase. Their explanation has been given in a brilliant paper by W. Helfrich [40]. He considers a cholesteric planar texture which is assumed to be *blocked* (i.e.: spatially immobile) because of certain anchoring effects at the walls, and assumes a uniform flow velocity $\mathbf{v}$, parallel to the helical axis $\mathbf{q}_0$. The dissipation per unit volume in this situation is easily derived from the general eqn 5.21 for the entropy source

$$T\dot{S} = (\boldsymbol{\sigma}:\mathbf{A}) + \mathbf{h}.\mathbf{N}. \tag{6.65}$$

For a uniform $\mathbf{v}$, the shear rate tensor $\mathbf{A}$ drops out. On the other hand, we must remember that

$$\mathbf{N} = \frac{d\mathbf{n}}{dt} - \boldsymbol{\omega} \times \mathbf{n}$$

involves the *total* derivative of the director **n**, as experienced by a flowing molecule. In the present case, we have

$$\frac{d\mathbf{n}}{dt} = v\frac{\partial \mathbf{n}}{\partial z} = \boldsymbol{\Omega} \times \mathbf{n} \tag{6.66}$$

where z is an axis parallel to **v**, and:

$$\boldsymbol{\Omega} = v\mathbf{q}_0. \tag{6.67}$$

Also, since v is uniform, the local rotation velocity $\boldsymbol{\omega} = \frac{1}{2}$ curl **v** vanishes. Finally we may write the entropy source in the form:

$$T\dot{S} = \boldsymbol{\Gamma}.\boldsymbol{\Omega}, \tag{6.68}$$

where the rotation velocity $\boldsymbol{\Omega}$ is given by (6.67), and the torque $\boldsymbol{\Gamma} = \mathbf{n} \times \mathbf{h}$ is given by eqn (5.32):

$$\boldsymbol{\Gamma} = \mathbf{n} \times (\gamma_1 \mathbf{N} + \gamma_2 \mathbf{A}.\mathbf{n})$$

$$= \gamma_1 \boldsymbol{\Omega}. \tag{6.69}$$

Inserting this into eqn (6.68), and equating the result to the work done by the pressure gradient p', we get

$$p'v - \gamma_1 q_0^2 v^2$$

$$p' = \gamma_1 q_0^2 v. \tag{6.70}$$

Eqn (6.70) corresponds to an apparent viscosity

$$\eta_{\mathrm{app}} = \tfrac{1}{8}\gamma_1(q_0 R)^2. \tag{6.71}$$

Typical capillary radii R used in viscosity measurements are of the order of 300 μm while $q_0 = 2\pi/P$ is of order 10^5 cm^{-1}. Thus $\eta_{\mathrm{app}}/\gamma_1$ may reach values of order 10^6; the dissipative process is not, as usual, a friction between neighbouring fluid regions flowing at slightly different velocities, but rather a friction between the individual molecules (in uniform flow) and a blocked cholesteric texture. The word 'permeation' has been introduced in this connection by Helfrich [40].

Of course, in practice, there is no reason for **v** to be exactly parallel to $\mathbf{q}_0$ in an unoriented sample; the details of the cholesteric texture achieved in the capillary will influence η_{app}. Conversely, the flow will react on the orientation of the domains, and this is probably the source of the non-newtonian properties which are observed.

The Helfrich assumption ($v =$ constant in all the capillary section) may appear surprising at first sight, since, at the capillary walls, v must, in fact, vanish. However, we shall now show that the effects of this boundary condition are important only up to a small distance (of order P) from the walls; if, as is usual, the capillary is much larger than P in diameter, the Helfrich assumption is correct.

For simplicity, we shall discuss this not for a three dimensional problem with a circular capillary, but rather for a two dimensional problem of flow between two walls separated by a gap $2R$. Qualitatively, we may write the dissipation in the form

$$T\dot{S} = \gamma_1 q_0^2 v^2 + \bar{\eta}\left(\frac{\partial v}{\partial x}\right)^2 \qquad (6.72)$$

where the x-axis is normal to the walls, while the velocity v is along the z-axis (in the plane of the walls). The viscosity $\bar{\eta}$ is a certain average of the Leslie coefficients. The force (per unit volume) along the z direction is then

$$-p' + \gamma_1 q_0^2 v - \bar{\eta}\,\frac{\partial^2 v}{\partial x^2} = 0. \qquad (6.73)$$

This equation for v, plus the boundary condition $v(x = \pm R) = 0$, specifies the problem entirely. The solution is

$$v = \frac{p'}{\gamma_1 q_0^2}\left\{1 - \frac{\cos h\ \kappa x}{\cos h\ \kappa R}\right\} \qquad (6.74)$$

where

$$\kappa^2 = \frac{\gamma_1}{\bar{\eta}}\,q_0^2. \qquad (6.75)$$

Equation (6.74) does show that the Helfrich value $v = p'/\gamma_1 q_0^2$ is obtained everywhere, except in a small region of thickness κ^{-1} near the walls. Since we expect $\bar{\eta}$ and γ_1 to be of comparable magnitude, we see from eqn (6.75) that κ^{-1} is roughly equal to $P/2\pi$: q.e.d.

6.3.2.2. *Future experiments*. What are the experiments suggested by the Helfrich idea? First one should check that the ratio p'/v is independent of R (provided that $\kappa R \gg 1$). To pin strongly the cholesteric planes, one might substitute *optical gratings* for the walls (with a grating interval comparable to P).

Clearly, we need measurements on single domain textures; this is hard to realize, but not entirely impossible, especially in cholesterics with negative local dielectric anisotropy ($\epsilon_\| < \epsilon_\perp$): as explained in Section 6.2, it is possible in principle to induce a planar texture in such a case by suitable low frequency a.c. electric fields [41]. It should then be possible to study viscous frictions in two very different situations:

(1) when the flow velocity is parallel to $\mathbf{q_0}$, where the Helfrich permeation dominates (if the texture is anchored); and

(2) when the flow is normal to $\mathbf{q_0}$; in this case the effective viscosity should be comparable in magnitude to the Leslie coefficients. This second case has been analysed theoretically by Leslie [33]. His calculation allows for local distortions of the helical pattern: these distortions turn out to be of order

$$\frac{\eta}{K q_0^2} A,$$

where A is the shear rate. They could lead to some amusing mechano-optical effects.

Finally, the possible role of *dislocations* in the non-newtonian viscosity should be studied (in analogy with the plastic properties of solid crystals).

6.3.3. Convective instabilities

A remarkable electro-optic effect has been observed in nematic–cholesteric mixtures by Hellmeier and his coworkers [42]. A thin slab, with the conventional planar texture, is driven by a static field E_0 parallel to the helical axis. When E_0 exceeds a certain threshold E_c, the texture breaks up into small domains (of typical size 10 μm) which have different orientations (different $\mathbf{q_0}$) and give rise to a strong scattering of light.

If the field E_0 is turned off, the domains persist (the planar texture 'heals' very slowly): for this reason the process is referred to as the '*storage mode*' of cholesterics. However it is possible to erase the information (i.e., to accelerate the recovery of the planar texture) by an a.c. field (of frequency in the kilohertz range): this last effect is based simply on the dielectric anisotropy and has been discussed in Section 6.2.2.2. It requires $\epsilon_\perp > \epsilon_\parallel$.

How do we explain the onset of the polydomain texture in increasing fields E_0? The initial model proposed by Hellmeier assumed that the field produced an emulsion of cholesteric particles in a nematic matrix. Two observations apparently supported this idea:

(1) no effect was observed on pure cholesterol derivatives; and

(2) the slow recovery of the planar texture is ruled by a kinetic law reminiscent of the Smoluchovski equation for the association of droplets.

However these facts were slightly misleading. We now tend to believe that the Carr–Helfrich process of charge accumulation governs the storage mode: nematic–cholesteric mixtures are helpful mainly because the nematic component is usually of low purity and high conductance (while pure cholesterol derivatives are better insulators). Also the second property (2) may occur with many textural transformations independently of a phase separation.

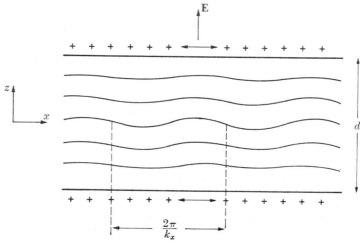

FIG. 6.16. The Helfrich distortion mode for a cholesteric planar texture, under an electric field ($\epsilon_a > 0$). The square-lattice distortion of Fig. 6.17 is the superposition of two such distortions, oriented at right angles (along x and y).

An elegant interpretation of the instability has been worked out by Helfrich [43]. The material is described by two conductances $\sigma_{\parallel}$ (along the local director **n**) and $\sigma_{\perp}$ (normal to **n**). The case of interest here is $\sigma_{\parallel} > \sigma_{\perp}$. Similarly there are two dielectric constants $\epsilon_{\parallel}$ and $\epsilon_{\perp}$. We start with a planar texture and investigate the effects of a small distortion, as shown on Fig. 6.17. The spatial wavelengths involved are assumed to be much larger than the pitch. Then we can use a coarse-grained description, as in the problem at the end of Section 6.2. We take as our variable the displacement $u(\mathbf{r})$ of the cholesteric planes (the z-axis is put parallel to the unperturbed helical axis, and u is measured along z). The type of distortion discussed by Helfrich corresponds to

$$u = u_0 \exp(ik_x x)\sin(k_z z), \tag{6.76}$$

with $k_z = \pi/D$. This choice corresponds to $u = 0$ on both glass plates ($z = 0$ and $z = D$), D being the sample thickness. The wave vector k_x

will be chosen later in order to maximize the instability. The local state of affairs may also be described by a unit vector **d** normal to the layers. Here we have

$$d_z \simeq 1$$

$$d_x \simeq -\frac{\partial u}{\partial x}. \tag{6.77}$$

At this coarse-grained level, the cholesteric medium is uniaxial with conductances

$$\sigma_{\|\mathbf{d}} = \sigma_\perp \qquad \text{(along } \mathbf{d}\text{)}$$

$$\sigma_{\perp\mathbf{d}} = \tfrac{1}{2}(\sigma_\| + \sigma_\perp) \qquad \text{(normal to } \mathbf{d}\text{)}. \tag{6.78}$$

Note that $\sigma_{\perp\mathbf{d}} > \sigma_{\|\mathbf{d}}$ for the case of interest ($\sigma_\| > \sigma_\perp$). We may write, in dyadic notation:

$$\boldsymbol{\sigma} = \sigma_{\perp\mathbf{d}} + (\sigma_{\|\mathbf{d}} - \sigma_{\perp\mathbf{d}})\mathbf{d}:\mathbf{d} \tag{6.79}$$

Similar equations hold for the dielectric tensor.

Let us now write down the current **J** present in the distorted structure. **J** is due to the total field E, the sum of the external field E_0 plus the fields caused by the Carr-Helfrich charges

$$\mathbf{J} = \boldsymbol{\sigma} \cdot \mathbf{E} \tag{6.80}$$

Both vectors **E** and **d** have small components along x. Treating these components to first order we find that J_z is a constant, while

$$J_x = \sigma_{\perp\mathbf{d}} E_x + (\sigma_{\|\mathbf{d}} - \sigma_{\perp\mathbf{d}}) E_0 d_x. \tag{6.81}$$

The condition of charge conservation div $\mathbf{J} = 0$ leads to $(\partial/\partial x)J_x = 0$ or finally to $J_x = 0$. This fixes the lateral field component

$$E_x = +E_0 d_x \frac{\sigma_{\perp\mathbf{d}} - \sigma_{\|\mathbf{d}}}{\sigma_{\perp\mathbf{d}}} = -E_0 \frac{\partial u}{\partial x} \frac{\sigma_{\perp\mathbf{d}} - \sigma_{\|\mathbf{d}}}{\sigma_{\perp\mathbf{d}}}. \tag{6.82}$$

Knowing the field distribution we can derive the corresponding charge densities:

(1) The density of the mobile carriers ρ_c is given by

$$\rho_c = \frac{1}{4\pi} \operatorname{div}(\boldsymbol{\varepsilon} \cdot \mathbf{E}) = \frac{\epsilon_{\perp\mathbf{d}} E_0}{4\pi} \frac{\partial^2 u}{\partial x^2} \left\{ \frac{\sigma_{\|\mathbf{d}}}{\sigma_{\perp\mathbf{d}}} - \frac{\epsilon_{\|\mathbf{d}}}{\epsilon_{\perp\mathbf{d}}} \right\}. \tag{6.83}$$

(2) The total charge density (mobile carriers plus dielectric polarization) is

$$\rho = \frac{1}{4\pi} \operatorname{div} \mathbf{E} = -\frac{E_c}{4\pi} \frac{\partial^2 u}{\partial x^2} \frac{\sigma_{\perp\mathbf{d}} - \sigma_{\|\mathbf{d}}}{\sigma_{\perp\mathbf{d}}}. \tag{6.84}$$

The vertical electric force due to the mobile carriers is

$$\phi_c = \rho_c E_0. \tag{6.85}$$

This is not the total electric force. There is another contribution ϕ_d, due to the following fact: the distortion of the cholesteric structure (described by u) imposes some changes on the dielectric tensor, and thus on the electrostatic energy. We can find ϕ_d from the dielectric torque Γ_d (directed along the y axis) acting on the molecules

$$\Gamma d = \frac{\epsilon_{\|d} - \epsilon_{\perp d}}{4\pi} E_0^2 \left(\frac{E_x}{E_0} - d_x \right). \tag{6.86}$$

In an infinitesimal transformation $u(x) \to u(x) + \delta u(x)$ the work done (per unit length along z) is

$$\int \Gamma_d \, \delta d_x \, dx = - \int \Gamma_d \frac{\partial}{\partial x} \delta u(x) \, dx$$

$$= \int \frac{\partial \Gamma_d}{\partial x} \delta u(x) \, dx.$$

This means that

$$\phi_d = \frac{\partial \Gamma_d}{\partial x} = \frac{\epsilon_{\|d} - \epsilon_{\perp d}}{4\pi} E_0^2 \frac{\sigma_{\|d}}{\sigma_{\perp d}} \frac{\partial^2 u}{\partial x^2} \tag{6.87}$$

(where we have made use of eqn 6.82). The total electric force acting on the layers is

$$\phi_z = \phi_c + \phi_d = \frac{\epsilon_\|}{4\pi} \frac{\sigma_\| - \sigma_\perp}{\sigma_\perp} E_0^2 \frac{\partial^2 u}{\partial x^2}. \tag{6.88}$$

It is interesting (although not obvious) to note that ϕ_z is the product of the total charge density ρ and the displacement field $D_0 = \epsilon_{\|d} E_0$.

We must now balance ϕ_z against an elastic restoring force ϕ_{el}. We derive ϕ_{el} from the results of the problem on coarse-grained continuum theory (see end of Section 6.2). The elastic energy density is

$$F_{cg} = \tfrac{1}{2} K_2 q_0^2 k_z^2 u^2 + \tfrac{1}{2} \tilde{K} k_x^4 u^2, \tag{6.89}$$

where $\tilde{K} = \tfrac{3}{8} K_3$.
Thus

$$\phi_{el.z} = -(K_2 q_0^2 k_z^2 + \tilde{K} k_x^4) u \tag{6.90}$$

The condition defining the threshold is

$$\phi_z = -\phi_{el.z}. \tag{6.91}$$

The ratio

$$-\frac{\phi_{el.z}}{\phi_z} = (K_2 q_0^2 k_z^2 / k_x^2 + \tilde{K} k_x^2) \left(\frac{4\pi}{\epsilon_{\|d} E_0^2} \right) \left(\frac{\sigma_{\perp d}}{\sigma_{\perp d} - \sigma_{\|d}} \right) \tag{6.92}$$

is minimum when

$$k_x^2 = q_0 k_z (K_2 / \tilde{K})^{\frac{1}{2}}. \tag{6.93}$$

This defines the wavelength of the perturbation which becomes unstable first in increasing fields. Since K_2 and $\tilde{K}$ are comparable in magnitude, this wavelength is proportional to the geometric mean of the pitch $P = 2\pi/q_0$ and of the sample thickness $D = \pi/k_z$. Our analysis requires $k_z \ll q_0$; this condition is indeed satisfied if $D \gg P$. The Helfrich formula for the threshold field is then

$$E_c^2 = \frac{8\pi}{\epsilon_{\|d}} \left(\frac{\sigma_{\perp d}}{\sigma_{\perp d} - \sigma_{\|d}} \right) (K_2 \tilde{K})^{\frac{1}{2}} q_0 \frac{\pi}{D}. \qquad (6.94)$$

Thus the threshold field is proportional to $(PD)^{-\frac{1}{2}}$. In the original Hellmeier experiments, the pitch P was small. The voltage DE_c was high, and, most important, the periodicity of the spatial pattern, $(PD)^{\frac{1}{2}}$, was too small for direct optical observations. More recent experiments have used nematic–cholesteric mixtures, where P is large and the instability can be studied more accurately [44]. As shown on Fig. 6.17 (see between pp. 148–9), above the threshold, the undulations build up a square pattern (the superposition of two undulation waves at right angles). The dependence of E_c on both P and D, and also the dependence of the spatial wavelength on D (eqn 6.93) have been verified in detail [44]: this gives a very strong confirmation of the Helfrich idea.

6.3.4. Torques induced by a heat flux

In any chiral liquid the symmetry of the physical laws is unusual, as stressed in particular by Pomeau [45]: for instance an electric field E may induce a magnetic moment $\mathbf{M} = \alpha\mathbf{E}$! But these effects are usually very small. In a cholesteric, however, because of the long-range helical order, the orders of magnitude are much more favourable. An interesting class of such effects has been analysed by Leslie [46] and will be briefly summarized here.

6.3.4.1. Transport equations.

Let us consider a certain transport current $\mathbf{J}$, which may describe an electric current, a heat current, or a diffusion current. With this current is associated a conjugate field $\mathbf{E}$: for the three cases discussed above we would have respectively

$$\mathbf{E} = -\nabla V \qquad (V = \text{electrical potential})$$

$$\mathbf{E} = -\frac{\nabla T}{T} \qquad\qquad\qquad\qquad (6.95)$$

$$\mathbf{E} = -\nabla\mu \qquad (\mu = \text{chemical potential of the diffusing species}).$$

The entropy source including flow, rotation of the director, and transport is of the form

$$T\dot{S} = \mathbf{A}{:}\boldsymbol{\sigma}' + \mathbf{h}{.}\mathbf{N} + \mathbf{J}{.}\mathbf{E}. \tag{6.96}$$

We shall define as fluxes the quantities $\mathbf{A}$, $\mathbf{N}$, and E and as forces $\boldsymbol{\sigma}'$, $\mathbf{h}$, and $\mathbf{J}$. This choice is somewhat unsymmetrical because E is even under time reversal, while $\mathbf{A}$ and $\mathbf{N}$ are odd. But it is convenient in practice.

The phenomenological equations between fluxes and forces may then be written in the form

$$\sigma'_{\alpha\beta} = \sigma^H_{\alpha\beta} + \mu_1 n_\alpha (\mathbf{E}\times\mathbf{n})_\beta + \mu_2 n_\beta (\mathbf{E}\times\mathbf{n})_\alpha \tag{6.97}$$

$$\mathbf{h} = h^H + \nu\mathbf{n}\times\mathbf{E} \tag{6.98}$$

$$\mathbf{J} = \sigma_\perp E + (\sigma_\| - \sigma_\perp)(\mathbf{n}{.}\mathbf{E})\mathbf{n} + \nu\mathbf{n}\times\mathbf{N} - (\mu_1 + \mu_2)\mathbf{n}\times(\mathbf{A}{:}\mathbf{n}). \tag{6.99}$$

In these equations $\boldsymbol{\sigma}^H$ and $\mathbf{h}^H$ are the standard contributions from nematodynamics† (they are linear functions of $\mathbf{A}$ and $\mathbf{N}$ given explicitly in eqns (5.31) and (5.32)) and $\sigma_\perp$ and $\sigma_\|$ are the usual conductivities. There are three new coefficients μ_1, μ_2, and ν. Note the signs occurring in eqn (6.99); they express the fact that $\mathbf{J}$ is odd in time reversal, while $\boldsymbol{\sigma}'$ and h are even. Finally the torque balance condition gives

$$\Gamma_z = \sigma'_{yx} - \sigma'_{xy} = (\mathbf{n}\times\mathbf{h})_z$$

$$\Gamma = \mathbf{n}\times(\gamma_1\mathbf{N} + \gamma_2\mathbf{A}{.}\mathbf{n}) - \nu\mathbf{E}_\perp \tag{6.100}$$

where $\mathbf{E}_\perp$ is the component of $\mathbf{E}$ normal to $\mathbf{n}$, and ν is related to μ_1 and μ_2 by the identity

$$\nu = \mu_1 - \mu_2. \tag{6.101}$$

Equation (6.100) shows that a field $\mathbf{E}$ (which is a polar vector) may induce a torque (an axial vector) in a cholesteric: this is possible only because the cholesteric differs from its mirror image. Some consequences of the cross terms have been discussed by Leslie (ref. 46, and *Proceedings 5th Faraday Symp.* 1972) and by Prost (*Solid St. Commun.* **11**, 183, 1972). We shall select here one of them.

6.3.4.2. The Lehmann effect. A cholesteric droplet, when submitted to a thermal gradient parallel (or nearly parallel) to its helical axis, shows a uniform rotation of the local molecular axes. This effect was observed by Lehmann [47] and discussed later by Oseen [48]. It can be understood from the Leslie eqns (6.97), (6.98), (6.99). Consider for simplicity a flat slab with planar texture (helical axis parallel to the normal Oz to the slab) under a field $E = -\nabla T/T$ parallel to Oz. It is

† The symbol H stands for 'hydrodynamics'.

easily verified that in this situation there is no hydrodynamic flow ($\mathbf{A} \equiv 0$). The equation for the angle $\varphi(z)$ between the director and a fixed axis (x) in the plane of the slab is

$$\gamma_1 \frac{\partial \varphi}{\partial t} = K_2 \frac{\partial^2 \varphi}{\partial z^2} - \nu E. \tag{6.102}$$

The boundary conditions depend on the nature of the two limiting surfaces:

(1) if we have a freely-suspended film (a situation which is probably not too different from the Lehmann droplets) the condition is one of zero surface torque: this may be shown to give

$$\left.\frac{\partial \varphi}{\partial z}\right|_{z=0} = \left.\frac{\partial \varphi}{\partial z}\right|_{z=D} = q_0, \tag{6.103}$$

where the sample surfaces are located at $z = 0$ and $z = D$. Then the solution of (6.102) is

$$\varphi = q_0 z - \frac{\nu E}{\gamma_1} + \text{constant}, \tag{6.104}$$

and the molecules rotate at a uniform speed $\nu E/\gamma_1$.

(2) if we have one anchored surface (at $z = 0$) and one free surface (at $z = D$) the boundary conditions may be taken as

$$\varphi(0) = 0$$
$$\left.\frac{\partial \varphi}{\partial z}\right|_D = q_0 \tag{6.105}$$

and the solution has the time independent form

$$\varphi = qz + \frac{1}{2}\frac{\nu E}{K_2} z^2.$$

Here q is slightly different from the unperturbed value q_0. To satisfy eqn (6.105) we require

$$q = q_0 - \frac{\nu E}{K_2} D.$$

The quantity which would be most easily measured in such an experiment is the angle at the free surface

$$\varphi(D) = q_0 D - \frac{1}{2}\frac{\nu E}{K_2} D^2. \tag{6.106}$$

Let us guess the order of magnitude of ν by a dimensional argument. When $E = -\nabla T/T$, ν has the dimension of energy per unit area.

Furthermore ν must vanish when q_0 vanishes, and must be odd in q_0. This suggests

$$\nu = xK_2q_0 \tag{6.107}$$

where x is an unknown numerical coefficient. The estimate (6.107) gives

$$\frac{\varphi(D, E) - \varphi(D, O)}{q_0 D} \sim x \frac{T(D) - T(0)}{T}.$$

Thus the distortion due to the heat gradient may be quite sizeable. Of course the experiment should be done with a suitable mixture where the pitch is essentially independent of temperature ($dq_0/dT = 0$).

6.4. Textures and defects in cholesterics

The ideal spiral arrangement which we discussed up to now is easily distorted into more complicated textures: a lucid review of the main textures is given in the review article by G. Friedel [49]. Associated with the textures are various singular lines. The structure of these lines is often amazingly complex; in the present section, we shall present only the simplest types of textures and of lines.

6.4.1. Textures

6.4.1.1. Focal conics. If a common cholesteric liquid is cooled from the isotropic phase, between two glass plates, one does not obtain usually the simple planar texture (with helical axis normal to the plates); what is found is a different arrangement, where the planes of equal phase are distorted into curved surfaces (Fig. 6.18; see between pp. 148–9).

This texture is most easily obtained in rather thick samples, using cholesteric–nematic mixtures with pitches P in the 5 μm range. It is essentially identical to the 'focal conic' texture of smectics A, which will be described in more detail in Chapter 7. Both in cholesterics and in smectics we have lamellar structures, which can be deformed easily provided that the thickness of the layers is not altered. Of course this thickness is much larger in cholesterics ($\sim 1\ \mu$m) than in smectics (~ 20 Å) but the geometrical consequences are similar. A detailed discussion of these textures, exhibiting clearly the similarities and differences with smectics A, has been given by Bouligand (ref. below Fig. 6.18). The similarity in textures between smectics A and cholesterics was a source of confusion in the early literature. One of the great achievements of G. Friedel was to show that the similarity is restricted to certain macroscopic features, and that cholesterics resemble nematics much more than smectics on the molecular scale.

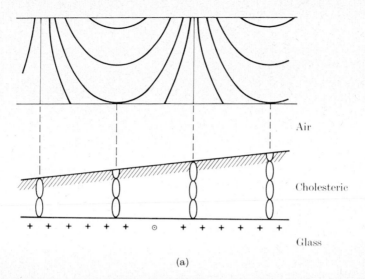

Air

Cholesteric

Glass

(a)

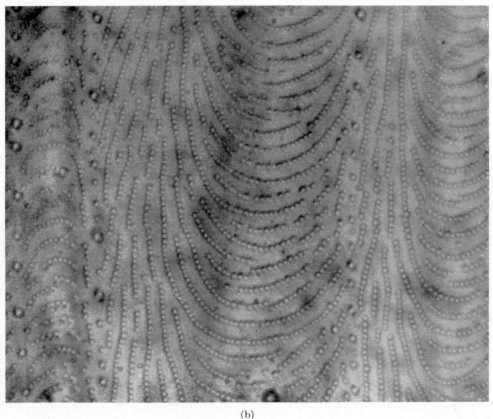

(b)

FIG. 6.19. (a) Molecular arrangement in a cholesteric droplet with tangential boundary conditions. The lower part of the Figure shows a small part of the droplet, where the free surface makes a nearly constant angle with the horizontal plane. The upper part of the Figure shows the arrangements of the molecules at the free surface, as seen from above the droplet. The molecules at the free surface are disposed in 'arceaux'. (b) Display of the arceaux by a decoration technique [after P. Cladis, M. Kleman, P. Pieranski, *C.r. Acad. Sci.* (*Paris*), **273**, 275 (1971)].

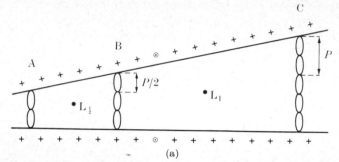

(a)

FIG. 6.20. (a) The Cano wedge. Region A with two half terms is separated from region B (three half turns) by a line $L_{\frac{1}{2}}$ (of strength $\frac{1}{2}$). Region B is separated from region C (five half-turns) by a line L_1 (strength 1). (b) Aspect of the lines in the wedge between a cylindrical lens and a flat plate. (Courtesy Orsay group). The image refers to a field-free situation. The thin lines are of strength $\frac{1}{2}$. The thick ones are of strength one. (c) On the lower image a field $H \sim 8000$ Gauss has been applied normal to the lines. The lines $L_{\frac{1}{2}}$ are unaffected. The lines L_1 are deformed into zig-zags.

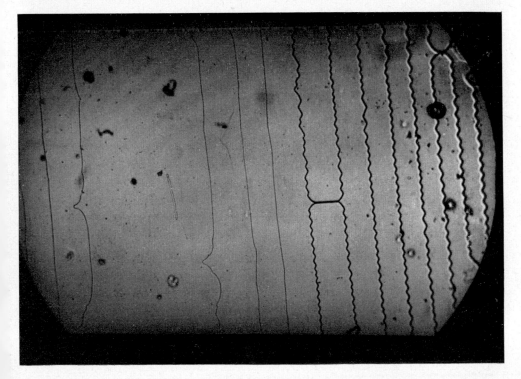

Fig. 6.20 (b)

Fig. 6.20 (c)

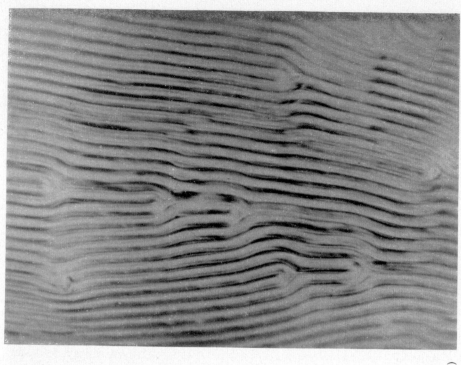

(b)

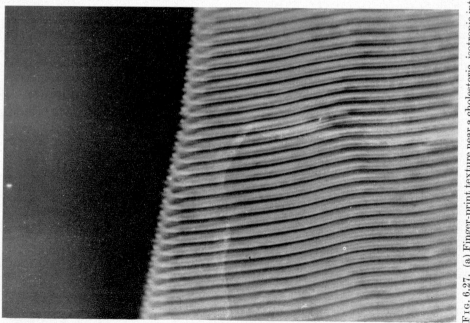

(a)

Fɪɢ. 6.27. (a) Finger-print texture near a cholesteric–isotropic interface. (b) Defects in a finger-print texture. Note that all the defects shown involve λ lines only (no core energy) (courtesy P. Cladis and J. Rault).

FIG. 7.2. Typical focal conics. The ellipses are just at the interface between the smectic sample and the cover glass. (The image is found on this plane.) The hyperbolas are normal to the plane of the sheet. (Courtesy C. Williams). In the present example, the various ellipses are grouped in a superstructure (polygons): for a discussion of this superstructure, see ref. 3.

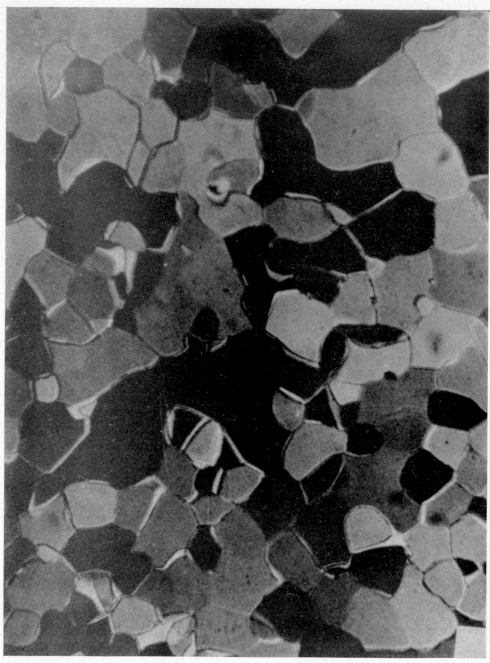

FIG. 7.5. Typical mosaic texture of a smectic B.
(after J. Doucet, *These 3ᵉ Cycle*, Orsay 1972)

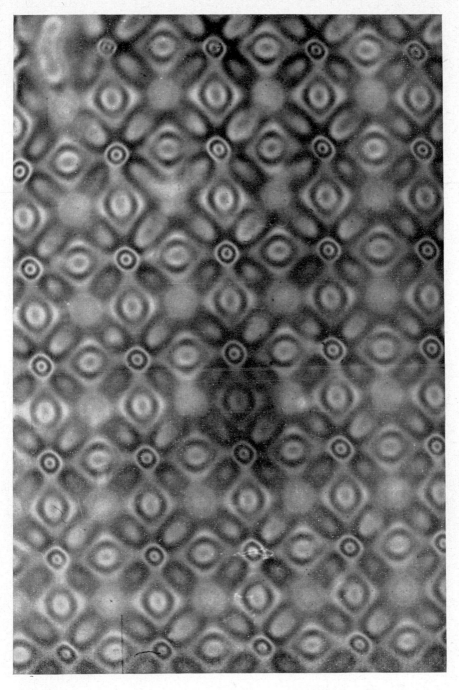

FIG. 7.9. Square pattern of undulation induced by mechanical tension in a cholesteric (courtesy F. Rondelez). Similar patterns are generated by mechanical tension in smectics, but they are much smaller and less regular.

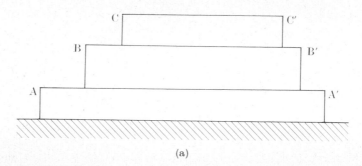

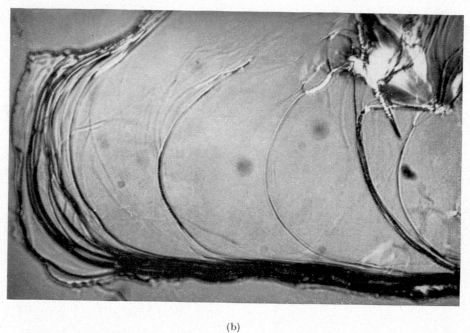

(b)

FIG. 17.12. 'Gouttes à gradin'. (a) Lateral view = what the observer O detects is the 'cliffs' located in C, C', etc. The 'mesas' (i.e., the flat regions between C and C', etc) are homeotropic and transparent. (b) Typical view from above: the cliffs appear as dark lines (courtesy J. Billard).

To eliminate the focal conics, it is often sufficient to displace slightly one of the glass plates [49].

6.4.1.2. Cholesterics wedges. To probe the helical arrangement, it is interesting to insert a cholesteric, not between parallel plates, but rather in a small-angle wedge. Three cases must be distinguished:

(1) Droplet with one free surface, and tangential boundary conditions Fig. 6.19. On the bottom plate the molecules are anchored in one orientation, but at the free surface their orientation is arbitrary. The helix can then build up its natural pitch and is undistorted at all points. The orientation of the molecules at the free surface may be determined either by optical techniques or by 'surface labelling' with microprecipitates or bubbles, as explained in Chapter 4. One then sees a pattern of 'arceaux,' as shown in Fig. 6.19 (see facing p. 260).†

(2) Cholesteric wedge between two oriented solid surfaces (Fig. 6.20). In the pioneering observations by Grandjean [50] this was obtained with a cleavage gap inside a sheet of mica. In more recent studies [6] the wedge is made with two polished glass surfaces. It is important to realise that in case (2) the number of turns allowed for the helix is quantized: the pitch of the spiral is modified, as is apparent in Fig. 6.20a (see between pp. 260–1). One can probe the local pitch through the optical rotatory power, using the de Vries equations (e.g. eqn 6.36). By this technique, Cano [6] has been able to show that the representation in Fig. 6.20 is correct: the local half-pitch $P/2$ is related to the local thickness d by

$$\frac{P}{2} = \frac{d}{n},$$

where n is an integer, chosen in such a way that P differs as little as possible from its unperturbed value P. A domain with n half-turns and a domain with $n+1$ half-turns are separated by a sharp discontinuity: the nature of this discontinuity will be discussed later in this section.

Remark: As first shown by J. Rault [52] by applying a horizontal field H to a droplet with a free surface, it is possible to 'quench' the orientation at this surface,‡ and to go from case (1) to case (2):

---

† The same pattern is found in electron micrographs of oblique cuts in the cuticle of certain crabs, where the building blocks are also arranged in a helix: see Y. Bouligand, *Journal de Physique* **30**, *Suppt.* C4, 90 (1969).

‡ The direction of alignment at the surface is not along H, but at 45° from H. This can be understood by an elementary calculation of the magnetic energy for a spiral of finite length [52].

at a certain threshold field H_2 a set of discontinuities appears.

(3) A situation very similar to (2) is obtained between two *parallel* polished plates (d fixed) if we use a cholesteric with a concentration gradient (say along X): the unperturbed pitch P_0 is then a function of X. The real pitch P is quantized, but tends to remain as close as possible from $P_0(X)$: again this leads to a succession of domains, separated by the same sharp discontinuities [51].

6.4.1.3. 'Fingerprint' texture. With a cholesteric of large pitch observed just below the clearing point T_c, one often finds a texture where the helical axis is *in the plane of the plates.* This situation allows for a 'side view' of the helix, and for very simple measurements of the pitch. There are some minor complications:

(1) Sometimes the helical axis deviates from the plane of the preparation.

(2) Near the glass slides which limit the sample at the bottom and at the top, the helix must distort to adjust to the boundary conditions (see ref. 53).

But, on the whole, the fingerprint texture is extremely useful for the observation of certain defects (see Fig. 6.27).

6.4.2. Singular lines

Just as in nematics, we find in cholesterics certain singular lines where the director field is discontinuous. But, because the unperturbed structure is a helix, the lines are much more complex in a cholesteric: for this reason, we shall discuss their geometry first, and come to the experiments at a later stage.

6.4.2.1. The Volterra process. A process generating singular lines inside an ordered medium was invented long ago by Volterra [54]. The application of this theoretical method to cholesterics is due to J. Friedel and M. Kleman [55]; we shall follow their presentation here.

Let us start from an ideal cholesteric helix, and carry out the following operations:

(1) We 'freeze' the cholesteric into a solid body,

(2) We cut this solid along a certain surface S, limited by a line L.

(3) We displace one side (say S_1) of the cut with respect to the other side S_2 by a certain translation **b**. We also rotate S_1 by an angle Ω around a certain axis **ν**.

We want to ensure that, after these operations, the displaced part S_1 is again 'in register' with S_2 along the entire cut: this imposes the condition that the translation **b** and the rotation $(\Omega, \mathbf{v})$ belong to the allowed symmetry operations of the unperturbed helix. (We shall list the possibilities below)

(4) We fill any voids present with some extra cholesteric matter.† (Alternatively, if we had an overlap of the two parts, we remove the excess.)

(5) We 'defreeze' the object; the director **n(r)** may adjust itself to minimize the Frank energy. At the same time, if the solid object had local dilations or compressions, we can relax them by displacing the molecules themselves.

Since the two sides S_1 and S_2 were 'in register', the director field **n(r)** is not discontinuous at S. But it will (usually) be discontinuous at L. Thus the process generates a *singular line* (which we call a disclination).

For the specific case of cholesterics, what are the allowed operations $(\mathbf{b}, \Omega, \mathbf{v})$? Here is a list of possibilities:

(1) A translation **b** normal to the helical axis has no effect on the director. It may create some local density changes in the frozen system: but, when we defreeze, these compressions will relax to zero by viscous flow; no effect is obtained. Thus this type of translation may be omitted.

(2) A translation **b** along the helical axis: here, to satisfy the 'in register' condition we must have

$$b = mP \qquad (m = \text{integer or half integer})$$

(3) A rotation $\Omega = 2m\pi$ around an axis **v** parallel to the helical axis. However, because of the screw symmetry of the helix, it is easily seen that this operation is not different from (2).

(4) A rotation $\Omega = 2m\pi$ with an axis **v** normal to the helical axis. This will be an allowed symmetry operation if (and only if) **v** is either parallel or perpendicular to the local director.

Of course we can also use any combination of the possibilities (1–4).

It may be helpful to compare this set of allowed operations with the allowed operations for a *nematic*: in the latter case all translations **b** become allowed and trivial; the only useful operation is a rotation $\Omega = 2m\pi$ around an axis **v** normal to the unperturbed director. The position of the rotation axis **v** is irrelevant, since two rotations (Ω) around two parallel axes (**v** and **v′**) differ only by a translation. This observation led to the simpler Volterra process described in Chapter 4

---

† The extra material must be positioned and oriented as the initial spiral: then it will be 'in register' with S_1 and S_2.

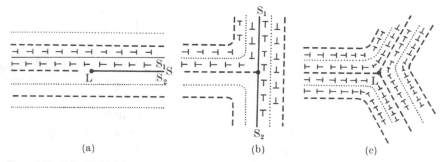

FIG. 6.21. The Volterra process. Example: generation of a τ^- line in a cholesteric. (a) The material is cut on a half plane S. (b) The two lips S_1S_2, of the cut are rotated around the axis L by a relative angle of π ($+\pi/2$ for S_2 and $-\pi/2$ for S_1). The empty space left on the right is then filled with cholesteric matter, 'in register' with the director field on S_1 and S_2. (c) The structure is relaxed: we are left with a singular line L. Dots = optical axis normal to the sheet. Broken lines = axis parallel to the sheet. Nail = tilted axis.

where each molecule is rotated around its own centre of gravity. For cholesterics, however, we must keep the full machinery.

6.4.2.2. Simple disclinations. Let us now go to some examples, and start with the Volterra process associated with operation (4) of the above list; it is enough to consider $\Omega = \pi$. Take first the case where the rotation axis $\mathbf{v}$ is normal to the molecules in the unperturbed helix. The process is shown in Fig. 6.21. The result is a line called τ^-. The ($-$) symbol recalls that, after the two lips S_1S_2 were separated, we had to fill in some voids. On Fig. 6.22 is shown a line λ^- (rotation Ω parallel to the local director).

If we move the two lips in the opposite sense, so that matter is subtracted rather than added we obtain two other lines, called λ^+ and τ^+

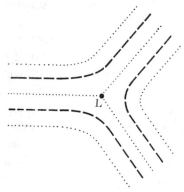

FIG. 6.22. Structure of a line λ^-. Notice that the director field $\mathbf{n(r)}$ is continuous on the line itself: a λ line has no core.

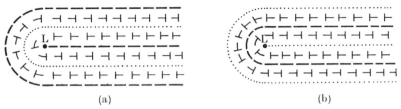

FIG. 6.23. (a) Line τ^+. (b) Line λ^+ (no core).

(Fig. 6.23). In practice there is an important difference between the λs and the τs. In a λ line the director is *continuous*: there is no core singularity.† On the other hand, a τ line has a core, and a higher line energy.

It is important to realize that an isolated λ (or τ) line is not easily deformed. This can be understood from the Volterra process: near the line L the relative displacements of the two lips S_1, S_2, must remain small, if we want to have a low line energy. To achieve this the rotation axis ν must coincide with L: this implies that L is a straight line.

Let us turn now to the lines generated by operation (2) or by its equivalent (3): these are called χ lines. To each χ line is associated a value of the strength (m). A χ line can be viewed in two different ways: as a *dislocation* in a system of layers (via operation 2); or as a *disclination* in a twisted nematic (via operation 3).

One schematic example is shown in Fig. 6.24. Note that a χ line is

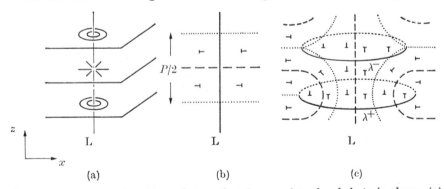

FIG. 6.24. Structure of a χ line, of strength $+1$, normal to the cholesteric planes. (a) Naive model with a strong discontinuity at the core. (b) Same model; arrangement in one diametral plane (xy). (c) Dissociated model: the line has no singularity at the core. It is surrounded by a periodic system of λ rings shown here in perspective. (After J. Rault, Doctoral Thesis, Orsay, 1972.)

† The situation is reminiscent of the 'escape in the third dimension' which has been described for disclinations of integral strength in nematics (Chapter 4).

more flexible than a λ or a τ: if we generate the χ by a rotation with $\mathbf{v}$ parallel to the helical axis (operation 3) we impose relative displacements to S_1 and S_2 which are normal to this axis, and which can be relaxed viscously: these displacements do not increase the line energy. Thus the χ line need not coincide with its rotation axis $\mathbf{v}$, and it may point in various directions.

6.4.2.3. *Dissociation of a χ line.* We have seen in Chapter 4 that, in nematic liquids, the disclinations *of integral* strength are generally unstable. A somewhat similar effect is found in cholesterics with the χ lines. Consider for instance a χ line of strength $m/2$, normal to the helical axis (Z) and parallel to the axis (Fig. 6.24). By analogy with eqn (4.1) we might, at first sight, expect an arrangement around the line of the form [56]

$$n_x = \cos(q_0 Z + m\psi + \alpha_0)$$
$$n_y = \sin(q_0 Z + m\psi + \alpha_0) \qquad (6.108)$$
$$n_Z = 0$$

where $\tan \psi = Z/X$ and α_0 is a constant. This corresponds to a strong singularity in the director field $\mathbf{n}(\mathbf{r})$ at the line $(X \to 0,\ Z \to 0)$. This singularity may possibly be meaningful for $m = \pm\frac{1}{2}$.† But for $m = \pm 1$ the χ line will tend to *dissociate* into a pair

$$\chi(1) \to \lambda^+ + \lambda^-,$$

as shown in Fig. 6.25. This is favourable, because both λ lines have no core singularity. The pair $(\lambda^+\lambda^-)$ is called a P pair because the material on one side $(X \to +\infty)$ has two extra cholesteric layers (or one full pitch P) when compared to the material on the other side $(X \to -\infty)$.

This important dissociation process was invented by Friedel and Kleman [55]. Similar (although more complex) dissociations occur probably for χ lines which are oblique with respect to the cholesteric planes [57].

We may ask whether dissociation also takes place for χ lines of half integral strength. It is indeed possible to consider the process

$$\chi(\tfrac{1}{2}) \to \lambda^+ + \tau^-$$

which is displayed in Fig. 6.26, or the analog $\chi(\tfrac{1}{2}) \to \lambda^- + \tau^+$. However, we now have a τ line in the final state, with a finite core energy: for lines of half-integral strength, it is not possible to eliminate the core singularity, as explained in Chapter 4. Thus it is not obvious that the

---

† However it must be pointed that eqn (6.108), although correct for a twisted nematic, is never rigorous for a cholesteric.

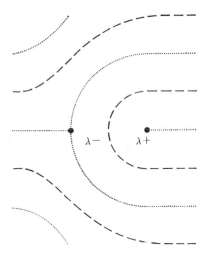

Fɪɢ. 6.25. Line of strength 1 (normal to the helical axis) dissociated into a $\lambda^-\lambda^+$ pair. This corresponds to the thick lines on Fig. 20(b).

dissociation into a $\lambda\tau$ pair will really lower the energy. (The $\lambda\tau$ pair is often called a $P/2$ pair.)

6.4.2.4. Observations on line defects.   The most direct evidence for the λ lines and the $\lambda^+\lambda^-$ pairs is obtained in the fingerprint texture (Fig. 6.27; see between pp. 260–1). But, for quantitative studies, observations of defects in a planar texture are more accurate. To generate lines *normal to the helical axis*, two methods are particularly useful: one static, and one dynamic.

(1) *Cano wedge* (Fig. 6.20.) In *thin* wedges, the discontinuities studied by Cano [6] were always found to separate domains with n and $(n+1)$

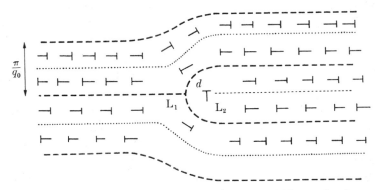

Fɪɢ. 6.26. Line of strength $\frac{1}{2}$ dissociated into a $\lambda^+\tau^-$ pair. This *may* be the structure of the thin lines of Fig. 6.20b.

half turns: Cano first interpreted them in terms of sheet singularities, but it soon became clear that the discontinuities must be singular lines (of strength $\frac{1}{2}$). In *thicker* wedges, the Orsay group later observed that lines of strength 1 (separating n from $n+2$) were systematically found [58]. At that time, the origin of this anomalous stability for 'double' lines was quite mysterious. Friedel and Kleman solved the paradox with the model of a $(\lambda^+\lambda^-)$ pair, devoid of any core singularity. If a magnetic field H is applied normal to the lines and to the helical axis, the double lines take on a zig-zag conformation (Fig. 6.20c) as soon as $H > H_Z$ where H_Z is a certain threshold field of order $H_c/2$ (H_c being the critical field for unwinding of the spiral). On the other hand, the simple lines are unsensitive to the field. This difference in behaviour has been explained by Kleman [55]: the experiments are made with nematic–cholesteric mixtures for which $\chi_a > 0$. Then, as seen earlier in this chapter, the helix prefers to have its axis normal to the field. The region between λ^+ and λ^- (Fig. 6.25) has a helical axis along x, and is thus in a high magnetic-energy conformation. The magnetic energy is lowered in the zig-zag conformation. Kleman showed that qualitatively $H_Z \sim H_c/m$ where $m/2$ is the strength of the line. Thus the lines of strength 1 have an observable $H_Z \sim H_c/2$.

The observations do not allow to conclude on the structure of the lines of strength $\frac{1}{2}$: for instance the absence of field effects on these lines is explained as well in a pure χ model than in a $\lambda\tau$ model (where $H_Z \sim H_c$ and thus H_Z would not be detectable). Certain qualitative optical features [58] suggest that both types (dissociated and undissociated) of lines of strength $\frac{1}{2}$ may be present, depending on the sample thickness. To clarify this point is difficult, but one can think of the following technique: polymerize the cholesteric inside a Cano wedge (using for instance cholesterol acrylate as one constituent [59]) and study the resulting frozen texture by electron microscopy.

(2) *Mechanical twist*: This method of line generation is the analogue of the Meyer technique for nematics (Chapter 4). The cholesteric planar texture (with n_0 half-terms) is realized between two polished glass plates. An abrupt change in the twist is then imposed, either by a rotation of one plate, or by a change in spacing, or by a combination of both processes. In the equilibrium state for these new conditions, the optimum number of turns N_t will be usually different from N_0. The transformation $N_0 \to N_t$ is realized by the migration of disclination loops. With suitable care one can generate lines of strength $\frac{1}{2}$ or of strength 1 [57] and measure their line tension or their mobility.

Up to now, we discussed only the generation of lines normal to the helical axis. To observe χ lines *parallel* to this axis, the following recipe is convenient [57]: one uses a cholesteric–mixture between a glass wall and a free surface (tangential boundary conditions). The transition from isotropic to cholesteric is obtained by slow evaporation of a passive solvent. One then obtains a planar texture with a few vertical χ lines present: the aspect is reminiscent of a schlieren texture. Each χ line has the optical appearance of a 'noyau.' Just as for nematics, by observations between crossed nicol prisms, one can determine the strength $m/2$ of the line.

The majority of the lines is found to have integral strength, and a rather thick core in optical observations. They are probably dissociated into pairs, but the geometry of these pairs is complex and not yet fully understood. Lines of half-integral strength are less frequent, and seem to have a thinner core. Detailed observations and plausible models for both are described in the doctoral work of Rault [57].

REFERENCES

[1] MATHIEU, J. P. *Bull. Soc. fr. Minér. Cristallogr.* **62**, 174 (1939).

[2] BERREMAN, D. W. and SCHEFFER, T. J. *Mol. Cryst. liquid Cryst.* **11**, 395 (1970).

[3] BILLARD, J. *Mol. Cryst.* **3**, 227 (1968).

[4] TAUPIN, D. *J. Phys. (Fr.) Colloq.* **69**, (Suppl. C4) 32 (1969).

[5] MAUGUIN, C. *Bull. Soc. fr. Minér. Cristallogr.* **34**, 3 (1911).

[6] CANO, R. *Bull. Soc. fr. Minér. Cristallogr.* **90**, 333 (1967); see also KASSUBECK, G. and MEIER, G. *Mol. Cryst. liquid Cryst.* **8**, 305 (1969).

[7] DE VRIES, H. *Acta Crystallogr.* **4**, 219 (1951). Earlier calculations on the same lines are found in [5] and in OSEEN, C. W. *Trans. Faraday Soc.* **29**, 833 (1933).

[8] See, for instance, LANDAU, L. D. and LIFSHITZ, E. M. *Quantum mechanics*, § 104, p. 384. Pergamon, London (1959).

[9] LEVELUT, A. and BILLARD, J. *Acta Crystallogr.* A **26**, 390 (1970).

[10] FERGASON, J. in *Liquid crystals (Proceedings of the second Kent conference)*, p. 89. Gordon and Breach, New York (1966); FERGASON, J., GOLDBERG, N., and NADALIN, R. *ibid.* p. 105. ADAMS, J. and HAAS, W. *Mol. Cryst.* **11**, 229 (1970).

[11] ALBEN, R. *Mol. Cryst. liquid Cryst.* **20**, 231 (1973).

[12] See, for instance, MAGNE, M. and PINARD, P. *J. Phys. (Fr.) Colloq.* **30**, (Suppl. C4) 117 (1960).

[13] FERGASON, J. *Scient. Am.* **211**, 77 (1964). *Mol. Cryst.* **1**, 309 (1966); ENNULAT, R. and FERGASON, J. *Mol. Cryst. liquid Cryst.* **13**, 149 (1971).

[14] CRISSEY, J., *et al.*, *J. invest. Derm.* **43**, 89 (1965); SELAWRY, O., *et al.*, *Mol. Cryst.* **1**, 495 (1966); GAUTHERIE, M. *J. Phys., Paris, Colloq.* **30**, (Suppl. C4) 122 (1969).

[15] HANSEN, J., FERGASON, J., and OKAYA, A., *Appl. Opt.* **3**, 987 (1964); KEILMANN, F. *Appl. Opt.* **9**, 1319 (1970).

[16] IIZUKA, K. *Electronics Lett.* **5**, 27 (1968).

[17] CANO, R. and CHATELAIN, P. *C.r. Acad Sci., Paris*, B **259**, 252 (1964). ADAMS, J., HAAS, W., and WYSOCKI, J. *Liquid crystals and ordered fluids*, p. 463. Plenum Press, New York (1970).

[18] BAESSLER, H. and LABES, M. *J. chem. Phys.* **52**, 631 (1970); ADAMS, J. and HAAS, W. *Mol. Cryst. liquid Cryst.* **15**, 27 (1971).

[19] ADAMS, J., HAAS, W., and WYSOCKI, J. *Phys. Rev. Lett.* **22**, 92 (1969).

[20] ADAMS, J., HAAS, W., and WYSOCKI, J. *Mol. Cryst. liquid Cryst.* **7**, 371 (1969).

[21] JENKINS, J. T., Thesis, Johns Hopkins University (1969).

[22] HAAS, W., ADAMS, J., and FLANNERY, J. B. *Phys. Rev. Lett.* **24**, 577 (1970); see also MULLER, J. H. *Mol. Cryst.* **2**, 167 (1966).

[23] SACKMANN, E., MEIBOOM, S., and SNYDER, L. C. *J. Am. chem. Soc.* **89**, 5982 (1967).

[24] WYSOCKI, J., ADAMS, J., and HAAS, W. *Phys. Rev. Lett.* **20**, 1025 (1968); **21**, 1791 (1968); *Mol. Cryst.* **8**, 471 (1969).

[25] DURAND, G., LEGER, L., RONDELEZ, F., and VEYSSIE, M. *Phys. Rev. Lett.* **22**, 227 (1969); MEYER, R. B. *Appl. Phys. Lett.* **14**, 208 (1969).

[26] KAHN, F. J. *Phys. Rev. Lett.* **24**, 209 (1970).

[27] HEILMEIER, G. and GOLDMACHER, J. *J. chem. Phys.* **51**, 1258 (1969).

[28] MEYER, R. B. (a) *Appl. Phys. Lett.* **14**, 208 (1968); (b) Thesis, Harvard University (1969).

[29] DE GENNES, P. G. *Solid State Commun.* **6**, 163 (1968).

[30] (a) LUCKHURST, G. and SMITH, H. *Mol. Cryst.* **20**, 319 (1973). (b) REGAYA, B., GASPAROUX, H., and PROST, J. *Rev. Phys. appl.* **7**, 83 (1972).

[31] (a) Magnetic field effects: RONDELEZ, F. and HULIN, J. P. *Solid State Commun.* **10**, 1009 (1972); (b) Electric field effects: GERRITSMA, G. and VAN ZANTEN, P. *Mol. Cryst. liquid Cryst.* **15**, 267 (1971); *Phys. Lett.* **37A**, 47 (1971); RONDELEZ, F., ARNOULD, H., *C.r. Acad. Sci. Paris* **B273**, 549 (1971).

[32] HELFRICH, W., *Appl. Phys. Lett.* **17**, 531 (1970); HURAULT, J. P. *J. chem. Phys.* (to be published).

[33] LESLIE, F. *Mol. Cryst. liquid Cryst.* **7**, 407 (1969).

[34] FAN, C., KRAMER, K., and STEPHEN, M. *Phys. Rev.* A **2**, 2482 (1970).

[35] GASPAROUX, H. and PROST, J. Private communication.

[36] (a) BROCHARD, F. *J. Phys.* (*Fr.*) **32**, 685 (1971). (b) MARTINOTY, E., and CANDAU, S., *Phys. Rev. Lett.* **28**, 1361 (1972).

[37] PINCUS, P. *C.r. hebd. Séanc. Acad. Sci., Paris,* B **267**, 1290 (1968).

[38] LANDAU, L. D. and LIFSHITZ, E. M. *Fluid mechanics,* Chapter 2, p. 17. Pergamon, London (1952).

[39] SAKAMOTO, K., PORTER, R., and JOHNSON, J. *Mol. Cryst. liquid Cryst.* **8**, **443** (1969). See also: POCHAN, J., MARCH, D., *J. Chem. Phys.* **57**, 1103 (1972)

[40] HELFRICH, W. *Phys. Rev. Lett.* **23**, 372 (1969).

[41] Only one experiment on the effect of electric fields on shear flow is available at present, and the conditions are not entirely clear: see WYSOCKI, J., *et al. J. appl. Phys.* **40**, 3865 (1969).

[42] HEILMEIER, G. H. and GOLDMACHER, J. E. *Proc. Inst. Elect. electron. Engrs.* **57**, 34 (1969).

[43] HELFRICH, W. (a) *Appl. Phys. Lett.* **17**, 531 (1970); (b) *J. Chem. Phys.* **55**, 839 (1971).

[44] GERRITSMA, G. Private communication.

[45] POMEAU, Y. *Phys. Lett.* A **34**, 143 (1971).

[46] LESLIE, F. *Proc. Soc.* A **307**, 359 (1968).

[47] LEHMANN, O. *Annln. Phys.* **2**, 649 (1900).

[48] OSEEN, C. W. *Trans. Faraday Soc.* **29**, 883 (1933).

[49] FRIEDEL, G. *Annales. Phys.* **18**, 273 (1922).

[50] GRANDJEAN, F. *C.r. Acad. Sci., Paris* **172**, 71 (1921).

[51] KELKER, H. *Mol. Cryst. liquid Cryst.* **15**, 347 (1972).

[52] RAULT, J. *Mol. Cryst. liquid Cryst.* **16**, 143 (1972).

[53] KLEMAN, M. and CLADYS, P. *Mol. Cryst. liquid Cryst.* **16**, 1 (1972).

[54] See, for instance, FRIEDEL, J. *Dislocations,* Pergamon, London (1964).

[55] KLEMAN, M. and FRIEDEL, J. *J. Phys.* (*Fr.*) **30**, (Suppl. C4) 43 (1969). See

also *Fundamental aspects of dislocation theory* (Simmons, deWit, and Bullough, eds.), p. 607. *Nat. Bur. Stand. spec. pub.* **317,** 1 (1970).

[56] DE GENNES, P. G. *C.r. hebd. Séanc. Acad. Sci., Paris* **266,** 571 (1968).

[57] RAULT, J. Thèse, Orsay (1971).

[58] ORSAY GROUP, *Phys. Lett.* A **28,** 687 (1969); *J. Phys. (Fr.)* **30,** (Suppl. C4) 38 (1969).

[59] STRELECKI, L. and LIEBERT, L. *Bull. Soc. chim. Paris,* no. 2, p. 597, 603, 605 (1973). *Mol. Cryst. liquid Cryst.* (to be published).

7

SMECTICS

' . . . où le mystère en fleurs s'offre à qui veut le cueillir.'
G. APOLLINAIRE

7.1. Symmetry of the main smectic phases

IN CHAPTER 1 we presented the main smectic phases A, C, and B. We emphasized that they are all characterized by a *layer structure*. In practice the layers are rather thick (20 to 30 Å) and they give rise to typical Bragg reflections at small angles in an X-ray experiment.

Inside this broad class of layered materials, we may distinguish two major subgroups [1a]:

Group 1: Each layer is a two-dimensional liquid ('ohne Ordnung' = without order)

Group 2: Each layer has some features of a two-dimensional solid ('mit Ordnung' = with order)

The two groups are extremely different, though their separate existence was recognized only rather late; G. Friedel was aware only of the first group, while Hermann and Alexander [1b] initially started from a picture where the second group alone was accepted. Later, Hermann [1a] included the two possibilities in his classification of smectics.

7.1.1. *Liquid layers* ('ohne Ordnung')

7.1.1.1. Basic properties. The main features associated with liquid layers are the following:

(1) The X-ray pattern at large angles is continuous, and liquid-like (no extra reflections).

(2) The layers can slip on each other, and their ideal planar stacking is easily deformed into more complicated arrangements where they are curved. (The only strict requirement is to keep the interlayer distance fixed Fig. 7.1.) This leads to the so-called 'focal conic textures' discussed in ref. [2] [3] and shown in Fig. 7.2 (see between pp. 260–1).

(3) The diffusion of the constituent molecules (or of solutes, and in particular of charge carriers) is easier inside one layer than in between the layers.

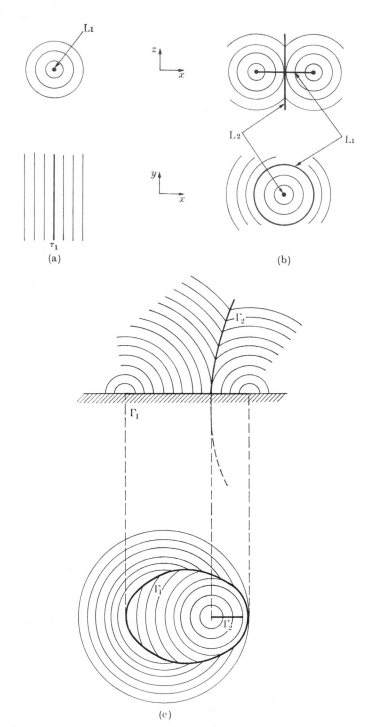

Fɪɢ. 7.1. Generation of the focal conic texture; (a) simple 'jelly-roll' or 'myelinic' arrangement generating a tube. (b) Tube closed into a torus: note the two singular curves (a circle and a straight line); (c) generalization: the circle becomes an ellipse, the straight line becomes a hyperbola. With tangential boundary conditions for the molecules, the ellipse is often stuck to the limiting surface.

All these features are found in both *smectics A and smectics C*. Let us now discuss the qualitative differences between A and C.

7.1.1.2. Biaxial versus uniaxial smectics. A uniaxial smectic must have its optical axis normal to the layers; this is what is called a *smectic A*. Biaxial smectics may be of various types: we shall classify them, assuming that the dielectric tensor (or any similar tensor) is enough to specify entirely the local symmetry.†

Following the notation of Chapter 2, we shall represent the anisotropic part of the tensor under study by the symbol $\mathbf{Q}_{\alpha\beta}$. The problem is to locate the three (orthogonal) principal axes of $\mathbf{Q}_{\alpha\beta}$ with respect to the smectic planes. There seem to be three possibilities:

(1) One of the axes may still coincide with the normal to the (z) to the smectic planes; but (excluding smectics A) the other two axes (ξ and η, both lying in the xy plane) must be inequivalent. This corresponds to:

$$\mathbf{Q}_{\xi\xi} \neq \mathbf{Q}_{\eta\eta}$$
$$\mathbf{Q}_{\xi\eta} = \mathbf{Q}_{\eta z} = \mathbf{Q}_{\xi z} = 0.$$

This possibility (first noticed by W. Mc Millan‡) might be favoured if, for instance, the benzene rings of the constituent molecules had a strong tendency to become parallel to each other. At the time of writing, the practical existence of smectics of this type has not been demonstrated, but they might show up in the future: we propose to give them the label C_M, where M stands for Mc Millan.

(2) If (z) is not a principal axis, we may still retain some symmetry, provided that one of the principal axes, (which we shall call η) remains in the layer plane. The other axes (ξ and ζ) are tilted: we shall call ω the angle between (z) and (ζ). The ($z - \zeta$) plane is a plane of symmetry of the structure. By a suitable choice it is always possible to have the axes (y) and (η) coincident. This gives the following structure to the $\mathbf{Q}$ tensor:

$$\mathbf{Q}_{zz}, \; \mathbf{Q}_{xz}, \; \mathbf{Q}_{xx} \neq 0$$
$$\mathbf{Q}_{xx} \neq \mathbf{Q}_{yy} \neq 0.$$

More physically, this phase is obtained with molecules aligned like in a nematic, but the director $\mathbf{n}$ being tilted (by an angle $\sim\omega$) with

† This omits certain possibilities, such as ferroelectric or antiferroelectric smectics, which have apparently not yet been found.

‡ To be published.

respect to the layer normal.† There is a certain amount of evidence
(from X-ray scattering data, and from more indirect sources)
favouring this picture for the usual biaxial smectics. We shall
retain the label 'smectic C' for this family.

The magnitude ω of the tilt angle is fixed (at a given temperature)
but the direction in which the tilt takes place may be chosen at will.

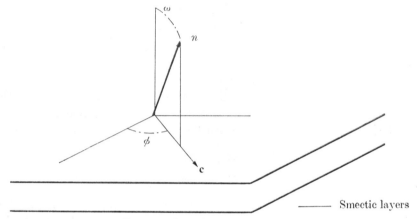

Fig. 7.3. Geometry of a smectic C.

It may be specified by a unit vector **c** in the plane of the layers (Fig.
7.3). We shall call **c** the *C-director* of the structure. Note that the
C-directors of successive layers are parallel: the C-director has long-
range order.‡

By suitable surface treatments, invented by Fergason and his coworkers [4], it is sometimes possible to prepare a smectic C between two
glass slides, with the layers parallel to the glass surfaces. In a texture of
this type, the smectic C behaves locally as a birefringent slab, the vector
c giving the axis of largest index. Thus one can map **c**, and one finds that
c behaves very much like the director in a nematic; it can vary progressively from point to point (to adjust to some lateral boundary
conditions) and it can show singular lines (disclinations). The overall
appearance is exactly that of a schlieren texture in a nematic.

We see that a smectic C may show two completely different textures:
focal conics or schlieren; for this reason, in the early days, certain

---

† In a strictly nematic phase, the tensor **Q** would have two equal principal values
(**Q**$_{\xi\xi}$ and **Q**$_{yy}$). But this degeneracy is removed by the presence of the smectic layers.
‡ One could think of smectics (in particular with lyotropic systems) where the C
directors of successive layers would be essentially uncorrelated: however this would
probably not correspond to thermal equilibrium.

smectics C were assigned incorrectly as being nematic! Note, on the other hand, that smectics A can show only one type of texture (focal conics).

The analogy between the C-director of smectics C and the director of nematics is important and covers many aspects, for instance:

(a) Addition of a chiral solute to a smectic C imposes a twist to the C-director and creates a helical structure [5] [6].

(b) The fluctuations of **c** are large (at long wavelengths) and give rise to a strong scattering of light [7]: smectics C are turbid like nematics.

On the other hand there are some significant differences between **c** and a nematic director: the most important one is that the states (**c**) and ($-$**c**) are not equivalent. This implies that, in a schlieren texture, we should observe only disclination lines with integral strength.

(3) In the third type of biaxial smectics the tensor $\mathbf{Q}_{\alpha\beta}$ may have all three axes tilted (none of them retaining a simple orientation with respect to the layers). We shall use the label C_G (where G stands for 'generalized') for this rather far-fetched possibility. Note that a C_G phase usually differs from its mirror image; if the constituent molecules were achiral, they would build up two enantiomers (like silicon oxide in quartz crystals).

7.1.1.3. Lack of continuous miscibility between A and C. Let us consider a smectic A and a smectic C, built up with molecules which are not too different, so that they tend to mix rather well. Let us start with the pure A phase, and add to it a small concentration x of the second compound. At first sight, we might think of two possible behaviours:

(1) As soon as some C is added, we find a finite tilt angle ω, increasing smoothly with the concentration x.

(2) At low x, the mixture remains non-tilted (A type). Above a certain x threshold, we have a phase transition, and then we get a tilted mixture.

Thermodynamics tell us that (2) is the correct answer: this point is crucial, because all the work of Sackmann and Demus [8] on the classification of mesophases is based on it (or on similar statements for other systems). Thus we shall try to explain it in more detail.

Let us start from the free enthalpy F of the A–C mixture, written as a function of the following variables: the temperature T, the concentration x of (C), and the tilt angle ω. By definition, the smectic A has

complete rotational symmetry around its optical axis. Thus F must be
an *even* function of ω and can be expanded in the form

$$F(T, x, \omega) = F_0(T, x) + \tfrac{1}{2}a(T, x)\omega^2 + \tfrac{1}{4}b(T, x)\omega^4 + \dots \qquad (7.1)$$

The equilibrium value of ω corresponds to the minimum value of F. To
simplify the discussion, let us assume first that $b(T, x)$ is positive
throughout the region of interest. Then we have two cases, depending
on the sign of a: if $a(T, x) > 0$, the optimum value is $\omega = 0$ (we have a
smectic A); if $a(T, x) < 0$, the optimum value is $\omega = -(a/b)^{\frac{1}{2}} \neq 0$, and
we have a smectic C.

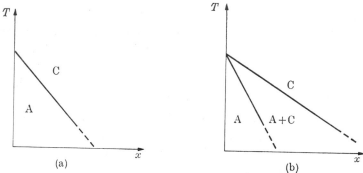

(a)   (b)

FIG. 7.4. Phase diagrams for mixtures of an A smectogenic compound with a C smectogenic compound: (a) limiting curve—second order transition; (b) limiting curve—first
order transition.

The two domains are thus separated by a certain critical line (in the
T, x plane) defined by

$$a(T, x) = 0. \qquad (7.2)$$

A typical aspect for the critical line is shown in Fig. 7.4a: this clearly
shows that we expect behaviour (2) as defined above, and not behaviour
(1). With our choice for the sign of b, the A $\rightleftharpoons$ C transition is of second
order. With $b < 0$, on the other hand, one must include terms up to
order ω^6 in the expression (7.1) to ensure that the minimum occurs for a
finite ω. When this is done, one now finds a first-order transition between
A and C: this is described in the (x, T) diagram by two lines $x_1(T)$ and
$x_2(T)$ giving the concentrations in the non-tilted, and in the tilted phase,
when they are in equilibrium (Fig. 7.4b). Again we find behaviour (2),
and not behaviour (1).

This property is in fact a particular example of a general rule, which
is usually stated as follows: *two phases of different symmetry must be
separated by at least one transition line.* For a more complete discussion of

this rule, the reader is referred to the book by Landau and Lifshitz.†
There is one point, however, which is not enough emphasized in this
reference, namely: the rule applies if, and only if, the molecules building
up the two species have the same internal symmetry. To make this
clearer, we shall now quote two examples, where this requirement is not
satisfied, and the rule does not hold.

A nematic and a cholesteric can (in many cases) be mixed continuously
without crossing any phase-transition line (no anomaly, for instance, in
the thermal properties). In this case, the free energy of the mixture, as a
function of the twist q, is of the form

$$F = \tfrac{1}{2}K_2 q^2 + \lambda q$$

where K_2 is a Frank constant and $\lambda(T, x)$ is a linear function of x at small
x. (For a microscopic discussion of λ, see the problem on page 236.) The
optimum twist is $q_0 = -\lambda/K_2$ and it varies smoothly with x.

Clearly, in this case, the rule breaks down because the solute molecules
do not have the mirror symmetry of the nematic solvent.

As a second example, in a 'gedanke experiment' consider two miscible
liquids, **1** and **2** consisting of elongated molecules. Component **1** has
centrosymmetric molecules. Component **2** has polar molecules. All
experiments are performed under a constant electric field E. Under E,
the pure fluid **1** has no vectorial order $\langle \cos \theta \rangle = 0$ (where θ is the angle
between the long axis of the molecule and **E**). On the other hand, fluid
2 has a non-zero $\langle \cos \theta \rangle$. As soon as we add a small concentration x of
2 to solvent **1**, we get a polarized phase. But no accident occurs in the
thermal properties.

Let us now return to the specific problem of smectics A and C, and
summarize our results: assume that we have two compounds X and Y,
with component molecules of the same symmetry. We know that X is a
smectic A in a certain temperature range. We want to know the symmetry of a certain phase of Y, through a study of mixtures of X+Y. If
the two phases of interest are continuously miscible, they are both
smectics A. If they are not, we cannot be sure: the Y phase may have a
different symmetry (e.g., it may be a smectic C) or the Y phase may also
be a smectic A, if the molecular sizes of X and Y are very different and
they do not mix readily. This type of argument, supplemented by
textural studies, was the main tool used by Sackmann and Demus [8] in
their classification of smectics.

---

† *Statistical mechanics*, ch. 14, Pergamon Press, Oxford, 1958.

7.1.2. 'Solid' layers ('mit Ordnung')

7.1.2.1. Facts. The group of smectics with solid layers contains principally the class B of Sackmann and Demus [8], plus some other, more exotic, types of which very little is known at present, and which we shall not discuss here. The basic properties of class B are the following.

(1) *X-ray scattering by polydomain samples:* Apart from the lines at small angles, due to reflexions on the smectic layers, one also finds lines at large angles, characteristic of an ordering inside each layer. In many cases the sequence of observed lines suggests a hexagonal arrangment in the layer—but other two dimensional lattices may also occur. The allowed symmetry groups for one layer have been listed in the PhD work of E. Alexander (see ref. [1b]). It may well be that the B class corresponds to hexagonal symmetry, (or deformed variants of it) while some of the more exotic smectics (F, G, etc) correspond to some other two dimensional lattices.†

(2) *X-ray scattering by single domain samples:* In the few cases where this type of study has been carried out [9], it appears that the scattering intensity is concentrated at some discrete *points* in reciprocal space. This means that the axes of successive layers are locked and parallel to each other.‡

Apart from these peaks in the scattering intensity, there is also a strong diffuse background, concentrated on planes in k space, and suggesting some 'chain like' correlations between molecules belonging to successive layers [9]. The meaning of these correlations is still unclear.

(3) *Textures and phase diagrams.* The large-scale deformations which lead to focal conic textures in smectics A and C, are not observed in smectics with solid-like layers. The characteristic texture here is a 'mosaic' (Fig. 7.5; see between pp. 260–1) in which the layers remain flat.

Of course we have to make a subdivision between tilted and non-tilted molecular arrangments: inside the B class we are thus led to distinguish two subclasses, which we call B_A (uniaxial) and B_C (biaxial).† Again, on general grounds, we would expect that a phase B_A should not be

---

† The so called H-phase defined by de Vries and Fishel, *Mol. Cryst.* **16**, 311, 1972 probably coincides with B_C.

‡ This result is opposite to the early guess of Hermann and Alexander [1b] who assumed that the orientations of two distant layers, inside one domain, would not be correlated.

continuously miscible with a phase B_C. However, in one case, the opposite behaviour has been found [10]! Of course, as pointed out by the authors of ref. [10] it may be that the species which they labelled as B_A is in fact a B_C with a small, and undetected, tilt angle. Or it may be also be that the line of separation between tilted and non-tilted corresponds to a second-order phase transitions, and is not easily detected within the accuracy of common optical techniques. The question remains open for the moment.

(4) *Electron spin resonance.* The case of a smectic B where the molecule carries two long aliphatic chains has been studied recently by C. Taupin and coworkers [11]. They prepared a molecule very similar to the smectic B under study, but carrying a paramagnetic spin (the latter being located essentially half-way along one aliphatic chain.). This molecule was then diluted in the host smectic B. A qualitative picture of the host and of the spin label as they stand in the solid phase is shown below:

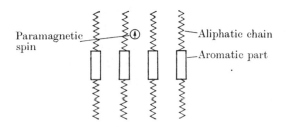

The e.s.r. spectrum of the mixture measures the degree of alignment S of the aliphatic chain of the label. In view of the great similarity between label and host, it is hoped that S is also a good measure of the alignment of the host aliphatic chains.

In the solid phase, S is found to be large: the chains are strongly aligned, as shown in the above picture. In the B phase, on the other hand, S is *practically equal to zero.* This suggests that the chains are then in a disordered, 'molten' state. This is to be contrasted with the aromatic portions, which must remain well ordered—at least locally—to account for the textural rigidity and for the X-ray data. These observations lead to a local picture for the B phase (of molecules with long terminal chains), which is shown at the top of p. 282.

Further evidence for the 'melting' of the chains is obtained from the transition enthalpy of the solid–smectic B transition:

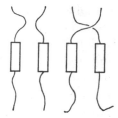

this is rather close to the enthalpy of melting for paraffins of length equal to the length of the terminal chains.

7.1.2.2. Discussion. The e.s.r. experiments described above have been carried out up to now for one system only, with long terminal chains. Let us restrict our attention to this type of molecule, and discuss the effective interactions between two of them in the B phase. The interactions between the rigid parts *inside* one same layer must be strong and tend to impose a two-dimensional ordering. The interactions between *consecutive* layers may be split into an average part (giving the overall attraction between layers) and a spatially-modulated part, which tends to stabilize a certain three-dimensional ordering of the successive layers. We are mainly interested in the latter part, which we shall call the *positional coupling* V_p between layers. We can think of two contributions to V_p;

Van der Waals interactions between the benzene rings, should be rather small, because of the large interlayer thickness.

Direct (steric, or other) interactions between the terminal chains of the two partners. This is a strong coupling when the chains are rigid, but only a weak coupling when the chains are molten, so that each chain is spread over a rather large area on the layer plane: the chains then act essentially as a liquid 'lubricant' between the two layers.

Depending on the size of V_p, we may try very different models for the long-range order of the B phase:

(1) If V_p is still large enough to impose three-dimensional order, the B phase is really a *crystal*. This means in particular that the shear modulus C_{44} for relative slip of two consecutive layers is finite. It also implies that the X-ray scattering intensity for a wave vector $\mathbf{q}$ close to a Bragg position $\boldsymbol{\tau}$ varies like $|\mathbf{q}-\boldsymbol{\tau}|^{-2}$. Of course, in this model, it must be expected that the crystal has a low yield stress for shear, because of the fluid state of the chains: it might be properly called a plastic crystal.

(2) If V_p is below a certain threshold, three-dimensional periodicity cannot be maintained: the layers can slip freely on each other, while still retaining locally an ordered two-dimensional structure. This possibility has been analysed by Sarma [12]. The model is illustrated on Fig. 7.6a. One of the consequences of the model

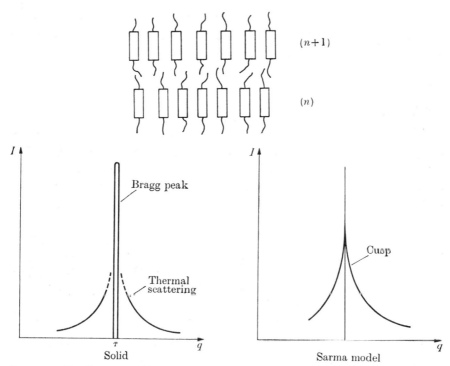

FIG. 7.6. The Sarma model for smectics B. (a) Arrangement of the molecules in two successive layers. The rigid, aromatic part of each molecule is represented by a rectangle. The terminal chains are pictured as flexible threads. The coupling between layers tending to create 3 dimensional order is weak: the result is a phase where successive layers are not necessarily in register. (b) Differences in the X-ray diagrams between the solid and the Sarma model.

is that the static shear modulus C_{44} vanishes. Also, the X-ray scattering near the nominal Bragg peaks τ of the structure is anomalous, because a two dimensional crystal has very pathological fluctuations. For most τ, there is no delta function peak, but only a weaker singularity like $|\mathbf{q} - \tau|^{-x}$, where $x(T)$ is a temperature dependent exponent.

At the time of writing, there are two experimental studies on smectics B which tend to favour model (1). One is an X-ray study of the diffuse scattering (M. Lambert and A. M. Levelut, to be published) which favours $I \sim |\mathbf{q} - \mathbf{\tau}|^{-2}$. The other experiment is a study of acoustic waves by Brillouin scattering (Y. Liao, N. Clark, and P. Pershan, *Phys. Rev. Lett.* **30**, 639, 1973). The results indicate that C_{44} does not vanish (although it is much smaller than the other moduli) at the Brillouin frequency $\omega/2\pi$ (of the order of 10^{10} H$_3$). However, as pointed out by the authors, this behaviour at high frequency does not entirely exclude the possibility of $C_{44} \to 0$ for $\omega \to 0$. Thus the question is still open.

7.2. Continuum description of smectics A and C

7.2.1. Statics of smectics A

7.2.1.1. Choice of variables.
Let us take an ideal single-domain sample of a smectic A, with parallel and equidistant layers (interval a, direction of the normal to the layers z). We want to impose some small, long wavelength distortions to this initial state. What variables must we include to provide a complete description of the distorted state? The main effects which come to mind are listed below.

The nth layer is displaced by a certain amount $u_n(x, y)$ (Fig. 7.7); u_n is the fundamental variable of the problem. The notation may be simplified somewhat, for the cases of slow spatial variations in which we are interested: we can substitute for the discrete index n the continuous variable $z = na$

$$u_n(x, y) \to u(x, y, z). \tag{7.1}$$

In the unperturbed state the molecules were normal to the layers. In the perturbed state, how will they be aligned? From a formal point of view, they do not remain exactly normal to the new plane of the layers: thus we might think of introducing certain tilt angles as extra variables in the theory. However, if the distortions take place over distances L which are much larger than a certain microscopic length $\lambda(T)$† it may be shown that these tilt effects are negligible. Thus, if we associate a unit director **n** to the optical axis, it will have the components

$$\left. \begin{aligned} n_x &= -\frac{\partial u}{\partial x} \ll 1 \\[2ex] n_y &= -\frac{\partial u}{\partial y} \ll 1 \end{aligned} \right\}. \tag{7.2}$$

† For a detailed discussion of λ, see ref. [43].

Equation (7.2) simply states that, at each point, **n** is normal to the layers. It has the interesting consequence that

$$\mathbf{n} \cdot \operatorname{curl} \mathbf{n} = 0. \tag{7.3}$$

Thus we see that the twist deformation which was allowed in nematics becomes forbidden in smectics A.

In the distorted material, the density ρ will usually differ from its unperturbed value ρ_0

$$\rho = \rho_0[1 - \theta(\mathbf{r})]. \tag{7.4}$$

However, for the static distortions of interest here, $\theta(\mathbf{r})$ will adjust itself to minimize the free energy in a given $u(\mathbf{r})$: thus (in the absence of external pressures which would change ρ) it need not be treated as an independent variable.† Similar arguments apply to all other internal variables.

In fact, the form of the relation between θ and u at equilibrium may be derived on pure symmetry grounds: we note first that a uniform displacement will not change ρ: thus θ must depend only on derivatives of u, the leading ones being first-order derivatives $\partial u/\partial x$, etc. A small uniform rotation along γ corresponds to $\partial u/\partial x \neq 0$ and will not change ρ. The only possibility left is then

$$\theta = m \frac{\partial u}{\partial z}, \tag{7.5}$$

where m is a dimensionless constant, characteristic of the material, which may in principle be positive or negative.‡

7.2.1.2. The distortion free-energy. The preceding discussion shows that, for *static* phenomena, a description in terms of the displacements $u(\mathbf{r})$ is enough. Let us now write the free energy F (per unit volume of the unperturbed system) as a function of the derivatives of $u(\mathbf{r})$. We shall constantly assume that grad u is small; physically, this means that we consider layers which are not very much tilted from the x, y plane. The assumption excludes certain interesting cases, such as the focal conic textures; but it makes the algebra much more transparent.

As regards symmetry, we simplify the discussion by postulating that, in the unperturbed smectic, the directions z and $-z$ are equivalent (no

---

† This statement applies only for static phenomena; to study dynamic effects, we must keep θ as an independent variable, because eqn (7.5), which is an equilibrium relation, breaks down at acoustic frequencies.

‡ Although the detailed definitions differ, there is a relation between m and the 'Poisson ratio' as it is defined in the elasticity of solid bodies [13].

ferroelectricity): this appears to be correct in all clear-cut cases. With all these provisos, we arrive at the following form for F [14].

$$F = F_0 + \tfrac{1}{2}\bar{B}\left(\frac{\partial u}{\partial z}\right)^2 + \tfrac{1}{2}\chi_\mathrm{a} H^2\left[\left(\frac{\partial u}{\partial x}\right)^2 + \left(\frac{\partial u}{\partial y}\right)^2\right] +$$
$$+ \tfrac{1}{2}K_1\left(\frac{\partial^2 u}{\partial x^2} + \frac{\partial^2 u}{\partial y^2}\right)^2 + \tfrac{1}{2}K'\left(\frac{\partial^2 u}{\partial z^2}\right)^2 + \tfrac{1}{2}K''\frac{\partial^2 u}{\partial z^2}\left(\frac{\partial^2 u}{\partial x^2} + \frac{\partial^2 u}{\partial y^2}\right). \quad (7.6)$$

F_0 is the unperturbed free energy. The second term represents an elastic energy for compression of the layers. The third term describes the coupling between a magnetic field H (parallel to z) and the molecular orientation: it can also be written as (see eqn 7.2)

$$\tfrac{1}{2}\chi_\mathrm{a} H^2(n_x^2 + n_y^2) = \mathrm{const} - \tfrac{1}{2}\chi_\mathrm{a}(\mathbf{n}\cdot\mathbf{H})^2. \quad (7.7)$$

The latter form is familiar from our discussion of nematics (Chapter 3). Usually the anisotropy χ_a will be comparable to what we found in that case (namely $\chi_\mathrm{a} > 0$ and $\chi_\mathrm{a} \sim 10^{-7}$ c.g.s. units).

It is important to understand that, in the absence of any orienting field ($H = 0$), there cannot be any term in eqn (7.6) proportional to $(\partial u/\partial x)^2$ or $(\partial u/\partial y)^2$. A constant (small) $\partial u/\partial x$ represents a uniform rotation around the y axis, and this does not alter the internal free energy.

It is thus necessary to include certain higher-order terms, and in particular

$$\tfrac{1}{2}K_1\left(\frac{\partial^2 u}{\partial x^2} + \frac{\partial^2 u}{\partial y^2}\right)^2 = \tfrac{1}{2}K_1(\mathrm{div}\ \mathbf{n})^2, \quad (7.8)$$

where again we make use of eqn (7.2). Equation (7.8) shows that this term is a splay energy, identical in form to what we have in a nematic.†

The last two terms in eqn (7.6) have been included for completeness but are in fact unobservable (with long-range distortions), because they contribute only if the displacement u varies along z. But in this case, they are dominated by the term $\tfrac{1}{2}\bar{B}(\partial u/\partial z)^2$ which is of lower order in the space variations. Thus, in all that follows, we shall omit K' and K''.

Note that the same argument cannot be applied to the splay term K_1: when the variations of u take place only in the layer plane ($\partial u/\partial z = 0$) the $\bar{B}$ term drops out and the elastic effects depend on K_1 alone.

To summarize: the static elastic properties of a smectic A and described by two constants: $\bar{B}$ (dimension energy length^{-3}) and K_1

---

† If the smectic A phase under study is followed, at higher temperatures, by a nematic phase, we expect that K_1 will have comparable magnitudes in both phases.

(dimension energy length^{-1}). It is often convenient to introduce the associated length

$$\lambda = \left(\frac{K_1}{\overline{B}}\right)^{\frac{1}{2}}. \tag{7.9}$$

Usually λ should be comparable to the layer thickness a.†

7.2.1.3. Boundary conditions. Having defined the relevant distortion (and magnetic) energies in eqn (7.6) we are able, in principle, to derive equations for the equilibrium distortion $u(\mathbf{r})$, which minimizes the overall free energy. However (just as in the case of nematics) not much insight is gained by writing down these equations in full generality; we shall directly attack some specific examples, where the algebra is comparatively simple. Most these examples are connected with the competition between wall alignment and field alignment. Thus we must begin by some statements concerning boundary conditions.

Studies on wall alignment effects are rather rare. But in many cases, when put into contact with a glass surface, or with a freshly cleaved solid crystal (e.g. mica) a smectic A will tend to put its layers parallel to the limiting surfaces. The molecules are then normal to the surface. This type of alignment could thus be called homeotropic. It can often be strengthened (as it is in nematics) by coating the solid surface with a detergent, if the aliphatic 'tails' of the detergent are normal to the wall.

Other types of boundary conditions may occur, with the molecules tangential to a glass wall, or even with oblique orientations at the surface.

Very little is known about the *strength* of these boundary effects. In the present discussion, we shall consider only the limit of *strong anchoring* as it is defined in Chapter 3.

Problem: A smectic A is put in contact with an undulating glass surface (Fig. 7.7). The smectic planes stay locally tangent to the surface. Study the distortions inside the smectic (G. Durand and N. Clark, 1972).

Solution: Let z be the local amplitude of the undulation, and consider the case of a sine wave.

$$z(x,y) = \alpha \cos kx \qquad (k\alpha \ll 1).$$

The smectic occupies the region above the surface ($z > 0$): The distortions in this region are of the form

$$u(x,y,z) = u_0(z)\cos kx$$

and the boundary condition imposes $u_0(0) = \alpha$.

---

† However, if we are in the vicinity of a second-order smectic $\rightleftharpoons$ nematic transition B will be small and λ will be larger than a.

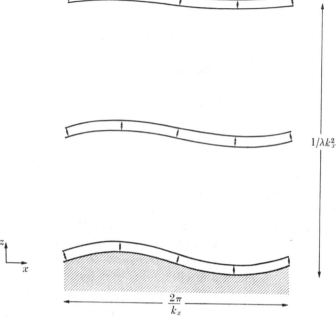

FIG. 7.7. A small undulation at the surface of a smectic has far-reaching effects inside the sample. The penetration length is much larger than for nematics (compare Fig. 3.9).

The elastic energy, described in eqn (7.6), is (after averaging in the x,y plane)

$$F = \tfrac{1}{2}\bar{B}\left(\frac{\mathrm{d}u_0}{\mathrm{d}z}\right)^2 \langle\cos^2 kx\rangle + \tfrac{1}{2}K_1 k^4 u_0^2 \langle\cos^2 kx\rangle$$

$$= \tfrac{1}{4}\bar{B}\left[\left(\frac{\mathrm{d}u_0}{\mathrm{d}z}\right)^2 + \lambda^2 k^4 u_0^2\right].$$

Minimizing the integral of F over the sample volume, we obtain an equation for u_0

$$-\frac{\mathrm{d}^2 u_0}{\mathrm{d}z^2} + \lambda^2 k^4 u^0 = 0.$$

The appropriate solution is†

$$u_0 = \alpha e^{-z/l}$$

where l gives us the thickness of the distorted region, and is given explicitly by

$$l = \frac{1}{k^2 \lambda}.$$

Note that l is much larger than the wavelength $(2\pi/k)$ of the undulation. For instance, if $2\pi/k = 10\ \mu$m, and $\lambda = 20$ Å, $l \simeq 1.4$ mm.

This remark is important from several different viewpoints: e.g. to prepare good single crystals of a smectic, it is clearly necessary to work with glass plates

† The solution is correct only for very weak modulations $\alpha \lesssim a$. For $\alpha > a$ non-linear effects must be included, and they reduce the amplitudes.

which are polished with great care. Also, the distortions react on the capillary properties of a smectic A; consider an interface between the smectic A and an isotropic fluid, the layers being always assumed to be locally parallel to the interface. The surface tension energy associated with an undulation $\zeta(x)$ is (per cm along y)

$$A_0 \int \sqrt{(dx^2 + d\zeta^2)} \simeq \text{const} + \tfrac{1}{2}A_0 \int \left(\frac{d\zeta}{dx}\right)^2 dx$$

(A_0 being the surface tension). For a sinusoidal distortion of amplitude α and wavevector k this gives (per cm^2)

$$\tfrac{1}{2}A_0 k^2 \alpha^2 \langle \sin^2 kx \rangle = \tfrac{1}{4}A_0 k^2 \alpha^2.$$

But, in a smectic A, we must add to this the terms discussed above, namely

$$\int_0^\infty dz \tfrac{1}{4}\bar{B}\left[\left(\frac{du_0}{dz}\right)^2 + \frac{u_0^2}{l^2}\right] = \frac{1}{4}\frac{\bar{B}\alpha^2}{l} = \tfrac{1}{4}l\bar{B}\lambda\alpha^2 k^2.$$

Thus the effective surface tension is increased by

$$\delta A = \bar{B}\lambda.$$

Even if we were able to make A_0 small by the use of suitable detergents, we would be left with a finite δA resulting from the bulk properties of the smectic.

7.2.1.4. Transitions induced by external forces: Helfrich–Hurault effect. Let us consider first a smectic A in a homeotropic texture between two glass plates; the unperturbed layers are parallel to the plane (x, y) of the plates; the molecules are aligned along z. We add on this system a magnetic field H in the x direction, and we assume that the diamagnetic anisotropy χ_a is positive. To minimize the magnetic energy, the system would like to rotate its optical axis. But the layers are strongly clamped at both walls. What will happen then?

If we were dealing with a nematic, we would expect to find, above a certain field threshold, a bend distortion of the Fredericks type, with the following structure for the director (see Chapter 3)

$$n_x(z) \neq 0 \qquad (n_x \ll \text{just above threshold})$$
$$n_y = 0$$
$$n_z \simeq 1.$$

According to eqn (7.2), this would correspond to

$$u(x, y, z) = \text{constant} - x n_x(z)$$

i.e., to layers which cross from one plate to the other, piling up at very high densities near both plates, and giving infinite elastic energies through the $\bar{B}$ term in eqn (7.6).

Thus we must invent another solution, also leading to a non-zero n_x—i.e. to some partial alignment along **H**.

A similar problem, for another layered system, has in fact been encountered in Chapter 6 in connection with cholesterics under electric fields (the Helfrich effect): the layers undergo a periodic distortion along X (Fig. 7.8). In terms of the displacements u, this corresponds to:

$$u(x, z) = u_0(z)\cos kx, \qquad (7.10)$$

where k is a certain wave vector, the optimal value of which will be derived later. Eqn (7.10) still contains a dependence in z, because the

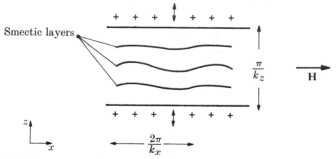

FIG. 7.8. The Helfrich–Hurault transition for smectics A: the transition can be induced in principle by a magnetic field **H**. In practice, it is more convenient to apply a mechanical tension to the limiting plates.

displacement must vanish on both plates. In fact, for small amplitude distortions (just above threshold) we can take $u_0(z)$ as a sine wave, vanishing both for $z = 0$ and $z = d$.

$$u_0(z) = u_0 \sin(k_z z)$$
$$k_z = \frac{\pi}{d}. \qquad (7.11)$$

This distortion of the layers corresponds to an optical axis locally defined by

$$n_x = -\frac{\partial u}{\partial x} = \epsilon \sin(k_z z)\sin kx$$
$$n_y = 0 \qquad (7.12)$$
$$n_z \simeq 1.$$

Let us now write down the elastic free energy $\langle F_{el} \rangle$ derived from eqn (7.6) (and averaged over the sample thickness):

$$\langle F_{el} \rangle = \tfrac{1}{2}\epsilon^2 \left[\bar{B}\left(\frac{k_z}{k}\right)^2 \langle\cos^2 k_z z\rangle\langle\cos^2 kx\rangle + K_1 k^2 \langle\sin^2 k_z z\rangle\langle\cos^2 kx\rangle \right]. \quad (7.13)$$

We must add to this the magnetic term: it differs from eqn (7.6) because, in the present example, the field is in the plane of the layers.

$$\langle F_{mag} \rangle = -\tfrac{1}{2}\chi_a H^2 \langle n_x^2 \rangle = -\tfrac{1}{2}\chi_a H^2 \epsilon^2 \langle \sin^2 k_z z \rangle \langle \cos^2 kx \rangle. \qquad (7.14)$$

All the averages of the form $\langle \sin^2\theta \rangle$ or $\langle \cos^2\theta \rangle$ are equal to $\tfrac{1}{2}$, and the total free energy becomes

$$\langle F_{el} + F_{mag} \rangle = \frac{\epsilon^2}{8}\bar{B}\left[\left(\frac{k_z}{k}\right)^2 + k^2\lambda^2\right] - \frac{\epsilon^2}{8}\chi_a H^2. \qquad (7.15)$$

When the overall coefficient of ϵ^2 in eqn (7.5) is positive for all k values, the unperturbed arrangement is stable. Instability will occur first for that value of k at which the first bracket in (7.15) is minimum. This corresponds to

$$k^2 = k_z/\lambda = \pi/\lambda d. \qquad (7.16)$$

Thus the optimal wavelength of the distortion $(2\pi/k)$ is (apart from numerical coefficients) equal to the geometric mean of the sample thickness d and of the microscopic length b. This result has been verified experimentally on cholesterics, where similar equations hold (λ being then related to the pitch) [15]. But no experiments have been performed, up to now, on smectics: the problem is to have very large samples, or very large fields. To understand this, let us now write down the threshold field H_{c1} obtained when the elastic and magnetic terms cancel exactly in eqn 7.15. We have

$$\chi_a H_c^2 = 2\bar{B}\left(\frac{k_z}{k}\right)^2 = 2\bar{B}k_z\lambda = 2\pi\frac{\bar{B}\lambda}{d}. \qquad (7.17)$$

Eqn (7.17) was derived by Hurault, inspired by an earlier calculation of Helfrich (ref. [32] of Chapter 6). It shows that H_c is proportional to $d^{-\frac{1}{2}}$. This behaviour is quite different from what we have in a conventional Fredericks transition, where $H_c \sim d^{-1}$. Here the field decreases more slowly with sample thickness. Writing eqn (7.17) in the form

$$\chi_a H_c^2 = 2\pi\frac{K_1}{\lambda d}, \qquad (7.18)$$

and using the estimates $K_1 = 10^{-6}$ dynes, $\lambda = 20$ Å, $d = 1$ mm, $\chi_a = 10^{-7}$, we arrive at $H_c \sim 60$ kG. To bring the value of H_c in a more convenient range (say 20 kG) we would need smectic monocrystals of thickness 1 cm. Samples of this size are in fact beginning to be prepared. However, there is one serious difficulty: above the threshold field H_c (say, for $H : 2H_c$) the distortion amplitude ϵ remains very small: the undulation of the layers is strongly limited by the requirement of (nearly) constant interlayer thickness: in such a situation we say that we have a 'ghost transition,' which is formally present but hard to observe.†

7.2.1.5. Transitions induced by external forces: undulation by mechanical tension.   When a homeotropic slab of a smectic A is put under mechanical tension, one observes a periodic undulation of the layers of the Helfrich Hurault type, with a wavelength $2\pi/k$ proportional to the square root of the sample thickness d, as in eqn (7.16). This remarkable deformation appears only under tension (not under compression). It has been found and studied by Delaye, Durand, and Ribotta [16a], through the diffraction of light on the periodic undulations. The effect was also predicted independently by the Harvard group [16b].

This 'mechano-optic effect' is easily understood, if we assume that, during the time under study, the number of layers present in the sample thickness is constant. (Changes in this number must involve dislocation motion, and will be discussed later.) Then, when the system is expanded, there are two possible behaviours: either the layers expand uniformly (and this costs a rather high energy, proportional to the elastic modulus $\bar{B}$) or they keep a nearly constant thickness, but manage to fill the available space by undulations as shown on Fig. 7.8 at high strains the latter solution is preferable.

The threshold may be analysed in terms of a simple generalization of the elastic energy [6] allowing for finite strains. This generalization is useful in many practical problems and will be described here in some detail. It is still sufficient to write the energy associated to a dilation γ of the layers in the form

$$F_{\text{dil}} = \tfrac{1}{2}\bar{B}\gamma^2. \tag{7.19}$$

But the linear form $\gamma = \partial u/\partial z$ must be improved. Consider first a case where the layers are tilted uniformly:

$$\frac{\partial u}{\partial z} = 0$$

$$\frac{\partial u}{\partial x} = -n_x \qquad (\ll 1).$$

Remember now that, in our problem, the total number of layers is constant: this means that the distance between them, *measured along z*

---

† Another geometry leading to a similar ghost transition for a smectic A under a magnetic field is discussed in more detail in ref. [28].

is still equal to the unperturbed interval a. But the true distance between them is a $\cos n_x \cong a(1 - \tfrac{1}{2}n_x^2) < a$.

Thus in this case, we have a small (second order) dilation

$$\gamma = -\frac{n_x^2}{2} = -\frac{1}{2}\left(\frac{\partial u}{\partial x}\right)^2.$$

Returning now to the general case (where u depends both on x and z, but the number of layers is still kept constant) we have

$$\gamma = \frac{\partial u}{\partial z} - \frac{n_x^2}{2}. \tag{7.20}$$

The elastic energy is still of the form

$$F = \tfrac{1}{2}\bar{B}\gamma^2 + \tfrac{1}{2}K_1\left(\frac{\partial^2 u}{\partial x^2}\right)^2 \tag{7.21}$$

γ being given by eqn (7.20). Let us now split the displacements u (or the resulting dilations) into two terms

$$u = u_0 + u_1(x, z)$$
$$\gamma = \gamma_0 + \gamma_1 \tag{7.22}$$

where $u_0 = \gamma_0 z$ is the uniform displacement associated with the imposed strain γ_0 ($\gamma_0 > 0$), while u_1 will describe the undulation. We are interested in the threshold for the onset of the undulations: thus γ_0 and u_0 are finite, while u_1 is infinitesimally small. Let us then expand the free energy (21) up to order u_2^1. We find:

$$F = F_0 + \bar{B}\,\gamma_0\frac{\partial u_1}{\partial z} - \bar{B}\,\gamma_0\frac{n_x^2}{2} + \tfrac{1}{2}K_1\left(\frac{\partial^2 u_1}{\partial x^2}\right)^2. \tag{7.23}$$

The term linear in $\partial u_1/\partial z$ gives 0 by integration on the z variable (since u_1 is defined as vanishing on both plates). The new term of interest is the third term. It is identical in form to the effect of a destabilizing magnetic field H, the correspondence being given by

$$-\tfrac{1}{2}\chi_a H^2 n_x^2 = -\tfrac{1}{2}\bar{B}\,\gamma_0 n_x^2. \tag{7.24}$$

Thus all our discussion of the Helfrich–Hurault threshold may be transposed to the present case: the spatial wavelength of the unstable mode is still given by eqn (7.16), and the threshold value of the strain γ_{oc} is obtained from eqns (7.17) and (7.24):

$$\gamma_{oc} = 2\pi\frac{\lambda}{d}. \tag{7.25}$$

This corresponds to an increase in sample thickness $2\pi\lambda \sim 150$ Å in typical cases. It is thus quite easy to achieve strains γ_0 which are much larger than γ_{oo}. In this case, the distortions become quite visible: the mechanical coupling is much more convenient than the magnetic coupling.

Theoretical predictions on the behaviour above threshold, based on non-linear elastic terms of the type (7.20), have been given recently by Debrieu (to be published). He shows in particular that the distortion will occur in the form of a square lattice (rather than a triangular, or hexagonal, lattice for instance). See Fig. 7.9 between pp. 260–1.

It must be emphasized that the distortions induced by tension are metastable; if we wait long enough, the number of layers present in the sample will tend to adjust itself through the motion of dislocations.† In practice, the stability is found to depend strongly on the amplitude of the perturbation γ_0: if γ_0 is not very large (although larger than γ_{oo}) the undulations persist only for a short time; while if γ_0 is large, the periodic distortions are not smooth any more (defects such as focal conics† may be present). They are then much harder to heal; we have here the analogue of the storage mode of cholesterics, discussed in Chapter 6.

7.2.1.6. Transitions induced by external forces: thermo-optic effect in smectics A. Another experimental set up, generating the Helfrich–Hurault undulations in smectics A, has been discovered by Kahn [16c].‡ A homeotropic sample is placed under a strong light beam, and is thus slightly heated because of the finite absorption of the material. The light intensity is then abruptly reduced: the sample cools down and the layers tend to contract. If the sample thickness d is fixed, this may be achieved by undulations; a distortion appears in the form of a very fine square lattice.

This Kahn effect suggests some interesting technical applications, since it allows us to store information on a very small scale in a smectic slab: the information can be written, read, and erased by light beams of suitable intensity [16c].

7.2.1.7. Fluctuations. Let us return to an unperturbed, single domain, smectic A and investigate the spontaneous fluctuations of the positions of the layers. These fluctuations control in particular the scattering of light, and the scattering of X-rays.

† A brief discussion of dislocations and focal conics in smectics is given in Section 7.2.1.6.

‡ The Kahn effect was found before the mechano-optic effect, but from a pedagogical point of view it is convenient to reverse the order.

To determine the amplitude of the fluctuations, we start from eqn (7.61) (putting $H = 0$ for simplicity) and go to the Fourier transform u_q of the displacement $u(\mathbf{r})$. In terms of the u_qs, the free energy F takes the form

$$F = \sum_q \tfrac{1}{2}(\bar{B}q_z^2 + K_\perp q_\perp^4)\,|u_q|^2. \tag{7.26}$$

where $q_\perp^2 = q_x^2 + q_y^2$. Applying the classical equipartition theorem to equation (26) we arrive at the following thermal averages

$$\langle |u_q|^2 \rangle = \frac{k_B T}{\bar{B}q_z^2 + K_1 q_\perp^4} = \frac{k_B T}{\bar{B}(q_z^2 + \lambda^2 q_\perp^4)}\,. \tag{7.27}$$

Let us now discuss how these fluctuations may be detected by light scattering [14]. In the unperturbed state, the dielectric tensor $\boldsymbol{\varepsilon}$ (at the light frequency on interest) has three non-vanishing components

$$\epsilon_{xx} = \epsilon_{yy} = \epsilon_\perp$$
$$\epsilon_{zz} = \epsilon_\parallel = \epsilon_\perp + \epsilon_a. \tag{7.28}$$

In the fluctuating system, we will first have some modifications in the magnitude of $\epsilon_\parallel$ and $\epsilon_\perp$, due to changes in the density and in the inter-layer distance; they will be similar to what is found in convention liquids, and will not give rise to any spectacular effect (see ref. [14] for more detailed discussion of this point). We also expect some deviations of the optical axis $\mathbf{n}$ from the normal to the layers, contributing to the depolarized scattering—but, as already mentioned, these deviations are not important at small wave-vectors q ($q\lambda \ll 1$).

Finally, the most interesting contribution is related to possible *rotations* of the layer system, the molecules remaining normal to the local layers. The fluctuations $\delta\boldsymbol{\varepsilon}$ of the dielectric tensor associated with such rotations may be obtained by writing (as was done in Chapter 3 for nematics)

$$\boldsymbol{\varepsilon} = \epsilon_\perp + \epsilon_a \;\; \mathbf{n}{:}\mathbf{n}$$
$$\delta\boldsymbol{\varepsilon} = \epsilon_a[\delta\mathbf{n}{:}\mathbf{n} + \mathbf{n}{:}\delta\mathbf{n}].$$

Using eqn 7.2 this gives two non-vanishing terms

$$\delta\epsilon_{xz} = -\epsilon_a \frac{\partial u}{\partial x}$$
$$\delta\epsilon_{yz} = -\epsilon_a \frac{\partial u}{\partial y}\,. \tag{7.29}$$

Assuming that ϵ_a is not too large, the scattered intensity (for a given scattering wave-vector $\mathbf{q}$) is given by the Born approximation formula

$$I(q) = \langle |\mathbf{i} \cdot \delta\boldsymbol{\varepsilon}(\mathbf{q}) \cdot \mathbf{f}|^2 \rangle \qquad (7.30)$$

Thus, to have a strong scattering, it is favorable to have the ingoing polarization vector $\mathbf{i}$ along z and the outgoing polarization $\mathbf{f}$ in the x, y plane (or the reverse). Let us assume that this condition is realized. Then, using eqn (7.27), (7.29), (7.30) we arrive at

$$I = \frac{k_B T}{\bar{B}} \, \epsilon_a^2 \, \frac{q^2}{q_z^2 + \lambda^2 q_\perp^4} \, . \qquad (7.31)$$

Eqn (7.31) predicts the existence of two widely different regimes:

(1) If $\mathbf{q}$ is oblique ($q_z \sim q_\perp \neq 0$), we may write

$$I \sim \epsilon_a^2 \frac{k_B T}{\bar{B}} \frac{q_\perp^2}{q_z^2} \, , \qquad (7.32)$$

since $q\lambda \ll 1$ for visible light. The intensity I is then of the order of $k_B T / \bar{E}$, i.e. comparable to what is found in a conventional liquid. Note that the 'oblique' situation corresponds to the majority of scattering events in transmission experiment; this eqn (7.32) indicates that a smectic A (in a single domain texture) should not be turbid like a nematic. This is indeed what is found.

(2) If $\mathbf{q}$ is in the plane of the layers ($q_z = 0$), the intensity becomes

$$I = \epsilon_a^2 \frac{k_B T}{\bar{B} \lambda^2 q^2} = \epsilon_a^2 \frac{k_B T}{K_1 q^2} \, . \qquad (7.33)$$

Eqn (7.33) has the form found in Chapter 3 for scattering by nematics, and predicts a very high intensity. Physically, the modes $q_z = 0$ are pure undulation modes (Fig. 7.10) for which the layers are deformed but the interlayer distance is not altered; fluctuations of this type require very little energy, and have large amplitudes.

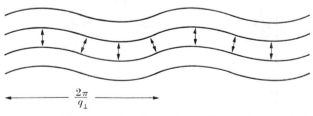

$$\frac{2\pi}{q_\perp}$$

FIG. 7.10. The pure undulation mode of smectics: in this mode the interlayer distance is unaffected.

The condition $q_z = 0$ restricts the intense scattering to a cone (Fig. 7.11). The angle of this cone is very sharply defined: from eqn (7.31) we see that I gets large when

$$q_z^2 < \lambda^2 q^4$$

$$\frac{q_z}{q} < \lambda q \sim 10^{-3}. \tag{7.34}$$

This implies that scattering experiments must be performed only on excellent single domain samples; also the thickness d of the sample must be large, to ensure that the undulation mode is not clamped

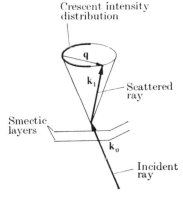

FIG. 7.11. Distribution of the light scattering intensity in an ideal smectic sample. For simplicity, the differences in refractive indices for the two polarizations i and f have been neglected: then $\mathbf{k_0}$ and $\mathbf{k_1}$ have the same length.

by the limiting surfaces. [The requirement on d is obtained by replacing $q_z \rightarrow \pi/d$ in eqn (7.34).] In practice, the light scattering due to thermal undulations has not yet been observed: it is masked by static undulations due to irregularities in the walls. See page 288 and also N. Clark, P. Pershan, *Phys Rev. Lett.* **30**, 3, 1973, R. Ribotta, G. Durand, J. D. Litster, *Sol. St. Comm.* **12**, 27, (1973).

Let us now discuss the effect of the fluctuations on the *scattering of X-rays* by smectics A. One important remark, in this connection, was first made by Peierls and Landau [17]: the mean square fluctuations $\langle u^2(\mathbf{r}) \rangle$ at one point $\mathbf{r}$, are *divergent* (in zero field). More generally, extending eqn (7.27) to the case of a finite field H (applied along the optical axis, and assuming $\chi_a > 0$) one finds [14]

$$\langle u^2(\mathbf{r}) \rangle = \frac{k_B T}{4\pi \bar{B} \lambda} \log(\xi_1/a) \tag{7.35}$$

where $\xi_1 = (K_1/\chi_a)^{\frac{1}{2}}H^{-1}$ diverges for $H \to 0$. In practice, the divergence of $\langle u^2 \rangle$ is weak, but conceptually it is important; it shows that (in zero field) the actual stacking of the layers cannot correspond to our usual notion of long-range order.

At this point it may be useful to review briefly what would be the consequences of conventional long-range order from the point of view of X-ray scattering. The main feature is that the intensity $I(\mathbf{q})$ when studied as a function of scattering wave vector $\mathbf{q}$, shows peaks which are δ functions in $\mathbf{q}$ space, at certain positions in the reciprocal lattice (Bragg peaks)

$$\mathbf{q} = \boldsymbol{\tau}.$$

For the case at hand $\tau_x = \tau_y = 0$, $\tau_z = n(2\pi/a)$ (n = integer). If we have thermal motion, but still retain long-range order, two new effects occur.

(1) We have some diffuse scattering at all q values.

(2) The intensity of the Bragg peak (τ) is reduced by a certain 'Debye–Waller factor' [18], related to the fluctuations of $\langle u^2(\mathbf{r}) \rangle$

$$\frac{I}{I_{\text{ideal}}} = \exp(-\tau^2 \langle u^2 \rangle). \tag{7.36}$$

Now let us turn to the actual situation in smectics A: $\langle u^2 \rangle$ is infinite, and the Debye–Waller factor vanishes; there cannot be δ function peaks in the scattering. There will remain, however, some *weaker singularities* at the nominal Bragg positions. These singularities have been studied theoretically in two very different cases:

In a two-dimensional model system built in analogy with the lamellar phase of soaps [19]. The model is strongly anharmonic, but is exactly soluble. The singularities in this case are found to be logarithmic.

On the three-dimensional problem, assuming a quadratic free energy of the form (7.26).† The calculation has been performed by A. Caillé [20], and leads to a singularity in $I(\mathbf{q})$ near $\mathbf{q} = \boldsymbol{\tau}$ which involves a characteristic exponent $x(T)$ varying with the temperature. This unusual form is also found theoretically in certain other 'quasi ordered' systems, but has not yet been checked experimentally.‡

7.2.1.8. Textures and defects. (1) *Terraced droplets.* We have seen that the smectics A often accept to be prepared in a homeotropic conformation between two glass plates. What happens if we use only one

---

† This neglects anharmonic effects, and is probably acceptable in most cases. However the assumption may break down in the vicinity of a smectic A ⇌ nematic second order transition.

‡ The experiments may be complicated by dislocation effects.

plate (made of glass or mica, etc) and leave one free surface? With due care one can obtain the 'terraced droplet shown in Fig. 7.12 (see facing page 261). This was discovered long ago by Grandjean and is known in the French literature as a *goutte à gradins*. The thickness of each terrace is macroscopic (typically 10 μm or more), but the whole figure is very suggestive of a layered material. Optical observations also show that the structure is quite fluid: successive terraces can slip on each other very easily.

Note that the lateral walls terminating one terrace are not simple vertical boundaries but have a more complex structure [3].

(2) *Focal conics.* The planar stacking of layers is easily deformed, *provided that the interlayer distance is maintained.*

Then the strong elastic term $\bar{B}$ (in eqn 7.6) does not contribute to the energies, and only the weak curvature term K_1 is involved. In terms of the director **n**, it may be useful to point out that, when the interlayer thickness a is unaltered, the only allowed distortion is a splay; bend and twist are forbidden. This may be shown as follows: for a system of curved layers, the curvilinear integral

$$\nu_{AB} = \int_A^B \mathbf{n} . d\mathbf{r} \, \frac{1}{a(\mathbf{r})}$$

measures the number of layers crossed when going from point A to point B on the integration path. In the absence of all defects (dislocations, etc) this number ν_{AB} must be independent of the path chosen between A and B. This imposes that the integral on any closed contour vanishes

$$\oint \mathbf{n} . d\mathbf{r} \, \frac{1}{a} = 0$$

or

$$\operatorname{curl}\left(\frac{1}{a}\,\mathbf{n}\right) = 0.$$

If a is independent of **r** (constant interlayer distance) this leads to curl **n** = 0. Then both bend (**n** $\times$ curl **n**) and twist (**n** . curl **n**) deformations are forbidden.

Assuming now that a keeps its unperturbed value, what will be the aspect of the allowed splay deformations?

(a) The simplest case is shown in Fig. 7.1a, where the layers are closed into cylinders around an axis Γ_1. This type of cylindrical arrangement is found in many cases, and in particular during the

growth of an A phase from an isotropic phase consisting of smecto-
gen + solvent. (The growth process has been studied recently in detail
by R. B. Meyer, to be published.) Textures showing such cylindrical
rods are commonly called 'myelinic', by analogy with the myelin
coating of certain nerves.

(b) We can bend the rod, and close it on itself, thus obtaining a
torus (Fig. 7.1b). Note that the arrangement now involves *two*
singular lines: a circle Γ_1 and the torus Γ_2. The detailed configuration
of the layers in the close vicinity of these lines is discussed in P. G. de
Gennes, *C.r.*, *Acad. Sci.*, Paris (1972), *275B*, 549.

(c) The most general possibility is achieved when one replaces
Γ_1 by an ellipse, and Γ_2 by a hyperbola (Fig. 7.1c). These are the
focal conics analysed first by G. Friedel [2]. A stacking of elements
such as the one shown in Fig. 7.1c gives the most common textures
'fans', 'polygons', etc. (see ref. [3] and [29]).

(3) *Dislocations*. The fundamental rule for distortions in smectics is
that twist and bend of the director are forbidden (see eqn 7.3): in an
ideal focal conic arrangement we have only splay. However, the
boundary conditions (or the action of external fields) may tend to
impose some twist or some bend. Is there a way to violate the rule in
such a case?

The answer is yes, if we insert in the structure a certain density of
dislocation lines. The nature of these lines in smectic systems has been
discussed by Friedel and Kleman (ref. 5 of Chapter 4). The form of the
long range distortion around one edge dislocation is shown on Fig.
7.13. See P. G. de Gennes, *C.r.*, *Acad. Sci.* (*Paris*), *275B*, 939 (1972).
For instance, Fig. 7.14 shows how a certain amount of bending can be

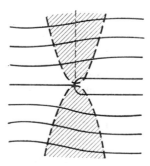

Fɪɢ. 7.13. An edge dislocation in a smectic A. The distortions induced by the dislocation
decrease rapidly along x, but only slowly (as $|Z|^{-\frac{1}{2}}$) along the optical axis. They are
practically confined to two parabolic regions. (Hatched areas.)

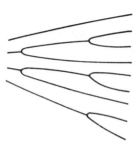

FIG. 7.14. A finite density of edge dislocations allows to introduce some bend in a smectic A.

obtained when edge dislocations are allowed. Evidence for some departures from ideality in focal conic textures, showing violations of the 'splay only' rule, is discussed by Bouligand in ref. [3].

7.2.2. Dynamics of smectics A

A simplified version of the dynamical equations for smectics A (neglecting damping and permeation effects) can be found in ref. [14]. A much more complete presentation has recently been constructed by Martin, Parodi, and Pershan [21]. Here we shall attempt to steer a middle course between these two extremes, and to present the basic features in a very low-brow style.

7.2.2.1. Principles. As explained earlier in Section 7.2.1, for dynamical studies we must treat the displacement u of the layers and the bulk dilatation as independent variables. To these two variables are associated two different sets of velocities:

The velocity of the molecules $\mathbf{v}$, which satisfies the conservation law

$$\text{div } \mathbf{v} + \dot{\theta} = 0. \tag{7.37}$$

The (vertical) velocity of the layers $\dot{u}$.

The acceleration equation for $\mathbf{v}$ has the form:

$$\rho \frac{d\mathbf{v}}{dt} = -\nabla p + \mathbf{g} + \nabla . \boldsymbol{\sigma}'. \tag{7.38}$$

Let us first explain the meaning of the various terms in eqn (7.38):

(1) The pressure p is derived from the energy density U by the equation†

$$p = -\frac{\partial U}{\partial \theta}. \tag{7.39}$$

---

† Eqn (7.39) omits the analogue of the Ericsen terms for nematics: they are quadratic in the amplitudes u and may be neglected for our study of small motions.

The explicit form of U is similar to what has been obtained for the free energy F, eqn (7.6). But we now have two types of dilatations: θ (bulk) and $\gamma = \partial u/\partial z$. Also the elastic coefficients are now defined for an adiabatic situation rather than for an isothermal one; this leads to the form

$$U = \tfrac{1}{2}A_0\theta^2 + C_0\theta\gamma + \tfrac{1}{2}B_0\gamma^2 + \tfrac{1}{2}\chi_a H^2\left[\left(\frac{\partial u}{\partial x}\right)^2 + \left(\frac{\partial u}{\partial y}\right)^2\right] +$$

$$+ \tfrac{1}{2}K_1\left(\frac{\partial^2 u}{\partial x^2} + \frac{\partial^2 u}{\partial y^2}\right)^2. \quad (7.40)$$

Thus, to first order in the amplitudes

$$p = -(A_0\theta + C_0\gamma). \quad (7.41)$$

(2) The force $\mathbf{g}$ in eqn (7.38) is parallel to Oz: it is the restoring force acting on the layers, and due to the energy U. Explicitly from (7.40) one finds

$$g = \frac{\partial}{\partial z}(C_0\theta + B_0\gamma) + \chi_a H^2\left(\frac{\partial^2 u}{\partial x^2} + \frac{\partial^2 u}{\partial y^2}\right) - K_1\left(\frac{\partial^2}{\partial x^2} + \frac{\partial^2}{\partial y^2}\right)^2 u, \quad (7.42)$$

(3) Finally, the tensor $\boldsymbol{\sigma}'$ in eqn (7.38) describes friction effects. Note that $\boldsymbol{\sigma}'$ may be taken as *symmetrical*: the external torques (due to the field H) being already taken into account *via* $\mathbf{g}$.
The entropy source may now be written as

$$T\dot{S} = \boldsymbol{\sigma}':\mathbf{A} + g(\dot{u} - v_z) + \mathbf{E}.\mathbf{J} \quad (7.43)$$

where $\mathbf{A}$ is the shear rate tensor. The first term is familiar. The second term describes *permeation* in the Helfrich sense; it vanishes for cases of uniform translation ($\dot{u} = v_Z = $ constant). The third term is associated with usual transport effects: to be specific, we shall say that $\mathbf{J}$ is a heat current, in which case $\mathbf{E} = -\nabla T/T$.

Taking as fluxes $\mathbf{A}$, g, and $\mathbf{E}$, and as forces the quantities $\boldsymbol{\sigma}'$, $\dot{u} - v_z$, and $\mathbf{J}$, we may then write the following phenomenological equations:

$$\sigma'_{\alpha\beta} = \alpha_0\,\delta_{\alpha\beta}A_{\mu\mu} + \alpha_1\,\delta_{\alpha z}\,\delta_{\beta z}A_{zz} + \alpha_4 A_{\alpha\beta} +$$

$$+ \bar{\alpha}_{56}(\delta_{\alpha z}A_{z\beta} + \delta_{\beta z}A_{z\alpha}) + \alpha_7\,\delta_{\alpha z}\,\delta_{\beta z}A_{\mu\mu}$$

$$\dot{u} - v_z = \lambda_p g + \mu E_z$$

$$J_z = \mu g + \sigma_\parallel E_z$$

$$J_\alpha = \sigma_\perp E_\alpha \quad (\alpha = x, y). \quad (7.44)$$

The equation for $\boldsymbol{\sigma}'$ introduces five viscosity coefficients (as usual for an anisotropic, compressible fluid with one preferred axis). Note that symmetry does not allow any coupling between $\boldsymbol{\sigma}'$ and g or $\mathbf{E}$. λ_p may be called the permeation coefficient. Finally μ describes coupling between material flow and the field E_z, which is familiar in membrane physics [22].

The equations (7.37), (7.38), (7.44) allow in particular to discuss the modes of small amplitude, in a smectic A, for an arbitrary wave vector $\mathbf{q}$ (provided that $\mathbf{q}$ is small on the molecular scale). The results depend critically on the orientation of $\mathbf{q}$: a few cases will be discussed below.

7.2.2.2. The undulation mode.   This type of mode has been shown on Fig. 7.10 and the light scattering intensities associated with it have been discussed in Section 7.2.1. It corresponds to a wave-vector $\mathbf{q}$ in the plane of the layers (in what follows we shall take $\mathbf{q}$ along the x-axis). For this particular mode there is no change in the interlayer spacing; the only restoring forces g are weak curvature forces, just as in a nematic. The mode is then very strongly damped, and so slow that the conditions are usually isothermal rather than adiabatic.

For simplicity we shall also assume $H = 0$. Then the equations of motion reduce to

$$g = -(K_1 q^2)q^2 u = \lambda_p^{-1}(\dot{u} - v_z) \tag{7.45}$$

$$\rho \dot{v}_z = -\eta_u q^2 v_z + g \tag{7.46}$$

where we have introduced an effective viscosity

$$\eta_u = \tfrac{1}{2}(\alpha_4 + \alpha_{56}). \tag{7.47}$$

The inertial term ρv_z in eqn (7.46) is negligible: eliminating v_z between (7.45) and (7.46) we thus arrive at

$$\dot{u} + \frac{K_1 q^2}{\eta_u}\, u + K_1 \lambda_p q^4 = 0. \tag{7.48}$$

The last term may be omitted for small q, and we see that the undulation relaxes with a rate

$$\frac{1}{\tau_q} = \frac{K_1 q^2}{\eta_u}\,. \tag{7.49}$$

It is also possible to investigate the amount of permeation from these equations: one finds

$$\frac{-\dot{u} + v_z}{\dot{u}} = \lambda_p \eta_u q_\perp^2 \ll 1.$$

Thus, permeation is very weak (in fact, in the original derivation of eqn (7.49) [14] permeation was omitted from the start). The relaxation properties of the undulation mode can in principle be studied (as in the case of nematics) by inelastic scattering of light (photon-beat techniques). The intensity is favourable; the main problem is to prepare a very good single crystal.

7.2.2.3. Acoustic waves. When the wave vector $\mathbf{q}$ is oblique with respect to the layers (e.g. $\mathbf{q}$ in the $[x, z]$ plane) the situation is very different: the displacement u must be associated with some modulation of the interlayer spacing. Similarly, the velocity v_x is linked to a bulk dilatation θ. For this reason, one expects *two types* of acoustic waves, describing coupled oscillations of the density and of the interlayer spacing [14]. The following features must be noted:

(1) Damping effects are weak: the frequencies are mainly real
(2) Permeation is again negligible ($v_z \cong \dot{u}$),
(3) The waves are adiabatic rather than isothermal.

These waves should be observable either by conventional acoustical methods, or by Brillouin scattering of light, and experiments are currently attempted in both directions. Preliminary data [23] suggest that the rigidity coefficient A_0 (eqn 7.40) is significantly larger than the two other coefficients B_0 and C_0. In this limit the oscillations of θ and of the layer system become nearly decoupled. One acoustic branch is associated with density fluctuations and has a velocity:

$$c_1 \simeq \left(\frac{A_0}{\rho_0}\right)^{\frac{1}{2}} \tag{7.50}$$

essentially independent of the direction of propagation. This is called a 'first sound'. The other acoustic branch describes changes in the interlayer distance, which take place at a nearly constant density. The velocity c_2 for this branch is much smaller than c_1, and strongly dependent on the angle f between $\mathbf{q}$ and z.

$$c_2 \simeq \left(\frac{B_0}{\rho}\right)^{\frac{1}{2}} \sin f \cos f. \tag{7.51}$$

This branch is called 'second sound': this name is reminiscent of superfluid helium, where two phonon branches also exist. In the latter case, second sound is associated with fluctuations of the phase of a complex order parameter. In smectics A, the displacement u is closely related to a phase—hence the analogy. (This point will be described in more detail in the next section.)

The calculation of c_2 in the limit of weak coupling to the first sound is instructive and comparatively simple. It will be briefly described here. In this limit, eqn (7.37) reduces to an incompressibility condition: putting the wave vector $\mathbf{q}$ in the xz plane, this has the form:

$$\frac{\partial v_z}{\partial z} + \frac{\partial v_x}{\partial x} = 0.$$

For the force g, eqn (7.42), all field and curvature effects may be dropped. Also the $C_0\theta$ term is negligible (since $\theta \to 0$). Thus $g \to B_0 \partial\gamma/\partial z$, and the acceleration eqn (7.38), neglecting all damping terms, reduces to:

$$\rho\dot{v}_z = -\frac{\partial p}{\partial z} + B_0 \frac{\partial \gamma}{\partial z}$$

$$\rho\dot{v}_x = -\frac{\partial p}{\partial x}.$$

The pressure may be eliminated using the incompressibility condition:

$$p = -\frac{1}{q^2} B_0 \frac{\partial^2 \gamma}{\partial z^2} = \frac{q_z^2}{q^2} B_0 \gamma$$

Finally, permeation is negligible ($v_z = \dot{u}$). This implies $\dot{\gamma} = \partial v_z/\partial z$. Inserting these results into the acceleration equations we get:

$$\rho\gamma = \left(1 - \frac{q_z^2}{q^2}\right) B_0 \frac{\partial^2 \gamma}{\partial z^2} = -B_0 \frac{q_x^2 q_\parallel^2}{q^2} \gamma$$

which leads finally to eqn (7.51).

The angular dependence of c_2 is remarkable. Eqn (7.51) shows that $c_2 = 0$ when $f = \frac{1}{2}\pi$. This case corresponds to the undulation mode discussed earlier. But c_2 also vanishes when $f = 0$. This feature was recognized, but left unexplained, in ref. [14]. It has been fully clarified by Martin, Parodi, and Pershan [21]: for $f \to 0$ permeation effects become very important. They are summarized in the next paragraph.

Apart from the acoustic waves (c_1 and c_2) which we have presented here, one also finds, for a given $\mathbf{q}$ vector, another mechanical mode associated to the velocity component v_y (normal to $\mathbf{q}$ and to the optical axis). This is a damped mode of hydrodynamic shear (uncoupled from u and θ), with a relaxation rate of the form $\eta_{\text{eff}} q^2/\rho$—where η_{eff} is a certain average of the friction coefficients α, and depends on the angle f.

7.2.2.4. Permeation effects.

The *apparent* viscosities of smectics, as measured by capillary methods [24], are enormous. This does *not* mean that the friction coefficients α introduced in eqn (7.44) are anomalously large. The effect can be understood in terms of a permeation process as first suggested by Helfrich [25].

The principle of this explanation has already been discussed in Chapter 6, in connection with cholesteric systems: it is fundamentally related to the existence of layers. If, in a capillary, we have a blocked texture, with fixed layers normal to the capillary axis (z), a certain pressure head $p' = -\partial p/\partial z$ will be required to pass the molecules through the layers. This can be related to our dynamical eqn (7.38) and (7.44). For a uniform flow (no gradients) in steady state we must have, from eqn (7.38)

$$-\nabla p + \mathbf{g} = 0. \qquad (7.52)$$

Let us now consider the second eqn (7.44), in an isothermal situation ($\mathbf{E} = 0$) and with blocked layers ($\dot{u} = 0$). We obtain:

$$-v_z = \lambda_p \nabla p. \qquad (7.53)$$

The flow rate is thus proportional to the pressure head and independent of the flow diameter: this is very different from the usual Poiseuille law and explains the huge apparent viscosities. The permeation coefficient λ_p can in principle, be extracted from flow studies in well-controlled textures.

Another possible approach to the permeation coefficient λ_p amounts to study the spontaneous fluctuation modes of the layers: in particular the modes of wave-vector $\mathbf{q}$ *collinear with the optical axis* ($q_\perp = 0$, or $f = 0$) are dominated by the permeation process.

To understand this, let us restrict our attention to one simple limiting case. We assume that:

(1) The changes in density are negligible [$\theta = 0$, and thus $v_z = 0$ as can be seen from eqn (7.37)].

(2) The temperature fluctuations relax faster than the fluctuations of u (no thermal gradients).

With these assumptions in mind, we return to the second equation (7.44) which now reads

$$\dot{u} = \lambda_p g. \qquad (7.54)$$

The explicit form of g is obtained from eqn (7.42), but using now isothermal rigidity coefficients B, C

$$g = \frac{\partial}{\partial z}[C\theta + B\gamma] \simeq B\frac{\partial\gamma}{\partial z}$$

$$= B\frac{\partial^2 u}{\partial z^2}. \qquad (7.55)$$

Comparing (7.54) and (7.55) we see that the displacement u obeys a simple diffusion equation [21]:

$$\dot{u} = \lambda_p B \frac{\partial^2 u}{\partial z^2}.$$ (7.56)

This shows that, when the angle f between $\mathbf{q}$ and the normal to the layers tends to zero, the second sound mode degenerates into a permeation mode, with a purely viscous behaviour, and a relaxation rate

$$\frac{1}{\tau_q} = \lambda_p B q^2.$$ (7.57)

In most cases we expect that $B\lambda_p$ will be comparable (although not identical) to the self-diffusion coefficient $D_{\parallel}$ of the constituent molecules (measured along the optical axis).†

7.2.3. Smectics C

The existence of tilted smectic phases has been accepted for a long time [1a]. However, it has been clearly demonstrated, by optical measurements on a single domain sample, only recently [4]. The symmetry is of the monoclinic type (Fig. 7.3). If the $\mathbf{c}$ director is along x, the (z, x) plane is a plane of symmetry.‡ There is also a centre of symmetry (at any mid-point in the layers).§ As discussed in section 7.1.1.2, a tensor such as the dielectric tensor has three non-equivalent axes: the first nearly parallel to the direction of alignment (in the symmetry plane), the second one normal to the first in this same plane, and the third along y.¶ In practice it appears that for many tensorial properties, the values measured along the second and third axis are nearly equal; the medium is nearly uniaxial, but with an axis tilted (by an angle ω) from the normal to the layer. Values of $\omega(T)$ for TBBA are shown on Fig. 7.15.

As already emphasized in Section 7.1, a smectic C has two distinct types of degrees of freedom.

(1) The 'C director' $\mathbf{c}$: in the case of Fig. 7.3 this is the unit vector of the x-axis. But of course it is possible to rotate the vector $\mathbf{c}$ around

---

† Eqn (7.57) holds only if the fluctuations of u are essentially isothermal: this requires that the effective diffusion coefficient $\lambda_p B$ be smaller than the thermal diffusivity D_t (= thermal conductivity/specific heat).

‡ We assume for the moment that the constituent molecules are non-chiral: thus planes of symmetry are allowed. We shall discuss briefly later the distortions which occur when chirality is introduced.

§ No case of ferroelectricity in smectics C is known for the moment.

¶ Note that the first axis need not be exactly the same for two different physical properties.

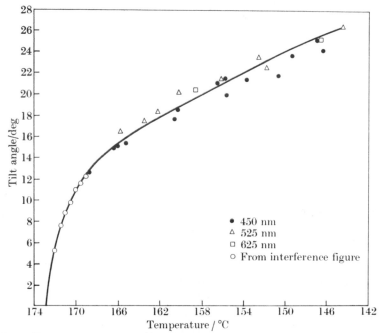

FIG. 7.15. Variation of the tilt angle with temperature in TBBA. After T. R. Taylor, J. Fergason, S. Arora, *Phys. Rev. Letts.* **25**, 722 (1970).

z, without changing the free energy; **c** is thus somewhat similar to the director in a nematic, and smectics C have many features of nematics. To describe situations where **c** is rotated, it will often be convenient to introduce the corresponding rotation angle, which we shall call Ω_z.

(2) the 'vertical' (along z) displacement of the layers $\mathbf{u}(r)$ is defined as in smectics A. Again, a uniform displacement does not change the energy. Thus the long wavelength fluctuations of u are inexpensive, and have large amplitudes; all the related effects which we discussed for smectics A have their counterparts here.

We shall now sketch briefly the static continuum theory for smectics C, in terms of the variables $\Omega_z(\mathbf{r})$ and $u(\mathbf{r})$. Since a constant $\partial u/\partial y$ represents a rotation Ω_x along the x-axis, it is sometimes convenient to use simultaneously the three (small) rotation angles [26]:

$$\Omega_x = \frac{\partial u}{\partial y}, \qquad \Omega_y = -\frac{\partial u}{\partial x}, \qquad \Omega_z. \tag{7.58}$$

Note that eqn (7.58) implies one relation:

$$\frac{\partial \Omega_x}{\partial x} + \frac{\partial \Omega_y}{\partial y} = 0. \tag{7.59}$$

Uniform rotations do not change the free energy F: thus the part F_d of F associated with elastic distortions must be a function of the gradients $\partial_\alpha \Omega_\beta$ (or more concisely $\nabla\Omega$). F_d will also depend on the changes in the interlayer distance, described by

$$\gamma = \frac{\partial u}{\partial z}.$$

To construct F explicitly, we note first that terms linear in $\nabla\Omega$ cannot occur, if the unperturbed structure has a centre of symmetry: in a reflection operation around this centre, a rotation vector (pseudo-vector) such as Ω is not changed, while the (∇) operator changes sign. Terms linear in γ will not occur either, if the unperturbed interlayer distance has its equilibrium value.

Finally, making use of the equivalence between terms which can be transformed into other by partial integration, and eliminating certain 'unobservable' terms [similar to K' and K'' in eqn 7.6)], one arrives at the following structure for F_d:

$$F_d = F_c + F_s + F_{cs}, \tag{7.60}$$

where F_c is associated with distortions of the C director, for fixed layers, while F_s describes distortions of the layers. Finally F_{cs} contains certain cross terms coupling the two effects. Explicitly one has

$$F_c = \tfrac{1}{2}B_1\left(\frac{\partial\Omega_z}{\partial x}\right)^2 + \tfrac{1}{2}B_2\left(\frac{\partial\Omega_z}{\partial y}\right)^2 + \tfrac{1}{2}B_3\left(\frac{\partial\Omega_z}{\partial z}\right)^2 + B_{13}\frac{\partial\Omega_z}{\partial x}\frac{\partial\Omega_z}{\partial z} \tag{7.61}$$

$$F_s = \tfrac{1}{2}A\left(\frac{\partial\Omega_x}{\partial x}\right)^2 + \tfrac{1}{2}A_{12}\left(\frac{\partial\Omega_y}{\partial x}\right)^2 + \tfrac{1}{2}A_{21}\left(\frac{\partial\Omega_x}{\partial y}\right)^2 + \tfrac{1}{2}\bar{B}\gamma^2 \tag{7.62}$$

$$F_{cs} = C_1\frac{\partial\Omega_x}{\partial x}\frac{\partial\Omega_z}{\partial x} + C_2\frac{\partial\Omega_x}{\partial y}\frac{\partial\Omega_z}{\partial y}. \tag{7.63}$$

The first group (F_c) has been introduced by Saupe [27]. The second and third group have been analysed in ref. [26].† In eqn (7.62) we find two types of terms associated with pure distortions of the layers:

(1) The terms A, A_{12}, A_{21} describe a curvature of the layers and are the analogue (for a system of monoclinic symmetry) of the splay term occurring in a smectic A (eqn 7.6). The reader will recall that this splay term had the form

$$\tfrac{1}{2}K_1\left(\frac{\partial^2 u}{\partial x^2} + \frac{\partial^2 u}{\partial y^2}\right)^2 = \tfrac{1}{2}K_1\left(\frac{\partial\Omega_x}{\partial y} - \frac{\partial\Omega_y}{\partial x}\right)^2.$$

† In this reference, the discussion was purposely restricted to cases of constant inter-layer thickness ($\gamma = 0$). In the present discussion, we allow for finite values of γ: this adds one term $\tfrac{1}{2}\bar{B}\gamma^2$ to equation (62).

21

(2) The term $\bar{B}$, associated with possible changes of the interlayer distance: just as in smectics A, one can show that these changes remain in fact small for most practical situations.

Finally the F_{cs} terms of eqn (7.63) describe some rather subtle effects according to which a deformation imposed on the C director reacts on the curvature of the smectic planes.

As regards dimensions and orders of magnitude, we may note that the Saupe coefficients (B_1, B_2, B_3, B_{13}), the A coefficients (A, A_{12}, A_{21}) and the C coefficients (C_1, C_2) have the dimensions of energy per unit length (dynes): they are probably comparable to the Frank elastic constants in a homologue nematic phase (i.e. of the order of 10^{-6} dynes). On the other hand $\bar{B}$ has the dimensions of energy per unit volume, as in smectics A.

At present, very little is known on all these elastic constants. However, they will be measurable in the future through various types of experiments.

7.2.3.1. Fredericks transitions under magnetic fields [28].

The idea is to start with a single domain of a smectic C, strongly anchored between two glass plates, and to impose a magnetic field **H** normal to the molecular axis **n**. We shall call z the normal to the layers, x the direction of the unperturbed **c**-director. The direction of the glass plates with respect to this reference frame will depend on the surface treatment. One finds theoretically two types of behaviour, depending on the detailed orientation of **H**:

(1) If **H** is in the symmetry plane (x, z) of the unperturbed material, (and **H** is normal to **n**), *no* transition of the Fredericks type is expected,† except for one particular case: if the layers are normal to the glass plates, a 'ghost' transition (as defined in Section 7.2) is allowed.

(2) If **H** is normal to the symmetry plane (along y) a conventional Fredericks transition should occur at a certain field H_c, inversely proportional to the sample thickness. The value of H_c gives one information on the Saupe coefficients defined in eqn 7.61.

7.2.3.2. Carr–Helfrich instabilities under electric fields.

Here, depending on the detailed geometry, one can expect effects of the nematic type (see Chapter 5) or of the smectic A type. No detailed prediction

---

† It must be emphasized that we discuss here only the 'second-order' transitions: it may be that other types of transitions occur, involving a discontinuous jump in the orientation of the layers, and leading to complex textures in the final state: see ref. [29].

on possible cross terms is available up to now. Preliminary experiments have been quoted in the literature [30].

7.2.3.3. Light scattering by smectics C. As usual, the intensity I of the scattered light is directly related to the elastic constants. Depending on the direction of the scattering wave-vector **q**, one expects two different types of behaviour:

(1) If the wave vector **q** is *oblique*, the fluctuations of the layers are weak (Ω_x and Ω_y negligible) but the fluctuations of the C-director still give rise to a strong scattering.† From angular plots of the intensity $I(\mathbf{q})$, one could determine the ratios between the various Saupe coefficients [26].

(2) If $q_z = 0$ the fluctuations of the C-director become strongly coupled to a certain undulation mode, and the intensity I involves all elastic coefficients.

At present, no data on the intensities I are available, but one study has been performed on the frequency dependence of the scattered light— at low frequency shifts; Rayleigh scattering, and for oblique **q** vectors [7]. In this situation, one finds a purely dissipative mode, very similar to what is known in nematics, and associated with fluctuations in the C-director. The detailed hydrodynamic theory which is adequate to account for these results is described in ref. [21]; but the number of unknown friction coefficients involved is so large that the situation is somewhat discouraging!

On the whole, we see that our knowledge of the experimental properties of smectics C is still extremely limited; this is largely due to the difficulty of preparing good single domain samples of reasonable size. However, by a suitable choice of surface treatments for the glass plates, this situation may improve rather fast.

Problem: A chiral solute is added to a smectic C. What will be the resulting distortions?

Solution: Let us construct the elastic free energy for this system, in terms of the local rotation vector $\boldsymbol{\Omega}$ defined as before. We always call (x, y, z) the local frame attached to the smectic, x being along the C-director. The symmetry of the material is lowered by the chiral solute: we lose the centre of symmetry. The only non-trivial symmetry operation which is left is a rotation by an angle of π around the y axis: we call this operation $\hat{R}$.

In the non-chiral smectic C_1 all terms linear in $\nabla\boldsymbol{\Omega}$ were prohibited in the elastic free energy, because they were incompatible with the inversion symmetry. Now, some of these terms become allowed: they are the analogue of the **n**.curl **n** term

† Smectics C are turbid like nematics.

in the Frank energy of cholesterics (Chapter 6). To list them exactly, let us write down first the effect of the rotation $\hat{R}$ on Ω and on the ∇ symbol

$$\hat{R}\Omega_x = -\Omega_x \qquad \hat{R}\Omega_y = \Omega_y \qquad \hat{R}\Omega_z = -\Omega_z$$

$$\hat{R}\,\frac{\partial}{\partial x} = -\frac{\partial}{\partial x} \qquad \hat{R}\,\frac{\partial}{\partial y} = \frac{\partial}{\partial y} \qquad \hat{R}\,\frac{\partial}{\partial z} = -\frac{\partial}{\partial z}.$$

We see that we may keep terms of the form

$$\frac{\partial}{\partial y}\,\Omega_y, \quad \frac{\partial}{\partial z}\,\Omega_z, \quad \frac{\partial}{\partial x}\,\Omega_x, \quad \frac{\partial}{\partial z}\,\Omega_x, \quad \frac{\partial}{\partial x}\,\Omega_z.$$

We must always remember the constraint

$$\frac{\partial}{\partial x}\,\Omega_x + \frac{\partial}{\partial y}\,\Omega_y = 0.$$

Finally we may simplify the result by considering that the layers are essentially incompressible ($\partial u/\partial z \to 0$). This implies that

$$\frac{\partial}{\partial z}\,\Omega_x = \frac{\partial^2}{\partial z\,\partial x}\,u = \frac{\partial}{\partial x}\left(\frac{\partial u}{\partial z}\right) = 0.$$

Then the first-order terms in the elastic energy reduce to

$$F_1 = D_1\frac{\partial}{\partial x}\,\Omega_z + D_2\frac{\partial}{\partial x}\,\Omega_x + D\,\frac{\partial \Omega_z}{\partial z}.$$

The second-order terms (of order $(\nabla\Omega)^2$) are essentially unaltered for a small concentration of solute.

Let us consider first the D_1 term: it would apparently tend to bend the C-director without deforming the layers, as shown on Fig. 7.16a. But in fact, for most practical situations, D_1 has no observable effect. To understand this, let us extend our notations to cover the case of finite rotations Ω_z: we always call (x, y, z) the local frame attracted to the smectic, and we introduce a laboratory frame (ξ, η, z); the two frames are related by a rotation Ω_z around z. The D_1 term may then be written, for finite Ω_z, in the form

$$\frac{\partial}{\partial x}\,\Omega_z = \cos\Omega_z\,\frac{\partial \Omega_z}{\partial \xi} + \sin\Omega_z\,\frac{\partial \Omega_z}{\partial \eta} = -\frac{\partial}{\partial \eta}(\cos\Omega_z) + \frac{\partial}{\partial \xi}(\sin\Omega_z) = (\text{curl } \mathbf{c})_z.$$

If $\mathbf{c}$ is continous (no regular lines) the volume integral of curl $\mathbf{c}$ may be transformed into a surface integral: thus the D_1 term does not contribute to the volume energy. If the molecules are strongly anchored at the limiting surfaces (non-degenerate case) the surface term D_1 is unobservable. The only possible effect of D_1 in such a case would be to generate a finite density of chiral defects, shown on Fig. 7.16b. These defects have not been found up to now. To discuss their stability one would have to know their core energies, i.e. to go beyond the continuum theory.

The term D_2, in the first-order free energy F_1, tends to transform a flat layer into a twisted ribbon. This type of twist is entirely acceptable for a single layer. For a many-layer system, it is still locally acceptable but it is not compatible with the rule of constant interlayer distance for a macroscopic sample; thus, in the absence of any defects, it will also be unobservable. (This feature is reminiscent of the ghost transitions discussed earlier for smectics A.) On the other hand, the D_2 term may become significant if dislocations of the smectic layers are allowed: one could realize the required twist with a small density of screw dislocations (all of the same chirality).

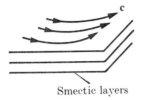

Smectic layers

FIG. 7.16. One type of distortion mode in a chiral smectic C. The C-director tends to bend. However, to gain a finite volume energy from this bend, one needs a bend of constant sign and finite magnitude, which cannot be achieved without singularities.

In practice, the only term which leads to strong, and observable, effects is the D term; this is coupled to a simple twist of the C-director

$$q = \frac{\partial \Omega_z}{\partial z}.$$

The terms involving q in the free energy are

$$Dq + \tfrac{1}{2} B_3 q^2,$$

and the optimum twist q_0 corresponds to the minimum of this expression

$$q = -D/B_3.$$

This leads to helical structures of period $2\pi/q$ (*not* π/q since the C-director does have a well-defined sense). These helices have been observed by various groups, either on mixtures or with pure chiral smectogenic compounds [5] [6]. The structure is shown in Fig. 1.11.

7.3. Phase transitions and precritical phenomena

In this section we shall be particularly concerned with the possible transitions between smectic A, smectic C, and nematic (N) phases; experimentally, many of these transitions are found to be of first order [32]. However, from the point of view of the Landau rules [17] all of them *may* be of second order. These second-order transitions, although less frequent, are of great interest because they are always announced by pretransitional anomalies.

From the point of view of statistical mechanics, the smectic phases give us a whole new class of critical phenomena, where the ideas developed in connection with simpler systems (ferromagnets, superfluids, etc.) can be transposed and subjected to new tests.

From the point of view of applications, in the nematic phase, the onset of quasi-smectic features may lead to a drastic change in certain important constants (elastic coefficients, transport properties, etc.). For instance, it has been shown recently by Rondelez [33] that the ratio of the parallel to the perpendicular electric conductivity $\sigma_\parallel/\sigma_\perp$ which is usually larger than unity in nematics, may become significantly smaller than unity in the vicinity of a smectic transition (because the charge carriers do not cross the layers easily). The electrohydrodynamic effects on which many display devices are based, depend critically on $(\sigma_\parallel/\sigma_\perp) - 1$: materials with positive dielectric anisotropy, which could not be used efficiently when $\sigma_\parallel/\sigma_\perp > 1$, become interesting.

7.3.1. The $C \rightleftharpoons A$ transition

The transition from tilted smectic C to uniaxial smectic A has been studied on a few typical materials such as TBBA [4]. This particular compound appears to have a continuous (or nearly continuous) transition of the type in which we are interested.

For a given set of layers, to describe the ordered (C) state, one must specify the magnitude ω of the tilt angle, and also the azimuthal direction of tilt, specified by an angle ϕ. Thus we can characterize the onset of order by two real parameters ω and ϕ, or equivalently by the complex number

$$\psi = \omega e^{i\phi}. \tag{7.64}$$

This brings in a remarkable analogy with superfluid helium [34]. In helium, what we have is a Bose condensation [35]. A macroscopic number of helium atoms are occupying *one* quantum state, described by a wave function $\psi(\mathbf{r})$. This is the order parameter and it is complex.

In both systems, an overall change of the phase ϕ does not modify the free energy F; in helium, this property is called gauge invariance. In smectics C, this simply means that two layer arrangement with the same tilt angle ω are equivalent. Thus the structure of F as a function of ψ will be essentially identical in both cases. This is discussed in ref. [34]†
and leads to the following predictions:

---

† The structure of the gradient terms of the free energy is oversimplified in this reference, but this does not alter the discussion.

(1) The transition $C \rightleftharpoons A$ *may* be continuous; if it is (as we shall assume in all that follows) the specific heat will show a logarithmic singularity at the transition point T_{CA}.

(2) Below T_{CA}, the tilt angle should vary according to the law

$$\omega = \text{constant} \, |\Delta T|^{\beta} \qquad (7.65)$$

where $\Delta T = T - T_{CA}$ and $\beta \simeq 0.35$. In TBBA a recent experimental value of β is $\beta = 0.40 \pm 0.05$ (Wise R., Smith D. and Doane W., *Phys. Rev. A7*, 1366, 1973).

(3) Above T_{CA}, if we can apply a magnetic field H which is *oblique* with respect to the optical axis (Oz) of the smectic phase (e.g. in the x,y-plane), we expect to induce a tilt angle

$$\omega = C \frac{\chi_a H_x H_z}{k_B T_{CA}} \left(\frac{T_{CA}}{\Delta T} \right)^{\gamma} \qquad (7.66)$$

where C is a numerical constant of order unity, χ_a is the diamagnetic anisotropy per molecule, and the critical exponent γ should be close to 1·30. The tilt predicted by eqn (7.66) is small, because the diamagnetic energy $\chi_a H^2$ is very weak when compared to $k_B T_{CA}$, but it could be measured by suitable optical methods, at least close to T_{CA}.

Similar experiments can be conceived with electric fields E. But, as explained in Chapter 2, the electric effects are usually much more complex than their magnetic counterparts.

(4) Again starting from the smectic A phase, and decreasing T towards T_{CA}, we expect to see the onset of a strong (depolarized) light scattering due to fluctuations in the tilt angle. In the (usual) situation where the wavelength of light is larger than the size ξ of the fluctuating domains, the scattered intensity I should be of the form

$$I = I_0 + I_1 \left(\frac{T_{AC}}{\Delta T} \right)^{\gamma} \qquad (7.67)$$

where I_0 is the normal scattering by the smectic A phase, and is small (except for very special circumstances as explained in Section 7.2). and I_1 is of comparable magnitude. Thus eqn 7.67 allows in principle another measurement of γ.

The determinations of β and γ are particularly interesting because these exponents are not accessible in helium (the amplitude of the condensate wave function ψ is not directly observable).

We must not end this discussion of critical exponents without a serious word of caution. The analogy between $\omega e^{i\phi}$ and the complex order parameter of superfluids is rigorous. However, we know two types of superfluids, with very different critical behaviours: the first type corresponds to helium-4, where the critical exponents show the non-trivial values quoted above. In helium, the coherence length far from T_c, which is usually called ξ_0, is comparable to the interparticle distance a. The second type of superfluids is found amongst the superconducting metals, and has $a/\xi_0 \ll 1$. In this case the anomalous critical region is restricted to an unobservably small temperature interval near T_c, and in all practical observations, it is the mean field exponents which are found. (For instance, $\beta = \frac{1}{2}$.)

We did invoke the existence of a 'small parameter' a/ξ_0 to explain the apparent 'mean field' behaviour of the nematic–isotropic transition in Chapter 2. Is there a similar small parameter (possibly based also on the width to length ratio of the constituent molecules) for the C–A transition? We do not know yet the complete answer to this question, but recent measurements of β suggest that we are rather dealing with a 'helium type' transition.

Apart from the static critical effects we can also think of various *dynamical* experiments which probe semi-slow motions somewhat similar to those which were described at the end of Chapter 5.

To summarize, we can say that the C $\rightleftharpoons$ A transition offers a very promising field of research. There may be some striking analogies between T_{CA} and the λ-point of helium. Certain critical properties are easier to probe in helium (e.g. the specific heat, because the high thermal conductivity allows for very precise temperature definitions). But other properties are accessible much more easily in smectics C (e.g. the exponents γ and β). Also, in spite of the analogy, there are some interesting differences, especially in the dynamical behaviour: below T_C the phase fluctuations in helium are associated with second sound, while in a smectic C they are expected to have a purely dissipative behaviour [36], very similar to what we have in a nematic.

7.3.2. The $A \rightleftharpoons N$ transition

The transition between smectic A and nematic is usually discontinuous, with a finite latent heat, etc. But this is not imposed by the symmetry of the problem. As first suggested by McMillan [37] on a specific model, with suitable values of the interaction constants one may have a second order transition. Using a slightly more general formulation we can present the basic ideas as follows.

In the nematic phase the alignment of the molecules around the optical axis (z) is described by the conventional parameter

$$S_0(T) = \langle \tfrac{1}{2}(3\cos^2\theta - 1)\rangle. \qquad (7.68)$$

The equilibrium value $S_0(T)$ is rather accurately given by the Maier–Saupe mean field calculation (see Chapter 2).

In the smectic phase we have a density wave:

$$\rho(\mathbf{R}) = \rho(z) = \rho_0[1 + 2^{-\frac{1}{2}}|\psi|\cos(q_s z - \Phi)] \qquad (7.69)$$

where ρ_0 is the average density, $|\psi|$ measures the strength of the smectic order (the factor $2^{-\frac{1}{2}}$ is for convenience), $q_s = 2\pi/d$ is the wave-vector of the density wave, and d the interlayer distance. Φ is an arbitrary phase.†

If the nematic alignment parameter is maintained fixed ($S = S_0(T)$) the free energy (per cm³) F can be expanded in powers of ψ as follows:

$$F_{\mathrm{S}} = \alpha_0|\psi|^2 + \beta_0|\psi|^4 + \cdots \qquad (7.70)$$

(only even powers of $|\psi|$ may come in). At a certain temperature T^*_{AN} the coefficient α_0 vanishes. Above this temperature it is positive. The coefficient β_0 is always positive. With these ingredients alone we could have a second-order transition at $T = T^*_{\mathrm{AN}}$.

However, there is some coupling between $|\psi|$ and S; if the alignment measured by S increases, the average attractions between neighbouring molecules in a smectic layer will increase. Because of this coupling; the optimal value of S need not coincide with $S_0(T)$. We shall put

$$\delta S = S - S_0(T). \qquad (7.71)$$

The coupling term, to lowest order in ψ and S, must have the form

$$F_1 = -C|\psi|^2\,\delta S. \qquad (7.72)$$

where C is a (positive) constant. Finally we must include in F the nematic free energy F_N, which is a minimum for $\delta S = 0$:

$$F_N = F_N(S_0) + \frac{1}{2\chi}\,\delta S^2. \qquad (7.73)$$

Here $\chi(T)$ is a response function, which is large (although finite) near the nematic isotropic transition point T_{NI}, but which is small for

† The order parameter is again a complex number $|\psi|e^{i\Phi}$: we find here a second analogy with superfluids, which has been noticed independently by McMillan [41] and by the present author [43].

$T < T_{\text{NI}}$ (since S_0 is then nearly saturated, and cannot fluctuate very much).

The overall free energy F, obtained by adding F_{S}, F_1, and F_{R}, must then be minimized with respect to δS, giving

$$\delta S = \chi C \psi^2 \tag{7.74}$$

$$F = F(S_0) + \alpha_0 \psi^2 + \beta \psi^4 \tag{7.75}$$

$$\beta = \beta_0 - \tfrac{1}{2} C^2 \chi. \tag{7.76}$$

The order of the transition depends critically on the sign of β (for $T \sim T^*_{\text{AN}}$)

(1) If $T^*_{\text{AN}} \sim T_{\text{NI}}$ $\chi(T^*_{\text{AN}})$ is large and β is negative. Then terms in ψ^6 must be added to (7.76) to ensure stability, and the resulting plots of $F(\psi)$ show that, for $T = T^*_{\text{AN}}$ the minimum of F is already at a non-zero $|\psi|$. The transition takes place at a higher temperature $T_{\text{c}} > T^*_{\text{AN}}$, and it is first order.

(2) If T^*_{AN} is significantly smaller than T_{NI}, the response function χ (T^*_{AN}) is small and $\beta \sim \beta_0 > 0$. Then looking at the plots of $F(\psi)$, one easily sees that the transition is of second order and the transition point is T^*_{AN}.

The change from second to first order is thus induced by a coupling between the order parameter (ψ) and an external variable (S). This is the exact counterpart of the so called 'Rodbell Bean' effect in magnetism [38]. In the magnetic case, the order parameter is the magnetization M, and the external variable is the density (or lattice parameter a). The coupling is due to the dependence of exchange interactions on a. If the magnetic crystal is strongly compressible (large χ) the transition becomes first order.

How can we, in practice, arrange for the transition to be second order? The discussion above suggests that this will occur if χ is small, i.e. when T^*_{AN} is significantly smaller than T_{NI}. We can act on these temperatures by changing the length l of the aliphatic chains which hang at the ends of the molecule at hand. Usually an increase in l decreases T_{NI} but does not change the limits of the smectic domain very much. Thus the most favourable situation should correspond to l small. One series of homologous compounds of variable l has been studied by Doane and his coworkers [39a] using n.m.r. measurements of S. They concluded that for the compound with the shortest l, the transition is of second order. This is confirmed by latent heat measurements [39b]. A completely different material ('NBOA') has also been found with a nearly second-order transition [39c] [41].

Let us now list a few remarkable pretransitional effects which will occur if the transition is of second order, or very nearly so:

7.3.2.1. Cybotactic clusters.   In the nematic phase, just above the transition point T_{AN}, we expect to see small domains with a local smectic organization (Fig. 7.17). This has been recognized by de Vries [40] who called these domains cybotactic clusters. From the point of view of X-ray scattering, one has two peaks of intensity in reciprocal space, concentrated near the points $\mathbf{q} = \pm q_s \mathbf{n}$. The width of these peaks gives some information on the size ξ of the clusters. Studies on the X-ray scattering intensities have been carried out, both below and above T_{AN}, on cholesterol derivatives [41a]; and more recently on simpler smectics [41b].

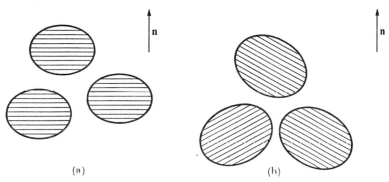

(a)   (h)

FIG. 7.17. Cybotactic groups above a smectic $\rightleftharpoons$ nematic transition: (a) smectic $A \rightleftharpoons$ nematic; (b) smectic $C \rightleftharpoons$ nematic.

7.3.2.2. Anomalies in the elastic constants.   In a smectic A phase, we have seen that both twist and bend deformations become forbidden; this suggests that just above T_{AN}, the Frank constants K_2 and K_3 become large. An effect for this sort was first found by the Freiburg group [42] on K_3. However, the systems at hand were smectics C rather than smectics A, and this tends to complicate the situation. On the theoretical side, a calculation of the anomaly has been described in ref. [43]. It predicts an increase $(\delta K_2, \delta K_3)$ proportional to the size ξ of the cluster. The calculation may be presented qualitatively as follows:

Consider for instance a *bend* deformation imposed on a cybotactic cluster (Fig. 7.18). The strains imposed to the layers are of order $\delta\theta$ where $\delta\theta$ is the angle between optical axis on both sides of the cluster

$$\delta\theta \sim \frac{\partial\theta}{\partial z} \xi. \tag{7.77}$$

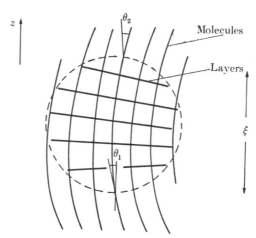

FIG. 7.18. Effect of a bending deformation on a cybotactic group, just above an A ⟷ N transition: the layers cannot remain equidistant.

The elastic energy due to layer distortion is (per cm³)

$$\tfrac{1}{2}\bar{B}(\delta\theta)^2$$

where $\bar{B}$ has been defined in eqn 7.6 below T_c, but is now used for a cluster above T_c. The correction to the bend energy is thus

$$\tfrac{1}{2}\bar{B}\xi^2\left(\frac{\partial\theta}{\partial z}\right)^2 = \tfrac{1}{2}\delta K_3\left(\frac{\partial\theta}{\partial z}\right)^2. \tag{7.78}$$

We must now estimate $\bar{B}$ for a cluster, i.e. in a temperature region where the order parameter ψ defined in eqn (7.72) is small. $\bar{B}$ vanishes with ψ and must be even in ψ: it is natural to assume that $\bar{B} \sim |\psi|^2$. The complete calculation shows that in fact, for the present problem, we must put

$$\bar{B} = \text{constant}\langle\psi^*(0)\psi(R)\rangle_{R\sim\xi}. \tag{7.79}$$

To a good approximation, the correlation function $\langle\psi\psi\rangle$ has the Ornstein–Zernike form

$$\langle\psi^*(0)\psi(R)\rangle = \frac{\text{constant}}{R}\exp(-R/\xi).$$

Then:

$$\bar{B} \sim \langle\psi^*(0)\psi(\xi)\rangle \sim 1/\xi$$

and

$$\delta K_3 \simeq \bar{B}\xi^2 \simeq \xi. \tag{7.80}$$

A similar argument holds the twist constant

$$K_2 = K_{20} + \delta K_2.$$

The pretransitional anomalies δK have very recently been measured by various groups. The results (to be published) can be summarized by an exponent $\delta K \sim (T - T_{AN})^{-x}$. The Orsay and Harvard results give $x \sim \frac{2}{3}$. The other results for x are slightly smaller. A value $x = \frac{2}{3}$ would be expected if the system is 'helium like'. On the other hand, if there is a 'small hidden parameter' a/ξ_0 (as defined for a similar case in Section 7.3.1) we would expect $x = \frac{1}{2}$.

The Harvard group has also compared the magnitude of the coherence length ξ, as extracted from δK_3, to a direct X-ray measurement by W. McMillan [41] on the same material, and has found some agreement.

7.3.2.3.   Pretransitional effects on the pitch of a cholesteric phase.   We may hope to obtain some indirect information on the twist constant K_2 through a study of the temperature dependence of the pitch in a cholesteric phase.† If t is the twist, the free energy for a cholesteric may be written as

$$F_{\text{chol}} = \tfrac{1}{2}K_2 t^2 - Dt + F_{\text{nem}},$$

and the equilibrium twist is $t_{\text{eq}} = D/K_2$. The onset of smectic short-range order modifies both K_2 and D. We have just seen that

$$\delta K_2 = K_2 - K_{20} \simeq \langle \psi^2 \rangle \xi^2.$$

The corrections to D can also be worked out† and are of order

$$\delta D \sim D \langle \psi^2 \rangle. \tag{7.81}$$

This effect on D is much smaller than the effect on K_2 (because ξ is large in the critical region). Thus in practice δD will often be negligible and we may write

$$t = t_0 \frac{K_{20}}{K_{20} + \delta K_2}, \tag{7.82}$$

where δK_2 is due to the clusters, and is a decreasing function of temperature; eqn (7.82) thus gives use to a twist $t(T)$ which is increasing with T, in agreement with what is observed in most cholesterics. The well-known experimental rise to $t(T)$ in cholesterol deviations may thus reflect simply the presence of cybotactic groups: this could be checked by comparing short-range order studies performed with X-rays [40] [41] (giving ξ and thus δK_2) to direct data on $t(T)$.

† See for instance R. Alben, *Molecular Crystals* **20**, 231, (1973).

7.3.2.4. *Dynamical effects.*   As seen in Section 7.2, in the smectic A phase we expect *two* branches of acoustical waves: this again shows a similarity with a superfluid such as helium where we have first and second sound. Going close to T_{AN}, and always assuming that the transition is of second order, we expect the second sound velocity c_2 to collapse. Another dynamical property of great interest is the Helfrich permeation coefficient, describing how the molecules can flow through the layers; clearly permeation becomes easier when ψ becomes small. These effects have been analysed recently by F. Brochard, using two distinct assumptions on dynamical scaling [44]: they should lead to a number of interesting experiments.

7.3.3. *The* C $\rightleftharpoons$ N *transition*

A direct transition from smectic C to nematic is observed in many mesomorphic systems—for instance with the longer homologs of PAA. This C $\rightleftharpoons$ N transition may also be of second order from the point of

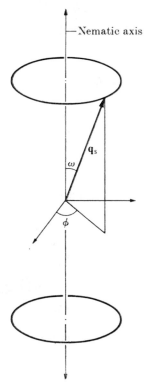

— Nematic axis

q_s

ω

ϕ

FIG. 7.19. Region of large X-ray scattering in a nematic having 'skewed cybotactic' groups.

view of the Landau rules [17]. Some instances where the $C \rightleftharpoons N$ heat of transition is particularly low can indeed be found in the literature. Thus, here again, we may expect interesting pretransitional phenomena.

However, from the point of view of statistical mechanics, $C \rightleftharpoons N$ is much more complex than $A \rightleftharpoons N$. Consider for instance the X-ray scattering by a single domain nematic containing some cybotactic groups as defined by de Vries. With 'normal' cybotactic groups (A type) the scattering is concentrated around two *points* in reciprocal space $\mathbf{k} = \pm\mathbf{q}_s$, $\mathbf{n} = \pm\mathbf{q}_s$. There are two and only two Fourier components of the density $\rho(\mathbf{q}_s)$ and $\rho(-\mathbf{q}_s) = \rho^*(\mathbf{q}_s)$ which show large fluctuations near T_c: the natural order parameter is a two component object $\psi = \text{const. } \rho(\mathbf{q}_s)$. On the other hand, with 'skewed' cybotactic groups (C-type) the scattering intensity is concentrated on two *rings* in reciprocal space, as shown in Fig. 7.19. All Fourier components $\rho(\mathbf{k})$ for which

$$k_z = q_v \quad (= q_s \cos \omega)$$

$$\sqrt{k_x^2 + k_y^2} = q_T \quad (= q_s \sin \omega)$$

have large fluctuations near T_c [45]. Thus the natural order parameter has an infinite number of components. This system has no obvious counterpart among superfluids, and a number of experimental data on the behaviour at T_c are urgently required.

REFERENCES

CHAPTER 7

[1] (a) HERRMANN, K. *Trans. Faraday Soc.* **29**, 972 (1933); *Z. Kristallogr. Kristallgeom.* **92**, 49 (1935); (b) ALEXANDER, E. and HERRMANN, K. *Z. Kristallogr. Kristallgeom.* **69**, 285 (1928).

[2] FRIEDEL, G. *Annls. Phys.* **18**, 273 (1922); BRAGG, W. H. *Nature* **133**, 445 (1934).

[3] BOULIGAND, Y. *J. Phys.* (*Fr.*) **33**, 525 (1972).

[4] TAYLOR, T. R., FERGASON, J. L., and ARORA, S. L. *Phys. Rev. Lett.* **24**, 359 (1970); **25**, 722 (1970).

[5] HELFRICH, W. and OH, C. S. *Mol. Cryst. liquid Cryst.* **14**, 289 (1971).

[6] BILLARD, J. and URBACH, W. Z. *C.r. Acad. Sci., Paris* **274B**, 1287 (1972).

[7] GALERNE, Y., MARTINAND, J. L., DURAND, G., and VEYSSIE, M. *Phys. Rev. Lett.* **29**, 561 (1971).

[8] (a) SACKMANN, H. and DEMUS, D. *Mol. Cryst.* **2**, 81 (1966); (b) DIELE, S., BRAND, P., and SACKMANN, H. *Mol. Cryst. liquid Cryst.* **16**, 105 (1972).

[9] LAMBERT, M. and LEVELUT, A. M. *C.r. Acad. Sci., Paris* **272B**, 1018 (1971); see also [8](b).

[10] URBACH, W. Z. *Thèse 3e Cycle*, Paris, 1973.

[11] DVOLAITSKY, M., POLDY, F., and TAUPIN, C. (to be published).

[12] DE GENNES, P. G. and SARMA, G. *Phys. Lett.* **A38**, 219 (1972).

[13] LANDAU, L. D. and LIFSCHITZ, E. M. *Theory of elasticity*, p. 13. Pergamon, London (1959).

[14] DE GENNES, P. G. *J. Phys.* **30**, *Colloq. C4*, Suppt. to No. 11–12, 65 (1969).

[15] RONDELEZ, F. and HULIN, J. P. *Solid State Commun.* **10**, 1009 (1972).

[16] (a) DELAYE, M., RIBOTTA, G., and DURAND, G. *Phys. Lett.* **A44**, 139 (1973); (b) CLARK, N., MEYER, R. B. *Appl. Phys. Lett.* **22**, 493 (1973); (c) KAHN F. *Appl. Phys. Lett.* **22**, 111 (1973).

[17] See LANDAU, L. D. and LIFCHITZ, E. M. *Statistical physics*, Chapter 13, Pergamon, London (1959).

[18] See, for instance, KITTEL, C. *Introduction to solid state physics* (3rd edn.), p. 68. Wiley, New York (1966).

[19] DE GENNES, P. G. *J. chem. Phys.* **48**, 2257 (1968).

[20] CAILLE, A. *C.r. Acad. Sci., Paris* **274**, 891 (1972). In eqn (9b) of this reference, the correct exponent should read $4 - 2x$ (not $2 - 2x$).

[21] MARTIN, P., PARODI, O., and PERSHAN, P. *Phys. Rev.* **A6**, 2401 (1972).

[22] See, for instance, KATCHALSKY, A. and CURRAN, P. *Non-equilibrium thermodynamics in biophysics*, Harvard University Press (1967).

[23] LORD, A. E. *Phys. Rev. Lett.* **29**, 1366 (1972); LIAO, Y., CLARK, N., and PERSHAN, P. S. *Phys. Rev. Lett.* **30**, 639 (1973).

[24] PORTER, R. S., BARRAL, E. M., and JOHNSON, J. F. *J. chem. Phys.* **45**, 1452 (1966); PORTER, R. S. and JOHNSON, J. F. *J. chem. Phys.* **66**, 1826 (1962).

[25] HELFRICH, W. (a) *Phys. Rev. Lett.* **23,** 372 (1969); (b) *Liquid crystals and ordered fluids* (J. F. Johnson, R. S. Porter, eds.), p. 405. Plenum Press, New York (1970).

[26] ORSAY GROUP on liquid crystals, *Solid State Commun.* **9,** 653 (1971).

[27] SAUPE, A. *Mol. Cryst. liquid Cryst.* **7,** 59 (1969).

[28] RAPINI, A. *J. Phys.* **33,** 237 (1972).

[29] BIDAUX, R., BOCCARA, N., SARMA, G., DE SEZE, L., DE GENNES, P. G., and PARODI, O. *J. Phys.* **34,** 661 (1973).

[30] CARR, E. F. *Mol. Cryst. liquid Cryst.* **8,** 247 (1969).

[31] JOHNSON, J. F., PORTER, R. S., and BARRAL, E. M. *Mol. Cryst. liquid Cryst.* **8,** 1 (1969).

[32] ENNULAT, R. D. *Mol. Cryst. liquid Cryst.* **13,** 27 (1971).

[33] RONDELEZ, F. *Solid State Commun.* **12,** 1675 (1972).

[34] DE GENNES, P. G. *C.r. hebd. Séanc. Acad. Sci., Paris, Ser. B* **274,** 758 (1972).

[35] For an introduction to Bose condensation, see LONDON, F. *Superfluids,* Part I; PENROSE, O. and ONSAGER, L. *Phys. Rev.* **104,** 576 (1956).

[36] See for instance DE GENNES, P. G., *Mol. Cryst. liquid Cryst.* **21,** 49 (1973).

[37] MCMILLAN, W. *Phys. Rev.* **A4,** 1238 (1971). An improved solution of the McMillan model has been given recently by S. Marčelja (private communication).

[38] BEAN, C. P. and RODBELL, D. *Phys. Rev.* **126,** 104 (1962).

[39] (a) DOANE, W., PARKER, R. S., CVIKL, B., JOHNSON, D., and FISHEL, D. *Phys. Rev. Lett.* **28,** 1694 (1972). (b) DUREK, D., BATURIC, J., MARCELJA, S., and DOANE, W. *Phys. Lett.* **A43,** 273 (1973). (c) CABANE, B. and CLARK, W. G. *Solid State Commun.* **13,** 129 (1973).

[40] DE VRIES, A. *Mol. Cryst. liquid Cryst.* **10,** 31 (1970); **10,** 219 (1970); **11,** 361 (1970).

[41] (a) MCMILLAN, W. L. *Phys. Rev.* **A6,** 936 (1972); (b) MCMILLAN, W. L. *Phys. Rev.* **A7,** 1673 (1973).

[42] GRULER, H. *Z. Naturf.* **28,** 474 (1972).

[43] DE GENNES, P. G. *Solid State Commun.* **10,** 753 (1972).

[44] BROCHARD, F. *J. Phys.* **34,** 411 (1973).

[45] The rings have been studied experimentally by MCMILLAN, W. L. (to be published).

AUTHOR INDEX

SUBJECT INDEX